# Lipid Oxidation Pathways
## Volume 2

# Lipid Oxidation Pathways
## Volume 2

**Editors**

**Afaf Kamal-Eldin**
Department of Food Science
Swedish University of Agricultural Sciences

**David B. Min**
Department of Food Science and Technology
Ohio State University

AOCS PRESS

URBANA, ILLINOIS

**AOCS Mission Statement**

To be a global forum to promote the exchange of ideas, information, and experience, to enhance personal excellence, and to provide high standards of quality among those with a professional interest in the science and technology of fats, oils, surfactants, and related materials.

**AOCS Books and Special Publications Committee**

M. Mossoba, Chairperson, U.S. Food and Drug Administration, College Park, Maryland
R. Adlof, USDA, ARS, NCAUR-Retired, Peoria, Illinois
M.L. Besemer, Besemer Consulting, Rancho Santa, Margarita, California
P. Dutta, Swedish University of Agricultural Sciences, Uppsala, Sweden
T. Foglia, ARS, USDA, ERRC, Wyndmoor, Pennsylvania
V. Huang, Yuanpei University of Science and Technology, Taiwan
L. Johnson, Iowa State University, Ames, Iowa
H. Knapp, DBC Research Center, Billings, Montana
D. Kodali, Global Agritech Inc., Minneapolis, Minnesota
G.R. List, USDA, NCAUR-Retired, Consulting, Peoria, Illinois
J.V. Makowski, Windsor Laboratories, Mechanicsburg, Pennsylvania
T. McKeon, USDA, ARS, WRRC, Albany, California
R. Moreau, USDA, ARS, ERRC, Wyndoor, Pennsylvania
A. Sinclair, RMIT University, Melbourne, Victoria, Australia
P. White, Iowa State University, Ames, Iowa
R. Wilson, USDA, REE, ARS, NPS, CPPVS-Retired, Beltsville, Maryland

AOCS Press, Urbana, IL 61802

©2008 by AOCS Press. All rights reserved. No part of this book may be reproduced or transmitted in any form or by any means without written permission of the publisher.

ISBN 978-1-893997-56-1

Library of Congress Cataloging-in-Publication Data

Lipid oxidation pathways / [edited by] Afaf Kamal-Eldin.
    p. cm
Includes bibliographical references and index.
  ISBN 1-893997-43X (acid-free paper)
  1. Lipids--Oxidation.   I. Kamal-Eldin, Afaf.

QP751.L5517 2003
572'.57--dc21

2003006200

CIP

Printed in the United States of America.
12 11 10 09 08  5 4 3 2 1

The paper used in this book is acid-free and falls within the guidelines established to ensure permanence and durability.

# Contents

Preface .................................................................................................................. vii

1. Chemistry and Reactions of Singlet and Triplet Oxygen in Lipid Oxidation
   *Hyun Jung Kim and David B. Min* ................................................................ 1
2. Chemistry and Reactions of Reactive Oxygen Species in Lipid Oxidation
   *Eunok Choe* .................................................................................................. 31
3. Oxidation of Long-Chain Polyunsaturated Fatty Acids
   *Kazuo Miyashita* .......................................................................................... 51
4. Oxidation of Conjugated Linoleic Acid
   *Taina I. Pajunen (née Hämäläinen) and Afaf Kamal-Eldin* ........................ 77
5. Oxidation of Cholesterol and Phytosterols
   *Afaf Kamal-Eldin and Anna-Maija Lampi* ................................................ 111
6. Tocopherol Concentrations and Antioxidant Efficacy
   *Afaf Kamal-Eldin, Hyun Jung Kim, Levon Tavadyan, and David B. Min* .... 127
7. Carotenoids and Lipid Oxidation Reactions
   *Afaf Kamal-Eldin* ........................................................................................ 143
8. Co-oxidation of Proteins by Oxidizing Lipids
   *Karen M. Schaich* ........................................................................................ 181
9. Lipid Oxidation in Food Dispersions
   *Eric A. Decker, Wilailuk Chaiyasit, Min Hu, Habibollah Faraji, and D. Julian McClements* ........................................................................ 273
10. Antioxidant Evaluation Strategies
    *Leif H. Skibsted* .......................................................................................... 291

Index ...................................................................................................................... 307

# Preface

The first volume of *Lipid Oxidation Pathways*, published in 2003, tried to highlight some of the anomalies and gaps in our current understanding of the lipid oxidation mechanism and kinetics. The different chapters discussed how lipid oxidation proceeds in different environmental surroundings and highlighted areas where further research is needed. A number of prominent scientists in academia and industry said they appreciate the book for its particular focus on the anomalies between observed and predicted behaviour and requested a new volume of the book that focuses on the oxidation kinetics and mechanisms governing the behaviour of different molecular species involved in lipid oxidation reactions. This second volume thus aims to complement the first volume and extend it to more detailed reviews of the reactions of lipid molecules other than conventional polyunsaturated fatty acids.

The first chapter discusses the basic chemistry of singlet oxygen and its involvement in lipid oxidation reactions with particular reference to reactions in foods such as reversion of flavour in soybean oil. The second chapter discusses the different reactive oxygen species and their particularities in lipid oxidation. Chapter 3 discusses in detail the oxidation of long chain polyunsaturated fatty acids, including eicosapentaenoic acid (EPA) and docosahexaenoic acid (DHA), and discusses the kinetic effects of different environments. Chapter 4 provides a detailed description of the oxidation of conjugated linoleic acid (CLA), particularly the fact that hydroperoxides are not the major primary oxidation products from this fatty acid and that conventional methods for estimating the degree of lipid oxidation (such as peroxide value and conjugated dienes) will lead to erroneous conclusions in this case. Chapter 5 describes the oxidation of sterols, which follow the same basic mechanism as for oleic acid. In chapter 6, possible reactions behind the paradoxical behaviour of tocopherols, i.e. increased rate of inhibited oxidation at high levels of these antioxidants or what has been known as tocopherol's pro-oxidant effect, has been delineated. Chapter 7 reviews the literature pertinent to the oxidation mechanisms and oxidation products of carotenoids. Like with the case of CLA, radical addition rather than hydrogen abstraction seem to be the most plausible mechanism. The co-oxidation of lipid and proteins is discussed comprehensively in Chapter 8 with emphasis on the reactions of proteins with free radicals and lipid hydroperoxides, epoxides, and carbonyl compounds. Chapter 9 discusses lipid oxidation in emulsions and how it is affected by interaction with proteins, type of emulsifier, antioxidants, and the structural organization of different molecular species. Chapter 10 discusses the complication in evaluating lipid antioxidants and inhibitors as it relates to the widely different modes of their action.

The editors sincerely thank the authors of the different chapters for their valuable contribution with timely literature reviews within this second volume of *Lipid Oxidation Pathways*. We also acknowledge, with sincere gratitude, Jodey Schonfeld, Brock Peoples, and their colleagues at the AOCS Press for their professional help in the editing and for the production of this volume.

Afaf Kamal-Eldin  
Uppsala, Sweden

David B. Min  
Columbus, OH

# 1

# Chemistry and Reactions of Singlet and Triplet Oxygen in Lipid Oxidation

Hyun Jung Kim and David B. Min
Department of Food Science and Technology, The Ohio State University, 2015 Fyffe Road, Columbus, OH 43210

## Introduction

Oxidation of food components can influence flavor quality, nutritional quality, consumer acceptability, and toxicity of food products (Min and Boff, 2002; Choe and Min 2006). Lipid oxidation products are especially responsible for the development of rancidity by the production of low-molecular-weight compounds that impart undesirable flavors. Oxidation occurs by both triplet oxygen and singlet oxygen. Atmospheric triplet oxygen contains two unpaired electrons, while singlet oxygen has no unpaired electrons. The electron arrangement of triplet oxygen does not allow for a direct reaction with compounds, such as unsaturated fatty acids, that is nonradical and in the singlet state. Triplet oxygen oxidation of foods has been extensively studied during the last 70 years (Labuza, 1971; Min and Lee 1996). Singlet oxygen formed in the presence of triplet oxygen by excitation is thought to be responsible for initiating lipid oxidation of food compounds, due to its ability to directly react with the electron-rich compounds. Singlet oxygen rapidly increases the oxidation rate of foods even at very low temperatures (Rawls and Van Santen, 1970). Singlet oxygen oxidation can produce new compounds, which are not found in ordinary triplet oxygen oxidation in foods (Frankel et al., 1981). During the last 30 years, the attention paid to singlet oxygen oxidation of foods has increased.

This chapter will discuss the basic information on the chemical properties, formation, inhibition, and detection of singlet oxygen oxidation as it relates to the oxidation of lipids.

## Electron Configuration of Triplet and Singlet Oxygen

The molecular orbital theory best describes the electron structure of molecular oxygen and its excited states (Korycka-Dahl and Richardson 1978). Molecular oxygen is composed of two oxygen atoms and has 10 molecular orbitals containing 12 valence electrons. Valence electrons are added sequentially to the orbitals in order of increasing energy to obtain molecular orbitals of triplet oxygen and singlet oxygen (Figures 1 and 2). According to Pauli's exclusion principle, only two electrons can occupy each orbital. Hund's rule states that one electron is placed into each orbital of equal energy one at a

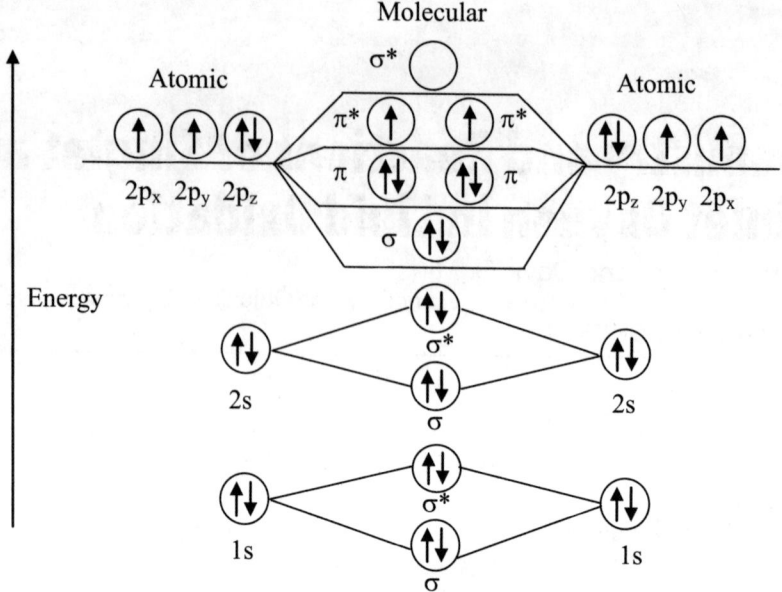

**Fig. 1.1.** Molecular orbital of triplet oxygen.

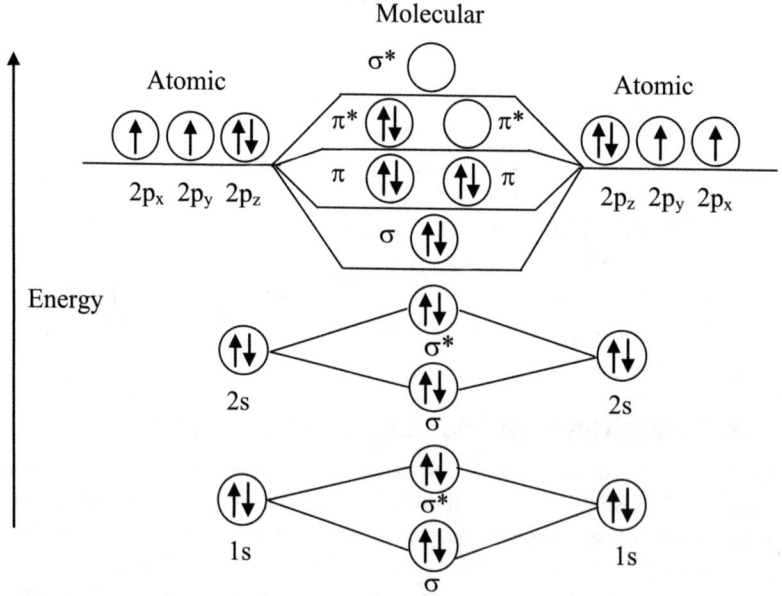

**Fig. 1.2.** Molecular orbital of singlet oxygen.

time before the addition of the second electron. Pauli's exclusion principle also states that electrons in a given orbital must have opposite spins. An electron in an atom or molecule generates two types of magnetic momentum and mechanical angular momentum. This behavior occurs from the motion of the electron around the nucleus and from the spin of the electron.

The spin multiplicity used to define spin states of molecules is defined as 2S + 1, where S is the total spin quantum number. The ground state of the molecule is singlet if the resultant spin (S) is 0, indicating the spin multiplicity to be 1. An excited state is formed by removing one of the electrons from the outermost filled orbital (bonding $\pi$) of the ground state to a vacant orbital (antibonding $\pi^*$) of the higher energy. The ground state of most stable molecules containing an even number of electrons is diamagnetic due to the arrangement of electrons into pairs with opposite spin magnetic moments. The ground state for most molecules is singlet state until a molecule is excited to the triplet state. However, molecular oxygen is exceptional among molecules with an even number of electrons. The electron configuration of triplet oxygen has two unpaired electrons occupying two degenerate antibonding $\pi^*$ orbitals (Fig. 1.1). The unpaired electrons in the antibonding $\pi^*$ orbitals can have the same spin, so the total spin number (S) is 1/2 + 1/2 = 1. The spin multiplicity (2S + 1) is 3, which is known as a triplet configuration because the spin has three possible alignments under magnetic field. Triplet oxygen is diradical and paramagnetic. Diradical triplet oxygen cannot react with food components, which are not radical compounds unless the food compounds become radical compounds. Triplet oxygen can react mostly with radicals.

Singlet oxygen having paired electrons is in violation of Hund's rule and is a highly energetic molecule. The resulting electronic repulsion can produce five excited state conformations. An activated $1\Delta$ state at 37.5 kcal above the ground state and an activated $1\Sigma$ state at 22.4 kcal above the ground state are two common states (Korycka-Dahl and Richardson, 1978; Girotti 1998). The $1\Sigma$ state of oxygen has two electrons with opposite spins in different orbitals this is very reactive and not able to survive relaxation to the ground state. The less energetic $1\Delta$ state of oxygen is sufficiently stable enough to react with other singlet state molecules. The $1\Delta$ state is responsible for most singlet oxygen oxidation and generally referred as singlet oxygen.

Singlet oxygen is not a radical compound and is very electrophilic. The energy of singlet oxygen is 22.4 kcal above the ground state of triplet oxygen and it exists long enough to react with other singlet state molecules (Korycka-Dahl and Richardson, 1978; Girotti, 1998). The lifetime of singlet oxygen is 50-700 microseconds depending on the solvent system of foods. Electrophilic singlet oxygen reacts with nonradical, singlet-state, and electron-rich compounds containing double bonds. Once singlet oxygen is formed, it is responsible for initiating singlet oxygen oxidation that rapidly produces free radicals that in turn can initiate a free-radical chain reaction. Table 1.1 shows a summary of chemical properties of triplet oxygen and singlet oxygen.

Table 1.1. Comparison of Triplet and Singlet Oxygen

|  | Triplet | Singlet |
|---|---|---|
| Energy Level | 0 | 22.4 kcal/mole |
| Nature | Diradical | Non-radical, Electrophilic |
| Reaction | Radical compound | Electron-rich compounds |

## Formation of Singlet Oxygen

Singlet oxygen can be formed chemically, enzymatically, and photochemically as shown in Fig. 1.3 (Choe and Min, 2005). Some mechanisms for singlet oxygen formation in Fig. 1.3 have not been unequivocally proven scientifically and have been questioned. Detailed studies of the many different mechanisms for the formation of singlet oxygen in foods should be studied further.

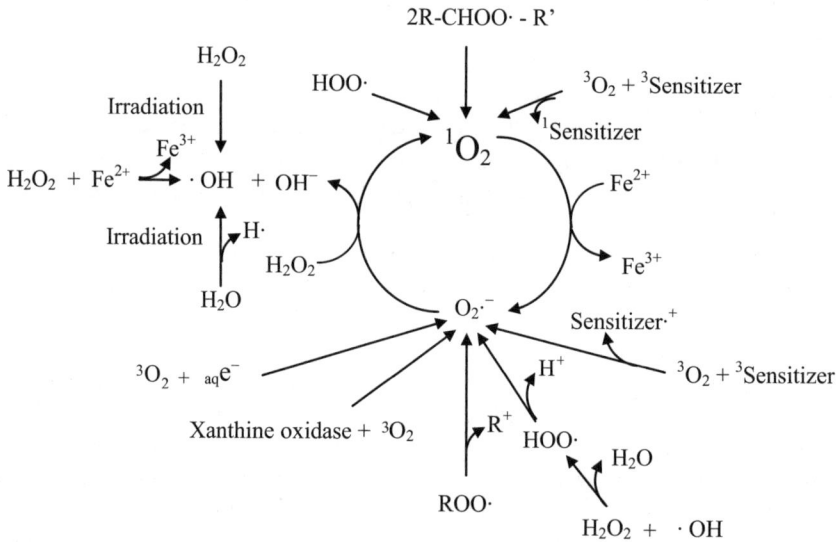

**Fig. 1.3.** Singlet oxygen formation by chemical, photochemical, and biological methods.

Singlet oxygen is produced from Haber-Weiss reaction (Kellogg and Fridovich, 1975). Superoxide anions formed from triplet oxygen produces hydrogen peroxide by spontaneous dismutation. Hydrogen peroxide reacts with a superoxide anion to form singlet oxygen by Haber-Weiss reaction (Halliwell and Gutteridge, 2001). Haber-Weiss reactions rarely occur in aqueous solution in the absence of transition metals, such as iron or copper, which catalyze the decomposition of hydrogen peroxide to the hydroxyl radical.

$$O_2^{\cdot -} + O_2^{\cdot -} + 2H^+ \xrightarrow{\text{Dismutation}} H_2O_2 + O_2$$

$$H_2O_2 + O_2^{\cdot -} \xrightarrow{\text{Haber-Weiss}} \cdot OH + OH^- + {}^1O_2$$

Singlet oxygen is also produced by the Russell mechanism from peroxy radicals (Halliwell and Gutteridge, 2001).

$$\underset{\underset{O\cdot}{\overset{|}{O}}}{\overset{|}{R\text{-}CH\text{-}R'}} + \underset{\underset{O\cdot}{\overset{|}{O}}}{\overset{|}{R\text{-}CH\text{-}R'}} \xrightarrow{\text{Russell}} \underset{\underset{H}{\overset{|}{O}}}{\overset{|}{R\text{-}CH\text{-}R'}} + \underset{\overset{\|}{O}}{\overset{|}{R\text{-}C\text{-}R'}} + {}^1O_2$$

The metastable phosphatidylcholine hydroperoxides produced singlet oxygen during hydroperoxide breakdown in the presence of $Cu^{2+}$ in the dark (Takayama et al., 2001). Retinal and retinyl palmitate in ethanol produced singlet oxygen by UV light (Delmelle, 1997).

The major pathway for the formation of singlet oxygen in foods is photosensitization. The interaction of light, photosensitizers, and triplet oxygen is the basis for the formation of singlet oxygen by photochemical mechanisms. The photosensitizers, such as chlorophyll, pheophytins, porphyrins, riboflavin, myoglobin, and synthetic colorants in foods, can absorb energy from light due to their conjugated double bonds and transfer it to triplet oxygen to form singlet oxygen (Foote et al., 1970; Afonso et al., 1999; Lledias and Hansberg, 2000). Once singlet oxygen is formed, it directly reacts with compounds that contain high densities of electrons, such as double bonds, and forms mixtures of conjugated and nonconjugated hydroperoxides that readily break down and produce undesirable compounds.

The effects of light on the flavor stability of food can be explained by both photolytic autoxidation and photosensitized oxidation. Photolytic autoxidation is the production of free radicals primarily from lipids during exposure to light. Direct interaction of UV or visible light and lipids in foods are minimal, and thus are not a primary concern. However, photosensitized oxidation occurs in the presence of photosensitizers and light very quickly. The photochemical mechanism begins with the absorption of light by photosensitizers, which depends on the arrangement of electrons around the atomic nuclei in its structure. As light energy is absorbed, an electron is boosted to a higher energy level and the ground singlet state of the sensitizer ($^1$Sensitizer) becomes an unstable excited singlet state ($^1$Sensitizer*). It takes only a few picoseconds for a chlorophyll molecule to absorb energy and become excited singlet chlorophyll (Foote, 1976). When light energy is removed, the electrons rapidly lose energy and return to the lower energy ground state. The excited singlet sensitizer may undergo three physical processes: internal conversion, emission of light, or intersystem crossing as illustrated by Jablonski's diagram (Fig. 1.4). Internal conversion involves the transformation of one excited state into another of the same spin state, resulting in

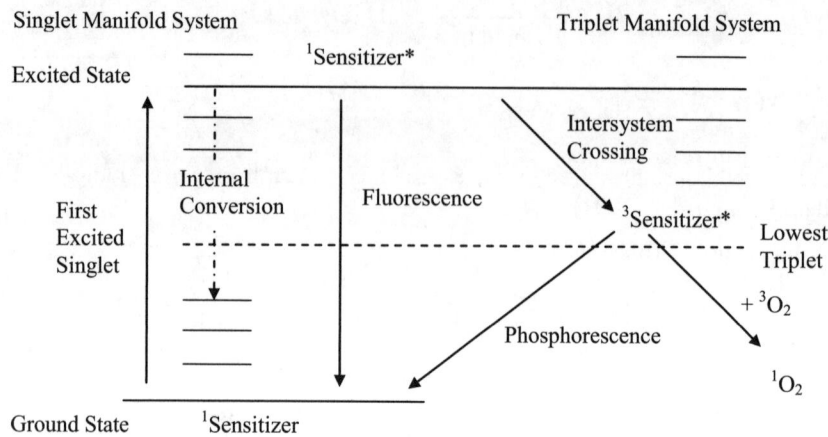

**Fig. 1.4.** Jablonski diagram.

loss of energy as heat. Not only can the excited singlet sensitizer lose energy as heat, but also it may emit light to the ground state. The nature of the emission light depends on the state multiplicity of the molecule. If the state from which the emission originates and terminates has the same state, the emission light is called fluorescence. Emission of fluorescence converts the excited singlet sensitizer to ground state singlet sensitizer, which are extremely fast processes.

The excited singlet sensitizer ($^1$Sensitizer*) becomes the excited triplet state ($^3$Sensitizer*) via intersystem crossing, which involves the conversion of the excited singlet state to the triplet state or vice versa. The excited triplet sensitizer is degraded to a lower, triplet-energy state that decays to the ground singlet state by the emission of light. Since the emission originates from the triplet state and terminates in a singlet state, the emitted light is called phosphorescence. The $^3$Sensitizer* is the reactive intermediate in photosensitized oxidation. The lifetime of the $^3$Sensitizer* is greater than the $^1$Sensitizer*. The excited triplet reacts with the same state of triplet oxygen ($^3O_2$) and forms singlet oxygen by triplet-triplet annihilation. The sensitizer returns to ground state and may begin the cycle again to generate singlet oxygen. Sensitizers may generate $10^3$-$10^5$ molecules of singlet oxygen before becoming inactive (Kochevar and Redmond, 2000).

## Type I and Type II Mechanisms

Once the $^3$Sensitizer* is formed, there are two major mechanisms it may undergo: Type I and Type II (Foote, 1976; Sharman et al., 2000). Type I is the direct interaction of the sensitizer with another molecule and is associated with the formation of free radicals. Type II is most commonly described as the transfer of energy from the

³Sensitizer* to atmospheric triplet oxygen to produce singlet oxygen.

The photosensitized oxidation process of the ³Sensitizer* is shown in Fig. 1.5. The Type I mechanism is characterized by hydrogen atom or electron transfer between an excited triplet sensitizer and a reducing substrate which results in the formation of free radicals (R·). The ³Sensitizer* acts as a photochemically activated free-radical initiator. The R· can react with radical triplet oxygen to form peroxy radicals which can abstract hydrogen from other compounds to produce oxidized compounds. The oxidized compounds readily break down to form free radicals that can initiate free-radical chain reactions. The ³Sensitizer* in the Type I mechanism can also react with $^3O_2$ to form superoxide anions by electron transfer to triplet oxygen. Less than 1% of the reaction of triplet sensitizer and triplet oxygen produces superoxide anion (Kepka and Grossweiner, 1972). The rate of Type I reactions is dependent on the type and concentration of the sensitizer and substrate. For example, the sensitizer benzophenone abstracts hydrogen from an alcohol 10,000 times faster than eosin; however, both eosin and benzophenone react at similar rates with more powerful reductants like N,N-dimethylaniline (Foote, 1976). Generally, the readily oxidizable compounds, such as phenols and amines or readily reducible compounds such as quinones, favor Type I mechanism (Korycka-Dahl and Richardson, 1978).

The ¹Sensitizer* usually does not have enough time to react with other molecules due to its short lifetime. The Type II mechanism is the energy transfer process in which the ³Sensitizer* reacts with triplet oxygen via triplet-triplet annihilation to generate singlet oxygen (Fig. 1.5). Energy is transferred from the high energy ³Sensitizer* to low energy triplet oxygen ($^3O_2$) to form high energy singlet oxygen ($^1O_2$) and low

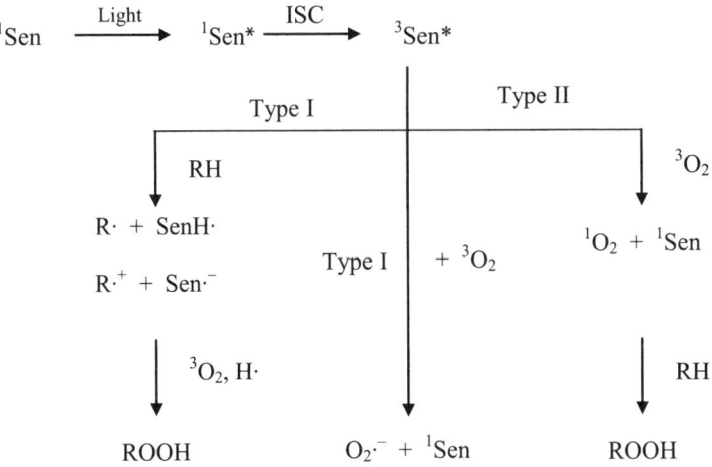

**Fig. 1.5.** Formation of excited triplet sensitizer (³sen*) and its reaction with substrate via Type I and Type II reaction.

energy ground state singlet sensitizer (Sharman et al., 2000). This reaction occurs very quickly and accounts for almost all of energy transfer from the $^3$Sensitizer* to triplet oxygen.

The competition between substrate and triplet oxygen for the $^3$Sensitizer* is a major determinant of whether the reaction mechanism is Type I or Type II. Electron-rich compounds, such as olefins and aromatic compounds, favor the Type II mechanism. The rate of Type II reaction mainly depends on the solubility and concentration of oxygen present in the food system. As the oxygen in a system becomes depleted, the shift from Type II to Type I mechanism is favored (He et al., 1998; Song et al., 1999). Oxygen is more soluble in lipids and nonpolar solvents than in water (Ke and Ackman, 1973). The trace amounts of chlorophyll as a sensitizer in vegetable oil tend to promote photosensitized oxidation by the Type II mechanism. In contrast, water-based food systems, such as milk, favor Type I mechanism due to the reduced availability of oxygen to interact with the $^3$Sensitizer*.

The natural decay rate of singlet oxygen to the ground state and the rate at which it reacts with a particular substrate must also be considered. The reactions of singlet oxygen with unsaturated fatty acids or other compounds depend on the lifetime of singlet oxygen. The lifetime of singlet oxygen is different depending on the type of solvent and ranges from 2 to 700 microseconds as shown in Table 1.2 (Adams and Wilkinson, 1972; Kearns, 1979). Temperature has little effect on the reaction rate of singlet oxygen oxidation, as indicated by negligible changes in the lifetime of singlet oxygen at a temperature range of -37.6 to 21.6°C (Min and Boff, 2002). This suggests that the lifetime of singlet oxygen in food systems is largely dependent on the type of matrix, water or fat, in foods.

## Detection of Singlet Oxygen

Direct demonstration of production and involvement of singlet oxygen in any reaction pathway is difficult due to its short lifetime. However, analytical techniques

Table 1.2. Lifetime of Singlet Oxygen in Solution

| Solvent | Lifetime (μs) |
| --- | --- |
| Protiated acetone | 46.5 |
| Deuterated acetone | 690 |
| Protiated acetonitrile | 54.4 |
| Deuterated acetonitrile | 600 |
| Protiated benzene | 26.7 |
| Deuterated benzene | 550 |
| Protiated water | 4 |
| Deuterated water | 68.1 |

have been developed to detect the singlet oxygen molecule. Singlet oxygen oxidation products are characteristically different from the products formed from a free-radical chain reaction. The chemical structures of photosensitized oxidized products can demonstrate the existence of singlet oxygen. For example, autoxidation only produces conjugated hydroperoxides, whereas singlet oxygen oxidation can produce nonconjugated and conjugated hydroperoxides. Singlet oxygen can be produced free of other reactive oxygen species and the reaction products of pure singlet oxygen oxidation can be compared with the experimental oxidation product. The following methods have been used to detect the singlet oxygen involvement in chemical reactions.

## Deuterium Oxide Effect

The physical and chemical properties of deuterium ($D_2O$ or $^2H_2O$) are similar to those of water ($H_2O$). The isotope effect of heavy hydrogen ($^2H$) is magnified further in biological systems, which are very sensitive to small changes in the solvent properties of water. The lifetime of singlet oxygen in $D_2O$ is about 13 times longer than that in $H_2O$ (Li, 1997). The longer lifetime of singlet oxygen in $D_2O$ allows for the detection of singlet oxygen in various reactions compared to $H_2O$. Therefore, the reaction rate of singlet oxygen in $D_2O$ should increase when a reaction in aqueous solution is dependent on singlet oxygen. The $D_2O$ effect is a result of the decay rate of singlet oxygen. The rate of photosensitized oxidation in $D_2O$ is 10 times higher than in $H_2O$ and would indicate the involvement of singlet oxygen in the reaction (Athar et al., 1988; Pecci et al., 2000). The effect on the ratio of the decay rate of singlet oxygen in $H_2O$ to $D_2O$ may actually be larger than 10 times.

Consider the following in order to study the effect of deuterium: the normal difference between pD and pH; the effect of H-D exchange on the active sites of the system under study; the possibility of different rates of free-radical reactions in $D_2O$ vs. $H_2O$; and different excited state lifetimes and intersystem crossing efficiencies in $D_2O$ vs. $H_2O$ (Foote, 1976; IUPAC, 2001). Simply observing an increase in the rate in $D_2O$ is not a specific test verifying the existence of singlet oxygen. This method has two important limitations: (1) if the isotope effect is to be observed, the lifetime of singlet oxygen must be limited by the solvent. If all the singlet oxygen is being scavenged by substrate, no effect will be observed. (2) The lifetime of the superoxide anion is also longer in $D_2O$ than in $H_2O$, so superoxide anion reactions should go more rapidly in $D_2O$ than in $H_2O$ (Foote, 1976).

## Chemical Traps

Chemical compounds can be used to trap singlet oxygen, forming a unique product that is indicative of the presence of singlet oxygen. *N,N,N',N'*-tetramethyl ethylenediamine and 1,3-diphenylisobenzofuran (DPBF) are known chemical trap agents of singlet oxygen (Kochevar and Redmond, 2000). The absorbance of DPBF in organic solvents measured at 410 nm decreases as the molecule reacts directly with singlet oxygen (Kochevar and Redmond, 2000).

Cholesterol has been used to distinguish between singlet oxygen oxidation and triplet oxygen oxidation. Cholesterol reacts with singlet oxygen to form specific oxidative products via the "ene" reaction. This specificity makes the use of cholesterol an effective indicator of singlet oxygen oxidation in situ, where the use of other detection techniques is difficult. The reaction of cholesterol with singlet oxygen produces 3β-hydroxy-5α-cholest-6-ene-5-hydroperoxide (5α-OOH), 3β-hydroxycholest-4-ene-6α-hydroperoxide (6α-OOH), and 3β-hydroxyc-holest-4-ene-6β-hydroperoxide (6β-OOH) as shown in Fig. 1.6. The reaction of cholesterol with triplet oxygen produces 3β-hydroxycholest-5-ene-7α-hydroperoxide (7β-OOH) and 3β-hydroxycholest-5-ene-7β-hydroperoxide (7β-OOH) (Girotti, 1998; Girotti and Korytowski, 2000). The specificity of cholesterol with triplet oxygen and singlet oxygen can differentiate between a singlet oxygen initiated reaction and a triplet oxygen initiated reaction. Girotti and Korytowski (2000) described in detail the use of cholesterol as an indicator of singlet oxygen.

## Quenchers

Inhibition of singlet oxygen reaction by quenchers, such as carotenoids, tertiary amines, tocopherols, histidine, 1,4-diazabicyclo[2,2,2]octane (DABCO), 2,5-dimethyl furan, and azide, has been used to detect singlet oxygen (Jung et al., 1991; Song et al., 1999; Bradley et al., 2006). All singlet oxygen quenchers have low ionization potential. Although none of the quenchers are specific for singlet oxygen, the quenchers physically or chemically interact with the singlet oxygen to inhibit oxidation. The calculation of a singlet oxygen quenching rate by quantitative analysis can effectively determine the characteristics of a quencher. The concentration that quenches half of the product can be given by the expression:

$$[Q] = (k_a[A] + k_d)/k_q$$

where $k_a$ is the rate of reaction with acceptor, [A] is the concentration of acceptor, $k_d$ is the decay rate of singlet oxygen in the medium, and $k_q$ is the quenching rate of the quencher. DABCO requires about 50 mM in aqueous solution to quench half of the singlet oxygen formed (Bradley et al., 2006).

Azide and histidine are commonly used to determine the singlet oxygen oxidation of compounds as these agents act as quenchers of singlet oxygen and greatly suppress the activity and the consumption of singlet oxygen (Song et al., 1999). Singlet oxygen quenching by azide is thought to be a charge transfer process in which molecular triplet oxygen is released after the reaction, therefore no oxygen is consumed. Further differentiation may be accomplished by using specific quenchers. Several quenching agents have specificity towards reactive oxygen species. DABCO is also an efficient scavenger of the hydroxyl radical. The azide ion is much more effective at quenching singlet oxygen than DABCO, but it is also known as hydroxyl radical scavenger.

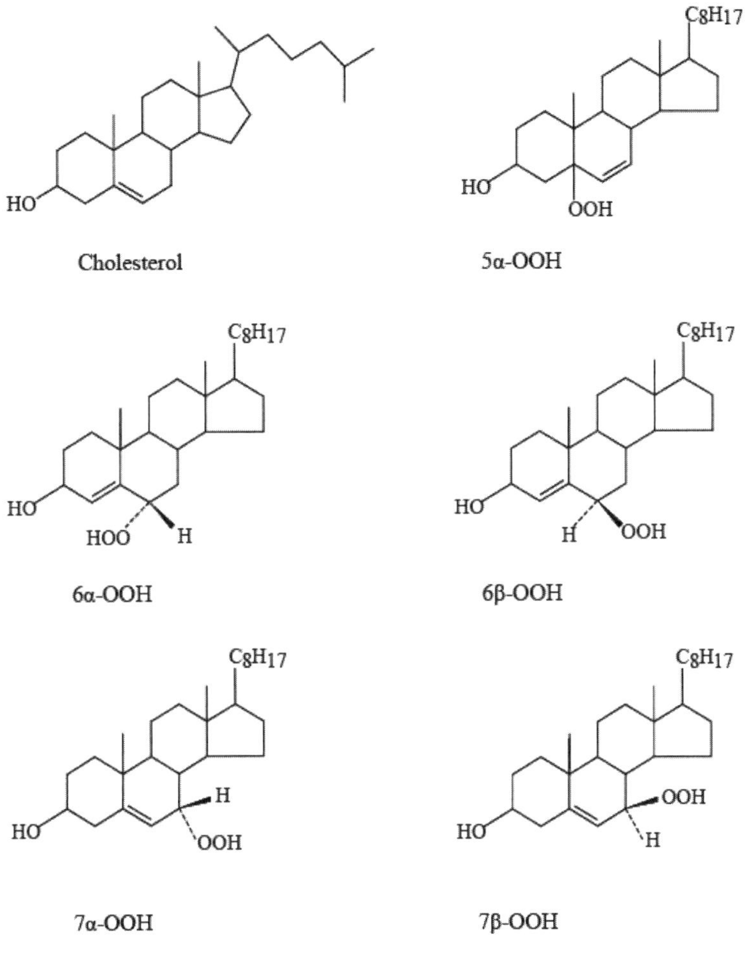

**Fig. 1.6.** Cholesterol oxidation products by singlet oxygen: 3β-hydroxy-5α-cholest-6-ene-5-hydroperoxide (5α-OOH), 3β-hydroxycholest-4-ene-6α-hydroperoxide (6α-OOH), and 3β-hydroxycholest-4-ene-6β-hydroperoxide (6β-OOH) and cholesterol oxidation products by triplet oxygen: 3β-hydroxycholest-5-ene-7α-hydroperoxide (7β-OOH) and 3β-hydroxycholest-5-ene-7β-hydroperoxide (7β-OOH) (Girotti 1998).

## Chemiluminence

Chemiluminenscence is produced when a chemical reaction yields an electronically excited species that emits light as it returns to its ground state. The chemiluminescence method measures very weak luminescence like emission excited by the reaction with singlet oxygen. Singlet oxygen is detected by its chemiluminescence at 1270 nm (Kanofsky and Axelrod, 1986; Macpherson et al., 1993; Darmanyan and Jenks, 1998). The energy differential between singlet oxygen and ground-state oxygen can be released as a 1270 nm photon, which is very specific to singlet oxygen. The weak emission from singlet oxygen is usually accompanied by light of other wavelengths in complex systems. Although these emissions are assigned to higher vibrational states of singlet oxygen, it seems likely that the extraneous emission comes from excited carbonyls or other species in the system. Precise wavelength determination is essential if this method is used for the detection of singlet oxygen.

## ESR Spectroscopy

Electron spin resonance (ESR) spectroscopy is a highly sensitive analytical method that detects the presence of free radicals. Although singlet oxygen is not a radical, various amines can interact with singlet oxygen forming nitroxide radicals that are readily detected by ESR. There is a high specificity of amines for singlet oxygen, and the existence of a charge-transfer complex between the amine and singlet oxygen has been proposed (Lion et al., 1976). The analytical method was developed in which stable nitroxide radicals were generated by reaction with singlet oxygen. Amine compounds, such as 2,2,6,6-tetramethly-4-piperidone (TMPD), can react with singlet oxygen to form a stable nitroxide radical adduct, 2,2,6,6-tetramethyl-4-piperidone-N-oxyl (TAN), which can be detected by ESR (Whang and Peng, 1988; Sharman et al., 2000). When photosensitized oxidation occurs in the presence of the amine, free nitroxide radicals will be produced. The reaction of TMPD with singlet oxygen for the formation of TAN is shown in Fig. 1.7 (Bradley et al., 2003). Although other reactive oxygen species, such as superoxide anion and hydroxy radical, can react with TMPD, they do not convert TMPD to TAN. This method is highly specific to singlet oxygen (Ando et al., 1997). Ando et al. (1997) further confirmed this specificity by observing an inhibition of TAN formation upon the addition of two known singlet oxygen scavengers, such as sodium azide and histidine, and not upon the addition of a hydroxy radical scavenger, such as dimethyl sulfoxide, or a superoxide anion scavenger, such as superoxide dismutase.

ESR spectroscopy detected the formation of singlet oxygen in meat (Whang and Peng, 1988) and milk (Bradley et al., 2003) using a spin-trapping technique. Effects of 0, 5, and 15 min illumination on an electron spin resonance spectrum of a TAN radical in a water solution of riboflavin and TMPD is shown in Fig. 1.8 (Min and Lee, 1996). Three hyperfine lines are the result of the coupling of the unpaired electron with an atom of nitrogen. The use of different spin-trapping compounds may be utilized to differentiate reactive oxygen species by ESR spectroscopy. He et al. (1997)

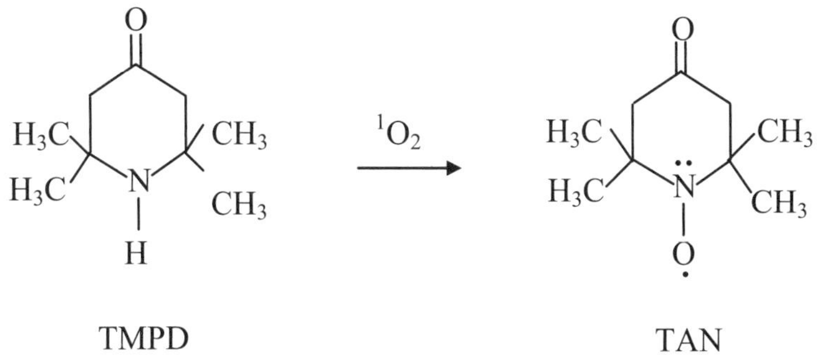

**Fig. 1.7.** Reaction of 2,2,6,6-tetramethyl-4-piperidone with singlet oxygen to form 2,2,6,6,4-peperidone-n-oxyl (Bradley et al. 2003).

used ESR to detect superoxide anion and hydroxyl radicals using 5,5-dimethyl-1-pyrroline-N-oxide as a spin-trapping agent.

The limitation of this method is that concentration of TAN should remain over $10^{-8}$ M for detection and over $10^{-6}$ M for good spectral resolution. Since the lifetime of singlet oxygen is less than 1 microsecond, steady-state concentrations greater than $10^{-7}$ M are rarely maintained. Coupling spin-trapping with ESR spectroscopy could improve this technique by diminishing the rate of disappearance.

Current singlet oxygen detection methods under investigation are the quantitative determination using laser deflection calorimetry (Schneider et al., 2000) and time-resolved singlet oxygen detection (Nonell and Braslavsky, 2000). Laser deflection calorimetry is very selective in the determination of singlet oxygen in heterogeneous systems (Schneider et al., 2000). Future advances in time-resolved singlet oxygen detection should include the detection of photosensitized and non-photosensitized production of singlet oxygen in heterogeneous systems such as foods (Nonell and Braslavsky, 2000). Andersen and Ogilby (2001) recorded the time-resolved absorption spectrum of singlet oxygen using transmission microscopy.

## Reaction of Singlet Oxygen with Fatty Acids

The oxidation rate of food depends on temperature, the presence of inhibitors or catalysts, the nature of the reaction environment, and the nature of the compounds. These factors are important in varying degrees to both singlet and triplet oxygen oxidation in foods. Temperature has little effect on singlet oxygen oxidation but has a significant effect on triplet oxygen oxidation, which requires high activation energy. Polyunsaturated fatty acids are more susceptible to radical-initiated triplet oxygen oxidation than monounsaturated fatty acids, a property that is primarily due to the lowered activation energy in the initiation of free-radical formation in polyunsaturated fatty acids compared to monounsaturated fatty acids. The type of polyunsaturated fatty acids is not important in singlet oxygen oxidation. The total numbers of double bonds is more important in singlet oxygen oxidation than the types of double bonds, such as nonconjugated or conjugated dienes or trienes, which are important in free-radical triplet oxygen oxidation.

Electrophilic singlet oxygen is seeking electrons to fill its highest degenerate vacant molecular orbital (Fig. 1.2). One of the most important reaction characteristics of singlet oxygen is that it can directly react with the electron-rich double bonds of unsaturated molecules (Adam, 1975; Beutner et al., 2000).

Singlet oxygen participates in reactions such as 1,4-cycloaddition to diene and heterocyclic compounds, "ene" reaction, and 1,2-cycloaddtion to olefins. All involve

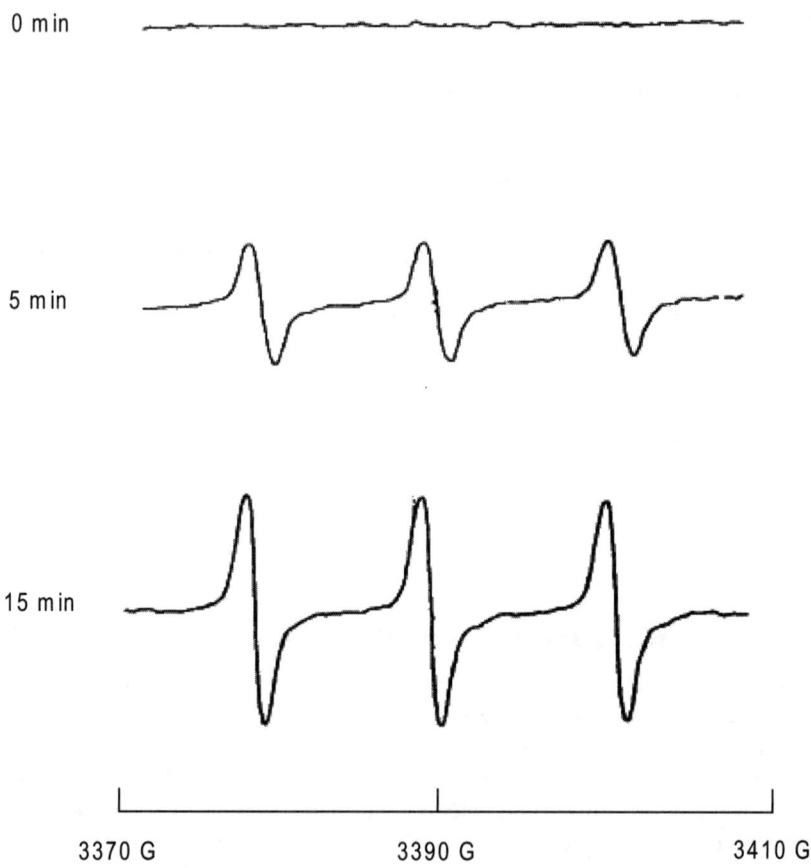

**Fig. 1.8.** Effects of 0, 5, and 15 minutes illumination on electron spin resonance spectrum of 2,2,6,6-4-piperidone-n-oxyl in water solution of riboflavin and 2,2,6,6-tetramethyl-4-piperidone (Min and Lee 1996).

direct reaction with double bonds as shown in Fig. 1.9. When singlet oxygen reacts with linoleic or linolenic acid, they form both conjugated and nonconjugated diene hydroperoxides (Fig. 1.10). The linoleic or linolenic acid reacts with triplet oxygen and produces only conjugated diene hydroperoxides (Fig. 1.10). The direct reaction of singlet oxygen with double bonds permits the formation of hydroperoxides at positions 10 and 12 in linoleic acid and 10 and 15 in linolenic acid, which do not form in triplet oxygen oxidation as shown in Table 1.3 (Frankel et al., 1979). These properties may be used to determine singlet oxygen activity in lipids (Stratton and Lieber, 1997) and produce compounds in the absence of triplet oxygen oxidation.

The reaction rates of singlet oxygen with oleic, linoleic, linolenic, and arachidonic acids are 0.74, 1.3, 1.9, and 2.4 × $10^5$ $M^{-1}S^{-1}$, respectively. The rate is relatively proportional to the number of double bonds in the fatty acid instead of the type of double bond, such as conjugated or nonconjugated double bonds, (Doleiden et al., 1974). The reactivity increases as the ionization energy decreases due to the presence of alkyl groups adjacent to the C=C double bond. However, steric hindrance to the double bond will lower the reactivity (Beutner et al., 2000).

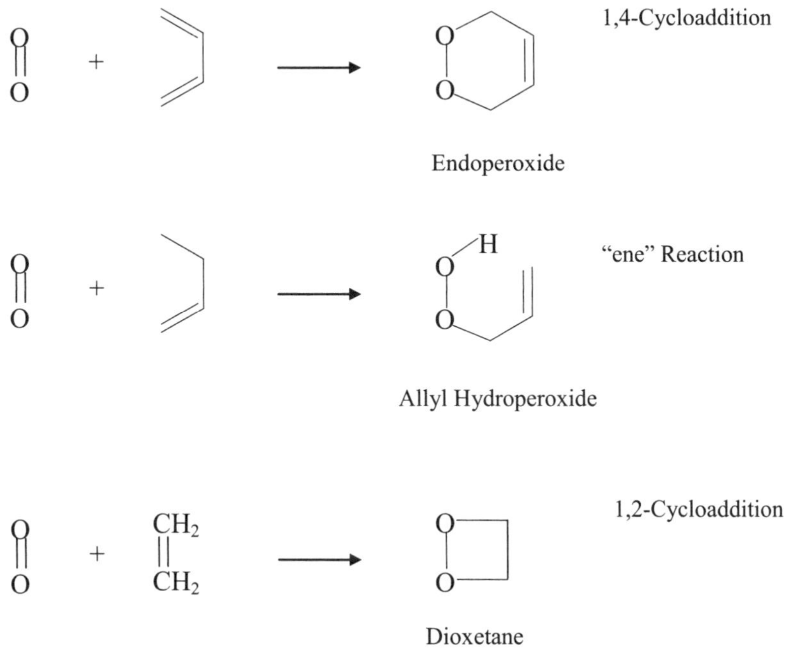

**Fig. 1.9.** Reactions of singlet oxygen with olefins by 1,4-cycloaddition, "ene" reaction, and 1,2-cycloaddtion (Min and Boff 2002).

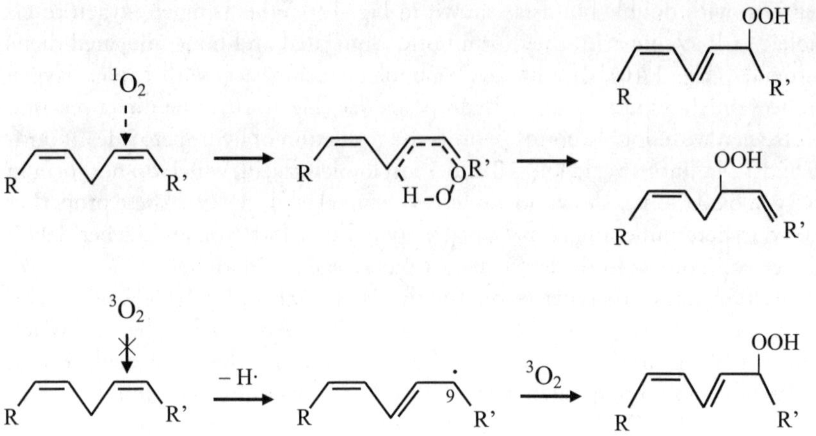

**Fig. 1.10.** Conjugated and nonconjugated hydroperoxide formation from a diene fatty acid by singlet oxygen and triplet oxygen.

Diradical triplet oxygen can react with radical food compounds. However, food compounds are not radical compounds and they should be in a radical state to react with triplet oxygen. The initiation of radical formation in food occurs at the carbon that requires the least energy to remove a hydrogen atom. The removal of hydrogen from a saturated fatty acid requires approximately 100 kcal/mol of energy. The carbon-hydrogen bond on carbon 8 or 14, which is α to the double bond of linoleic acid is about 75 kcal/mole. Hydrogen at position 11 of linoleic acid is most easily removed due to the presence of a double bond on both sides and requires only about 50 kcal/mol. The various strengths of hydrogen-carbon bond of fatty acids explain the differences of oxidation rates of stearic, oleic, linoleic, and linolenic acids during oxidation by triplet oxygen. Heat, light, metals, and reactive oxygen species facilitate the radical formation of food components. Once the hydrogen is removed, a pentadienyl radical intermediate between carbon 9 and carbon 12 of linoleic acid is formed. The pentadienyl radical provides an equal mixture of conjugated 9- and 13-diene radical and produces 9- and 13-conjugated diene hyroperoxides upon reaction with triplet oxygen (Table 1.3).

Triplet oxygen autoxidation produces only the conjugated diene hydroperoxides in linoleic and linolenic acids. The relative reaction ratios of triplet oxygen with oleic, linoleic, and linolenic acid for hydroperoxide formation are 1:12:25, which is dependent on the relative difficulty for the radical formation in the molecule (Min and Lee, 1996). The reaction rate of triplet oxygen with linolenic acid is about twice as fast as that of linoleic acid because linolenic acid has 2 pentadienyl groups in the molecule, compared to linoleic acid with 1 pentadienyl group.

Singlet oxygen oxidation produces hydroperoxides at the positions of double bonds without migration (Table 1.3). Gunston (1986) compared triplet oxygen oxidation and photosensitized singlet oxygen oxidation rate of oleic, linoleic, and linolenic acids as shown in Table 1.4. The reaction rate of singlet oxygen with oleic, linoleic, and linolenic acid is about 1,000~30,000 times faster than that of the triplet oxygen.

## Singlet Oxygen Oxidation in Foods

The effect of light energy has long been known to play a role in the flavor stability of vegetable oils and other fat-containing products, such as margarine, butter, and mayonnaise. Most food pigments and colorants are potential initiators of singlet oxygen oxidation in foods, due to their ability to absorb light in the visible light range, exhibit fluorescence and phosphorescence, reflecting both a singlet and triplet state, and have a high quantum yield of a long-lived triplet state. Photosensitizers include synthetic

**Table 1.3.** Hydroperoxides of Fatty Acids Formed by Triplet and Singlet Oxygen Oxidation

|  |  | Oleate | Linoleate | Linolenate |
|---|---|---|---|---|
| $^3O_2$ | | 8-OOH<br>9-OOH<br>10-OOH<br>11-OOH | | |
| | Conjugated hydroperoxides | | 9-OOH<br>13-OOH | 9-OOH<br>12-OOH<br>13-OOH<br>16-OOH |
| $^1O_2$ | | 9-OOH<br>10-OOH | | |
| | Conjugated hydroperoxides | | 9-OOH<br>13-OOH | 9-OOH<br>12-OOH<br>13-OOH<br>16-OOH |
| | Nonconjugated hydroperoxides | | 10-OOH<br>12-OOH | 10-OOH<br>15-OOH |

Frankel et al., (1979)

**Table 1.4.** Relative Oxidation Rates of Triplet and Singlet Oxygen with Oleate, Linoleate, and Linolenate

|  | Oleate | Linoleate | Linolenate |
|---|---|---|---|
| Triplet Oxygen | 1 | 27 | 77 |
| Singlet Oxygen | 30,000 | 40,000 | 70,000 |

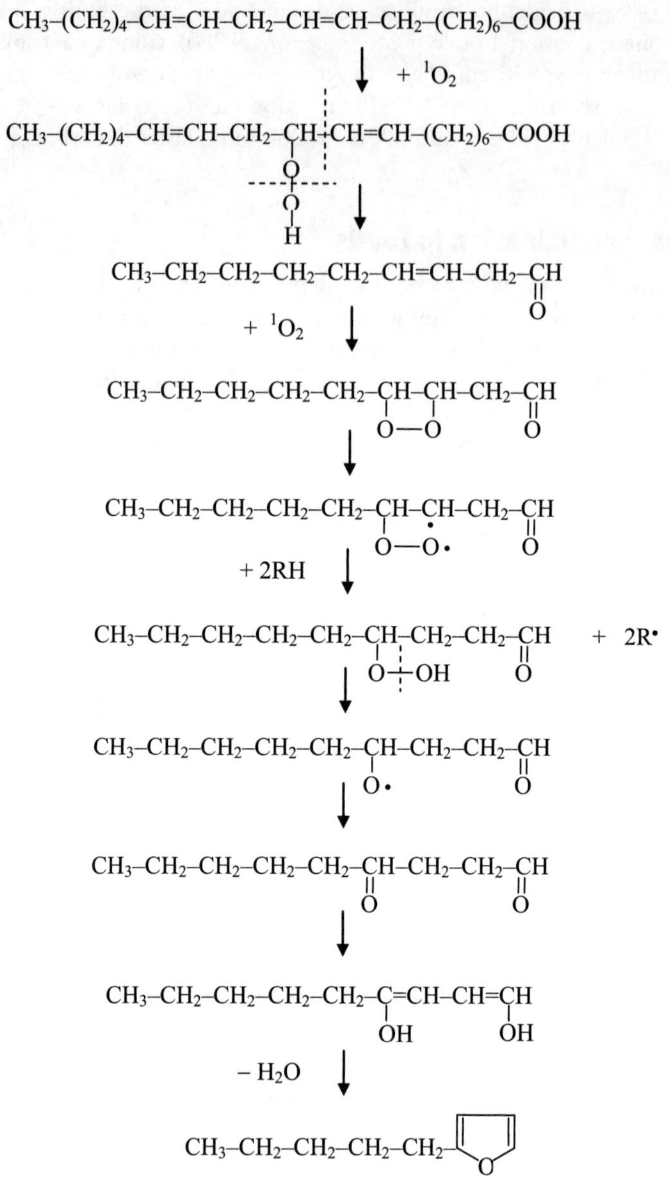

$CH_3-(CH_2)_4-CH=CH-CH_2-CH=CH-CH_2-(CH_2)_6-COOH$

$\downarrow + {}^1O_2$

$CH_3-(CH_2)_4-CH=CH-CH_2-CH-CH=CH-(CH_2)_6-COOH$
                              |
                              O
                              |
                              O
                              |
                              H

$\downarrow$

$CH_3-CH_2-CH_2-CH_2-CH_2-CH=CH-CH_2-CH=O$

$+ {}^1O_2 \downarrow$

$CH_3-CH_2-CH_2-CH_2-CH_2-CH-CH-CH_2-CH=O$
                         |   |
                         O—O

$\downarrow$

$CH_3-CH_2-CH_2-CH_2-CH_2-CH-CH-CH_2-CH=O$
                         |   |
                         O—O•

$+ 2RH \downarrow$

$CH_3-CH_2-CH_2-CH_2-CH_2-CH-CH_2-CH_2-CH=O \quad + \quad 2R^\bullet$
                         |
                         O—OH

$\downarrow$

$CH_3-CH_2-CH_2-CH_2-CH_2-CH-CH_2-CH_2-CH=O$
                         |
                         O•

$\downarrow$

$CH_3-CH_2-CH_2-CH_2-CH_2-CH-CH_2-CH_2-CH=O$
                         ||
                         O

$\downarrow$

$CH_3-CH_2-CH_2-CH_2-CH_2-C=CH-CH=CH$
                         |           |
                         OH          OH

$- H_2O \downarrow$

$CH_3-CH_2-CH_2-CH_2-CH_2-\underset{O}{\text{furan}}$

**Fig. 1.11.** Mechanism for the formation of 2-pentyl furan from linoleic acid by singlet oxygen (Min et al. 2003).

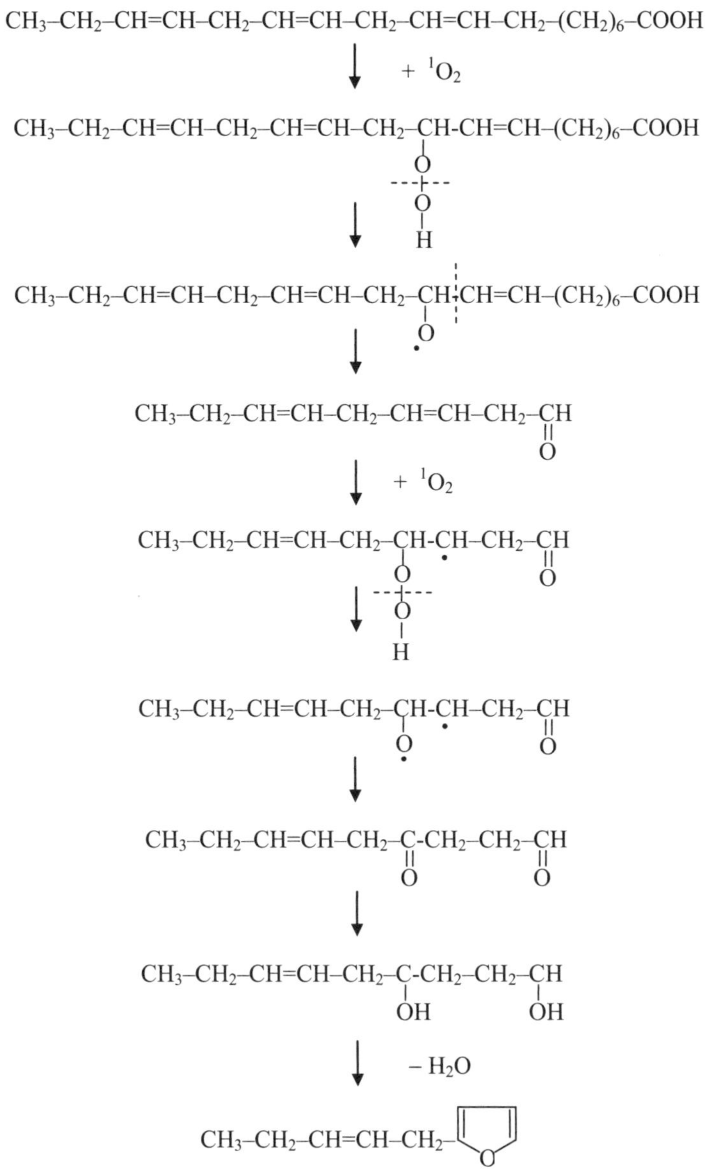

$$CH_3-CH_2-CH=CH-CH_2-CH=CH-CH_2-CH=CH-CH_2-(CH_2)_6-COOH$$

$$\downarrow + {}^1O_2$$

$$CH_3-CH_2-CH=CH-CH_2-CH=CH-CH_2-\underset{\underset{H}{\overset{|}{O}}}{\overset{\overset{O}{|}}{CH}}-CH=CH-(CH_2)_6-COOH$$

$$\downarrow$$

$$CH_3-CH_2-CH=CH-CH_2-CH=CH-CH_2-\underset{\overset{\bullet}{O}}{CH}\vdots CH=CH-(CH_2)_6-COOH$$

$$\downarrow$$

$$CH_3-CH_2-CH=CH-CH_2-CH=CH-CH_2-\underset{O}{\overset{\|}{CH}}$$

$$\downarrow + {}^1O_2$$

$$CH_3-CH_2-CH=CH-CH_2-\underset{\underset{H}{\overset{|}{O}}}{\overset{\overset{O}{|}}{CH}}-\overset{\bullet}{CH}-CH_2-\underset{O}{\overset{\|}{CH}}$$

$$\downarrow$$

$$CH_3-CH_2-CH=CH-CH_2-\underset{\overset{\bullet}{O}}{\overset{|}{CH}}-\overset{\bullet}{CH}-CH_2-\underset{O}{\overset{\|}{CH}}$$

$$\downarrow$$

$$CH_3-CH_2-CH=CH-CH_2-\underset{O}{\overset{\|}{C}}-CH_2-CH_2-\underset{O}{\overset{\|}{CH}}$$

$$\downarrow$$

$$CH_3-CH_2-CH=CH-CH_2-\underset{OH}{\overset{|}{C}}-CH_2-CH_2-\underset{OH}{\overset{|}{CH}}$$

$$\downarrow -H_2O$$

$$CH_3-CH_2-CH=CH-CH_2-\underset{O}{\bigcirc}$$

**Fig. 1.12.** Mechanism for the formation of 2-pentenyl furan from linolenic acid by singlet oxygen (Min et al. 2003).

dyes; naturally occurring pigments, such as chlorophyll, flavin, and porpyrin; coenzymes; and biochemical compounds, such as pyridoxals and psoralens; metallic salts; and transition metal complexes, such as ruthenium bipyridine.

Photooxidation of vegetable oils is a major concern of the food industry, in that they contain natural photosensitizers and are commercially sold under light. Salad dressings containing 30-40% vegetable oil account for 35% of the production of all dressings, mayonnaise, and sandwich spreads. Soybean oil has a 90% share of the prepared dressings market. Not only is soybean oil susceptible to oxidation due to the high concentration of linoleic acid, but it contains 1-5 ppm chlorophyll (Brekke, 1980), which is an excellent singlet oxygen sensitizer. Min and Lee (1988) reported that headspace volatile compounds of soybean oil increased as added chlorophyll increased from 0, 2, 4, and 6 ppm to 8 ppm. Soybean oil containing no chlorophyll, which was removed by silicic acid liquid column chromatography, did not produce headspace volatile compounds under light under identical experimental conditions at 10°C. However, the effects of 0, 2, 4, 6, and 8 ppm added chlorophyll did not have any effect on the formation of volatile compounds of soybean oil in dark storage. The formation of headspace volatile compounds in the soybean oil decreased inversely with the amount of added β-carotene, which quenches singlet oxygen (Lee and Min, 1990). The very rapid formation of volatile compounds in the soybean oil in the presence of chlorophyll, light, and oxygen was due to the singlet oxygen oxidation.

The oxidative deterioration of virgin olive oil sold as an unrefined green liquid is related to the amount of chlorophyll contained in the oil. Olive oil in its natural state contains chlorophyll, carotenes, tocopherols, and other phenolic compounds. The presence of 6 ppm chlorophyll acted as a photosensitizer, resulting in rapid oxidation of the oil during exposure to fluorescent light. The presence of β-carotene and nickel chelates both substantially inhibited oxidation in the first hours of illumination, thus supporting the concept that chlorophyll brings about formation of singlet oxygen, which is quenched by β-carotene and nickel chelates.

Chlorophylls and their decomposition products in vegetable oils are potential photosenstizers generating singlet oxygen in the presence of light and triplet oxygen. Singlet oxygen rapidly reacts with unsaturated fatty acids to produce a mixture of conjugated and nonconjugated hydroperoxides that rapidly decompose to produce undesirable flavor compounds. Most studies have been done in model systems that consist of one or two free fatty acids exposed to light in the presence of a sensitizer.

## Reversion Flavor in Soybean Oil

Soybean oil represents about 70% of all edible fats and oils consumed in the United States (Golbitz, 2000). Soybean oil is inexpensive and widely available compared to other edible oils. The development of reversion flavor, described as beany or grassy, is a unique defect to soybean oil and can be formed in soybean oils, which have low peroxide values (Ho et al., 1978). To improve the flavor stability and quality of soybean oil, reversion flavor has been extensively studied in soybean oil since 1936 (Ho et al., 1978).

Smouse and Chang (1967) identified 2-pentyl furan in reverted soybean oil and reported that it significantly contributed to the reversion flavor of soybean oil. Chang et al. (1983) isolated and identified all 4 2-pentenyl-furan isomers in reverted soybean oil.

Sensory evaluation showed that the addition of 2 ppm 2-pentyl furan to freshly deodorized and bland soybean oil produced the "reverted" soybean oil flavor. The addition of 2 ppm 2-pentyl furan to deodorized cottonseed oil and corn oil also produced reversion flavor found in reverted soybean oil (Chang et al., 1966). Ho et al. (1978) reported that 2-(1-pentenyl) furan contributed to reversion flavor. Smagula et al. (1978) reported that 2-(2-pentenyl) furan is also a contributor according to sensory evaluation. Flavor thresholds of the 2-pentenyl furan isomers were between 0.25 and 6 ppm.

Smouse and Chang (1967) and Ho et al. (1978) proposed the mechanisms for the formation of 2-pentyl furan from linoleic acid and 2-pentenyl furan isomers from linolenic acid using triplet oxygen, respectively. The proposed mechanisms for the formation of 2-pentyl furan from linoleic acid and isomers of 2-pentenyl furan from linolenic acid by triplet oxygen have been questioned. The formation of both the 2-pentyl furan and 2-pentenyl furan requires a hydroperoxide at carbon 10 of linoleic acid. The formation of 10-hydroperoxide in linoleic or linolenic acids by free-radical triplet oxygen oxidation is highly improbable but is very common in the singlet oxygen oxidation of linoleic or linolenic acids as shown in Table 1.2. Min et al. (2003) reported on the chemical mechanisms for the formations of 2-(2-pentenyl) furan from linoleic acid and 2-pentyl furan formation from linolenic acid using singlet oxygen as shown in Figs 1.11 and 1.12, respectively.

Min et al. (2003) identified 2-pentylfuran and 2-pentenyl furan in soybean oil containing 5 ppm chlorophyll b during storage under light for 96 hrs. The formation of 2-pentyl and 2-pentenyl furan increased with increasing storage time and concentration of chlorophyll. 2-Pentyl furan and 2-pentenyl furan were formed only in the presence of light and chlorophyll in soybean oil. Soybean oil containing 5 ppm chlorophyll did not produce 2-pentyl furan and 2-pentenyl furan during dark storage. The chlorophyll-free soybean oil obtained by silicic acid chromatography did not produce either 2-pentyl furan or 2-pentenyl furan during light storage. The singlet oxygen oxidation reaction is involved in the formation of 2-pentyl furan and 2-pentenyl furan, which have been reported to be mainly responsible for reversion flavor in soybean oil. The chlorophyll, which is an excellent photosensitizer, in soybean oil should be carefully removed during the oil processing to minimize the formation of reversion during storage.

## Singlet Oxygen Quenching Mechanisms

Other than exclusion of light and reduction of oxygen present, the use of quenching agents is the best way to reduce singlet oxygen oxidation. Natural food components, such as tocopherols, carotenoids, and ascorbic acid, can act as effective singlet oxygen quenchers (Lee et al., 2004). Quenching agents may be involved to minimize the development or activity of singlet oxygen at several stages in the oxidation of foods. Figure 1.13 shows the development of singlet oxygen and its subsequent reaction with substrate (A) to form the oxidized product ($AO_2$).

At every stage in this reaction, there is at least one alternate route, which, if taken, would minimize the oxidation of the compound (A). The first step represents

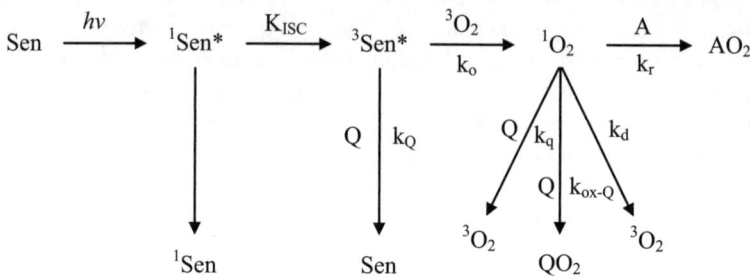

**Fig. 1.13.** Formation of singlet oxygen and its reaction with substrate a to produce the oxidized product $AO_2$.

the return of the $^1$Sensitizer* to $^1$Sensitizer without intersystem crossing which can form the $^3$Sensitizer*. The second represents a reaction with a quencher (Q) at a rate $K_Q$, returning the $^3$Sensitizer* to $^1$Sensitizer prior to reaction with triplet oxygen. The $^3$Sensitizer* may react with triplet oxygen ($^3O_2$) to form singlet oxygen ($^1O_2$). Following its creation, there are three fates for singlet oxygen in food: (1) it may naturally decay to the ground state at a rate $k_d$; (2) it may react with a singlet state compound (A) at a rate $k_r$ forming the oxidized product ($AO_2$); and (3) it may be destroyed by a quencher, either chemically at a rate $k_{ox-Q}$ to form the product $QO_2$, or physically at a rate $k_q$ to return to free triplet oxygen.

There are three points at which a quencher may act (Fig. 1.13): one is quenching of the $^3$Sensitizer*, and the other two are quenching of singlet oxygen by either chemically or physically. Chemical quenching involves the reaction of singlet oxygen with the quencher to produce an oxidized product ($QO_2$). Physical quenching results in the return of singlet oxygen to triplet oxygen without the consumption of oxygen or the formation of oxidized products by either energy transfer or charge transfer. Therefore, triplet oxygen quenchers must either be able to donate electrons or to accept 22.4 kcal above ground state. An example of the latter is β-carotene which has a low singlet energy state and can therefore accept the energy from singlet oxygen (Lee and Min, 1990). Ascorbic acid is an example of a chemical that can quench the excited sensitizer. Table 1.5 lists quenching rates of several antioxidants.

## Quenching Mechanism of Carotenoids

Carotenoids, which are responsible for the yellow and red colors of many plants and

**Table 1.5.** Singlet Oxygen Quenchers and Their Quenching Rates

| Quenching Compound | Quenching Rate (M$^{-1}$sec$^{-1}$) |
| --- | --- |
| β-Apo-8'-carotenal | 3.1 × 10$^9$ |
| β-Carotene | 4.6 × 10$^9$ |
| Lutein | 5.7 × 10$^9$ |
| Zeaxanthin | 6.8 × 10$^9$ |
| Lycopene | 6.9 × 10$^9$ |
| Isozeqaxanthin | 7.4 × 10$^9$ |
| Astaxanthin | 9.9 × 10$^9$ |
| Canthaxanthin | 11.2 × 10$^9$ |
| α-Tocopherol | 2.7 × 10$^7$ |
| 1,4-Diazabicyclo-(2,2,2)-octane | 1.5 × 10$^7$ |
| Dimethylfuran | 2.6 × 10$^7$ |
| *Bis*(di-n-butyldithiocarbamato)nickel chelate | 1.2 × 10$^9$ |
| {2,2'-Thiobis(4-1,1,3,3,-tetramethylbutyl) phenalto)}-*n*-butylamine)nickel chelate | 3.7 × 10$^7$ |

Min and Lee (1996), Li et al., (1998)

animal products, have been known to minimize singlet oxygen oxidation (Forse, 1979; Lee and Min, 1990). Carotenoids include a class of hydrocarbons called carotenes and their oxygenated derivatives called xanthophylls. Foote (1976) found that one molecule of β-carotene can quench 250 to 1000 molecules of singlet oxygen at a rate of $1.3 \times 10^{10}$ M$^{-1}$S$^{-1}$.

Energy transfer mechanism is responsible for the minimization of singlet oxygen oxidation of lipids by β-carotene (Forss, 1979). Electron excitation energy is transferred from singlet oxygen to singlet state carotenoid ($^1$CAR), producing triplet state carotenoid ($^3$CAR) and triplet oxygen, which is called singlet oxygen quenching. Energy is also transferred from $^3$Sensitizer* to the $^1$CAR, which is called triplet sensitizer quenching. The $^3$CAR can easily return to the $^1$CAR dissipating energy as a heat.

$$^1O_2 + {}^1CAR \rightarrow {}^3O_2 + {}^3CAR$$
$$^1CAR + {}^3Sensitizer^* \rightarrow {}^3CAR + {}^1Sensitizer$$
$$^3CAR \rightarrow {}^1CAR$$

The energy transfer from singlet oxygen (22.4 kcal/mole) to carotenoids with nine or more conjugated double bonds (<22.4 kcal/mole) is exothermic (Foote, 1970). Foote (1970) reported that the carotenoid with seven conjugated double bonds was effective at quenching triplet chlorophyll. Carotenoids with fewer than nine conjugated double bonds have energies above that of singlet oxygen and are less efficient singlet oxygen quenchers. Carotenoids with eleven or more conjugated double bonds quench at a diffusion-controlled rate of singlet oxygen.

The rate of singlet oxygen quenching by carotenoids is dependent on the number of conjugated double bonds and the type and number of functional groups on the ring portion of the molecule. This is important in the solubility of carotenoids. Kobayashi and Sakamoto (1999) compared the quenching activity of β-carotene to astaxanthin and found that the quenching activity of astaxanthin decreased with increasing hydrophobicity, while the quenching activity of β-carotene increased. Lee and Min (1990) evaluated the effectiveness of five carotenoids including lutein, zeazanthin, lycopene, isozeaxanthin, and astaxanthin in quenching chlorophyll-photosensitized oxidation of soybean oil and reported that the effectiveness increased with the number of double bonds in the carotenoid and the concentration of carotenoid added.

## Quenching Mechanisms of Tocopherols

Tocopherols are the most abundant and prevalent antioxidants in nature and are studied as free-radical scavengers. When present in systems that are vulnerable to singlet oxygen oxidation, tocopherols inhibit lipid peroxidation. Tocopherols were identified in soybean oil at about 1100 ppm and exist in α-, β-, γ-, and δ-tocopherol at approximately 4, 1, 67, and 29%, respectively (Jung et al., 1991). Jung et al. (1991) studied the effectiveness of α-, γ-, and δ-tocopherol in the chlorophyll-photosensitized oxidation of soybean oil and reported that α-tocopherol quenched singlet oxygen at the rate of $2.7 \times 10^7$ $M^{-1}S^{-1}$. Tocopherols involve charge transfer as singlet oxygen quenchers (Foote, 1970). Tocopherols can form a charge transfer complex with singlet oxygen by electron donation from tocopherol to singlet oxygen. The transfer complex undergoes an intersystem crossing to form triplet oxygen and tocopherol. Since this does not involve chemical reactions between tocopherol and singlet oxygen, it is called physical quenching:

$$T + {}^1O_2 \rightarrow [T^+ - {}^1O_2]_1 \rightarrow [T^+ - {}^1O_2]_3 \rightarrow T + {}^3O_2$$

The destruction of vitamin $D_2$ in a model system by singlet oxygen oxidation was reduced by the addition of α-tocopherol (King and Min, 1998). The rate of singlet oxygen quenching by α-tocopherol is similar to that of β-carotene.

## Determining Quenching Mechanisms

The quenching mechanism of photosensitized singlet oxygen oxidation can be determined by measuring the rate constant of total quenching, physical quenching, and chemical quenching. Quenchers work in numerous ways to inhibit the formation of oxidized products as has been previously described (Fig. 1.13).

The quantum yield of a photochemical reaction is defined as the ratio of the number of molecules of a product formed to the number of photons of light absorbed. This value is used to measure the relative efficiency of a photochemical reaction. The quantum yield of oxidized product formation (ØAO$_2$) can be defined by the equation:

$$\text{ØAO}_2 = A \times B \times C \quad (Eq.1)$$

where A represents the partitioning of singlet sensitizer for singlet oxygen oxidation; B, the partitioning of triplet sensitizer for singlet oxygen formation; and C, the formation of the oxidized product.

The concentration of quencher necessary to inhibit a substantial amount of the singlet oxygen sensitizer is particularly high since the lifetime of singlet oxygen sensitizer is very short. For these reasons, the singlet sensitizer quenching is not considered in the steady-state equation. Therefore, term A in Eq. 1 is a constant (K) that is equal to the quantum yield of intersystem crossing.
Term B in Eq. 1 represents the rate of singlet oxygen formation, which is dependent on the triplet sensitizer quenching rate and the rate of triplet-triplet annihilation. Therefore, B is:

$$B = \frac{k_o [\text{oxygen}]}{k_o [\text{oxygen}] + k_Q [\text{quencher}]} \quad (Eq.\ 2)$$

where $k_o$ is the reaction rate constant of triplet-triplet annihilation and $k_Q$ is the reaction rate constant of triplet sensitizer quenching (Fig. 1.13).

Term C in Eq. 1 represents the formation of oxidized product, which is dependent on the concentration and nature of the compound, the physical and chemical quenching of singlet oxygen, as well as the natural decay rate of singlet oxygen. The assemblage of these factors generates the following equation:

$$C = \frac{k_r [\text{substrate}]}{k_r [\text{substrate}] + (k_{ox-Q} + k_q)[\text{chemical} + \text{physical quencher}] + k_d} \quad (Eq.\ 3)$$

where $k_r$ is the reaction rate constant of the reaction between singlet oxygen and the substrate, $k_{ox-Q}$ is the reaction rate constant of chemical quenching, $k_q$ is the reaction rate constant of physical quenching, and $k_d$ is the decay rate constant of singlet oxygen.

If a quencher inhibits photosensitized oxidation by quenching singlet oxygen, the steady state equation can be written as:

$$\text{ØAO}_2 = K \frac{k_o [^3O_2]}{k_o [^3O_2] + k_Q [Q]} \times \frac{k_r [A]}{k_r [A] + (k_{ox-Q} + k_q)[Q] + k_d} \quad (Eq.\ 4)$$

where K is the quantum yield of intersystem crossing of the $^1$Sensitizer* (A from Eq. 1), and both B and C have been appropriately substituted with Eq. 2 and Eq. 3, respectively.

In a given system, if there is only singlet oxygen quenching such that $k_Q [Q] \ll k_o [^3O_2]$, then B is equal to 1. Therefore, the steady state equation becomes:

$$\emptyset AO_2 = K \frac{k_r [A]}{k_r [A] + (k_{ox-Q} + k_q)[Q] + k_d} \quad \text{(Eq. 5)}$$

This equation can be inverted to:

$$[\emptyset AO_2]^{-1} = K^{-1} \left[ 1 + \frac{(k_{ox-Q} + k_q)[Q] + k_d}{k_r} [A]^{-1} \right] \quad \text{(Eq. 6)}$$

so that it is in slope-intercept form.

Alternatively, if there is only triplet sensitizer quenching, such that $(k_{ox-Q} + k_q)[Q] \ll k_r [A] + k_d$, then the slope-intercept form of the equation is:

$$[\emptyset AO_2]^{-1} = K^{-1} \left[ 1 + \frac{k_Q [Q]}{k_o [^3O_2]} \right] \left[ 1 + \frac{k_d}{k_r [A]} \right] \quad \text{(Eq. 7)}$$

The significance of these two equations is the fact that one describes a system in which singlet oxygen quenching is dominant, and the other describes a system in which triplet sensitizer is dominant. A plot of $[AO_2]^{-1}$ against $[A]^{-1}$ at different $[Q]$ will appear in one of two ways, dependent on which mechanism dominates a system. If singlet oxygen quenching is dominant (Eq. 6), then the plots at various $[Q]$ will all have the same $y$-intercept but different slopes (Fig. 1.14). If triplet sensitizer quenching is dominant (Eq. 7), then both the intercept and the slope will vary (Fig. 1.15).

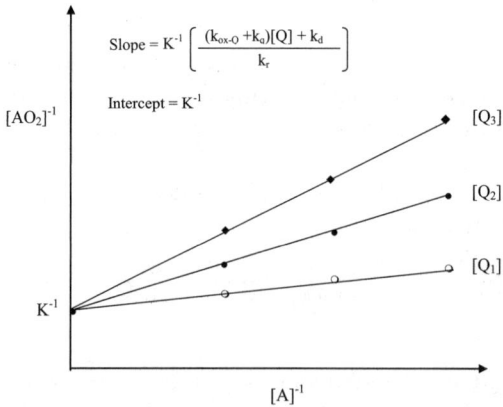

**Fig. 1.14.** Characteristics plot of a singlet oxygen quenching mechanism (Li et al. 2000).

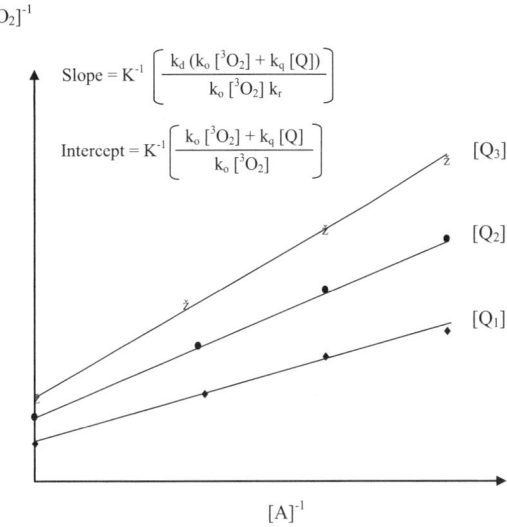

**Fig. 1.15.** Characteristic plot of a triplet sensitizer quenching mechanism (Li et al. 2000).

## References

Adams, D.R.; and Wilkinson, F. Lifetime of Singlet Oxygen in Solution. *J. Chem. Soc. Faraday Trans. 2,* **1972,** *68,* 686.

Adam, W. Singlet Molecular Oxygen and Its Role in Organic Peroxide Chemistry, *Chem. Ztg.* **1975,** *99,* 142.

Andersen, L.K.; and Ogilby, P.R. Time-Resolved Detection of Singlet Oxygen in a Transmission Microscope. *Photochem. Photobiol.* **2001,** *73,* 489.

Afonso, S.G.; Fnriquez de Salamanca, R.; and del C Bafle, A. M. The Photodynamic and Non-Photodynamic Actions of Porphyrins. *Braz. J. Med. Biol. Res.* **1999,** *32,* 255.

Ando, T.; Yoshikawa, F.; Tanigawa, T.; Kohno, M.; Yoshida, N.; and Motoharu, K. Quantification of Singlet Oxygen from Hematoporphyrin Derivative by Electron Spin Resonance. *Life Sci.* **1997,** *61,* 1953.

Athar, M.; Mukhtar, H.; and Bickers, D.R. Differential Role of Reactive Oxygen Intermediates in Photofrin-I-Initiated and Photofrin-II-Initiated Photocenhancement of Lipid Peroxidation in Epidermal Microsomal Membranes. *J. Invest. Derm.* **1988,** *90,* 652.

Beutner, S.; Bloedorn, B.; Hoffman, T.; and Martin, H.D. Synthetic Singlet Oxygen Quenchers. *Methods in Enzymology,* Packer, L., Sies, H. Eds., Academic Press: New York, 2000, Vol 319, 226-241.

Bradley, D.G.; Lee, H.O.; and Min, D.B. Singlet Oxygen Detection in Skim Milk by Electron Spin Resonance Spectroscopy. *J. Food Sci.* **2003,** *68,* 491.

Bradley, D.G.; Kim, H.J.; and Min, D.B. Effects, Quenching Mechanism, and Kinetics of

Water Soluble Compounds in Riboflavin Photosensitized Oxidation of Milk. *J. Agric. Food Chem.* **2006,** *54,* 6016-6020.

Brekket, O.L. In *Handbook of Soyoil Processing and Utilization,* Erickson, D.R., Pryde, E.H., Brekke, O.L., Mounts, T.L., Falb, R.A., Eds., American Oil Chemists' Society: Champaign, IL, 1980, 109.

Chang, S.S.; Shen, G.H.; Tang, H.; Jin, Q.Z.; Shi, H.; Carlin, J.T.; and Ho, C.T. Isolation and Identification of 2-Pentenylfurans in the Reversion Flavor of Soybean Oil. *J. Am. Oil Chem. Soc.* **1983,** *60,* 553.

Chang, S.S.; Smouse, T.H.; Krishnamurthy, R.G.; Mookherjee, B.D.; and Reddy, R.B. Isolation and Identification of 2-Pentyl-furan as Contributing to the Reversion Flavour of Soybean Oil. *Chem. Ind. (London),* 1966, 1926.

Choe, E.; and Min, D.B. Chemistry and Reactions of Reactive Oxygen Species in Foods. *J. Food Sci.* **2005,** *70,* 142.

Choe, E.; and Min, D.B. Mechanisms and Factors for Edible Oil Oxidation. *Comp. Rev. Food Sci. Food Saf.* **2006,** *5,* 169.

Darmanyan, A.P.; and Jenks, W.S. Charge-Transfer Quenching of Singlet Oxygen by Amines and Aromatic Hydrocarbons. *J. Phys. Chem.* **1990,** *109,* 7420-7426.

Delmelle, M. Retinol Damage by Light: Possible Implication of Singlet Oxygen. *Biophs. Struct. Mech.* **1997,** *3,* 195-198.

Doleiden, F.H.; Farenholtz, S.R.; Lamola, A.A.; and Rwozzolo, A.M. Reactivity of Cholesterol and Some Fatty Acids Toward Singlet Oxygen. *Photochem. Photobiol.* **1974,** *20,* 519.

Foote, C.S.; and Denny, R.W. Chemistry of Singlet Oxygen. VII Quenching by β-Carotene. *J. Am. Chem. Soc.* **1968,** *90,* 6233.

Foote, C.S.; Chang, Y.C.; and Denny, R.W. Chemistry of Singlet Oxygen. X. Carotenoid Quenching Parallels Biological Protection. *J. Am. Chem. Soc.* **1970,** *92,* 5216.

Foote, C.S. Photosensitized Oxidation and Singlet Oxygen: Consequences in Biological Systems. *Free Radicals in Biology,* Pryor, W.A., Ed., Academic Press: New York, 1976, 85.

Forss, D.A. Mechanism of Formation of Aroma Compounds in Milk and Milk Products. *J. Dairy Res.* **1979,** *46,* 691.

Frankel, E.N.; Neff, W.E.; and Bessler, T.R. Analysis of Autoxidized Fats by Gas Chromatography-Mass Spectrometry: V. Photosensitized Oxidation. *Lipids* **1979,** *14,* 961.

Frankel, E.N.; Neff, W.E.; and Selke, E. Analysis of Autoxidized Fats by Gas Chromatography-Mass Spectrometry: VII. Volatile Thermal Decomposition Products of Pure Hydroperoxides from Autoxidized and Photosensitized Oxidized Methyl Oleate, Linoleate, and Linolenate. *Lipids* **1981,** *16,* 279.

Frankel, E.N. Chemistry of Autoxidation: Mechanism, Products and Flavor Significance. *Flavor Chemistry of Fats and Oils,* Min, D.B., and Smouse, T.H., Eds., American Oil Chemists' Society: Champaign, IL, 1985, 1.

Girotti, A.W. Lipid Hydroperoxide Generation, Turnover, and Effector Action in Biological Systems. *J. Lipid Res.* **1998,** *39,* 1529.

Golbitz, P. *Soya and Oilseed Bluebook.* Soyatech, Inc., Bay Harbor, Maine, 2000, 2001.

Gunston, F.D. Chemical Properties. *The Lipid Handbook,* Gunston, F.D., Harwood, J.L., Padley, F.B., Eds., Chapman and Hall, New York, 1986 449-484.

Halliwell, B.; and Gutteridge, J.M.C. *Free Radicals in Biology and Medicine.* 3rd ed.; Oxford University Press: New York, 2001.

He, Y.Y.; An, J.Y.; and Jiang, L.J. EPR and Spectrophotometric Studies on Free Radicals ($O_2^*$, Cysa-HB*) and Singlet Oxygen ($^1O_2$) Generated by Irradiation of Cysteamine Substituted Hypocrellin B. *Int. J. Rad. Biol.* **1998**, *74,* 647.

Ho, C.T.; Smagula, M.S.; and Chang, S.S. The Synthesis of 2-(1-Pentenyl) furan and Its Relationship to the Reversion Flavor of Soybean Oil. *J. Am. Oil Chem. Soc.* **1978**, *55,* 233.

IUPAC Commission on Nomenclature of Inorganic Chemistry, Names for Muonium and Hydrogen Atoms and their Ions. *Pure Applied Chem.* **2001**, *73, 377–380.*

Jung, Y.J.; Lee, E.; and Min, D.B. α-, γ-, and δ-Tocopherol Effects on Chlorophyll Photosensitized Oxdiation of Soybean Oil. *J. Food Sci.* **1991**, *45,* 183.

Kanofsky, J.R.; and Axelrod, B. Singlet Oxygen Production by Soybean Lipoxygenase Isozymes. *J. Biol. Chem.* **1986**, *261,* 1099.

Ke, P.J.; and Ackman, R.G. Bunsen Coefficient for Oxygen in Marine Oils at Various Temperatures Determined by Exponential Dilution Method with a Polarographic Oxygen Electrode. *J. Am. Oil Chem. Soc.* **1973**, *50,* 429.

Kearns, D.R. Solvent and Solvent Isotope Effects on the Lifetime of Singlet Oxygen. *Singlet Oxygen,* Wasserman, H.H., Murray, R.W. Eds., Academic Press: New York, 1979, 115.

Kellogg, E.W.; and Fridovich, I. Suoperoxide, Hydrogen Peroxide, and Singlet Oxygen in Lipid Peroxidation by a Xanthin Oxidase System. *J. Biol. Chem.* **1975**, *250,* 8812-8817.

Kepka, A.; and Grossweiner, L.I. Photodynamic Oxidation of Iodide Ion and Aromatic Amino Acids by Eosine. *Photochem. Photobiol.* **1972**, *14,* 621.

King, J.M.; and Min, D.B. Riboflavin Photosensitized Singlet Oxygen Oxidation of Vitamin D. *J. Food Sci.* **1998**, *63,* 31.

Kobayashi, M.; and Sakamoto, Y. Singlet Oxygen Quenching Ability of Astaxanthan Esters from the Green Alga *Haematococcus pluvialis, Biotech. Lett.* **1999**, *21,* 265.

Kochevar, I.E.; and Redmond, R.W., Photosensitized Production of Singlet Oxygen. *Methods in Enzymology,* Packer, L., and Sies, H., Eds., Academy Press: New York, 2000, 20.

Korycka-Dahl, M.B.; and Richardson, T. Activated Oxygen Species and Oxidation of Food Constituents. *Crit. Rev. Food Sci. Nutr.* **1978**, *10,* 209.

Labuza, T.P. Kinetics of Lipid Oxidation in Foods, *CRC Crit. Rev. Food Sci. Nutr.* **1971**, *2,* 355.

Lledias, F.; and Hansberg, W. Catalase Modification as a Marker for Singlet Oxygen. *Methods in Enzymology,* Parks, O.W. and Sies, H., Eds., Academic Press: New York, 2000, 110.

Lee, S.H.; and Min, D.B. Effects, Quenching Mechanisms, and Kinetics of Carotenoids in Chlorophyll-Sensitized Photooxidation of Soybean Oil. *J. Agric. Food Chem.* **1990**, *38,* 1630.

Lee, J.; Koo, N.; and Min, D.B. Reactive Oxygen Species, Aging, and Antioxidative Netraceuticals. *Comp. Rev. Food Sci. Food Saf.* **2004**, *3,* 21-33.

Lledias, F.; and Hansberg, W. Catalase Modification as a Marker for Singlet Oxygen. *Methods in Enzymology,* Parks, O.W. and Sies, H., Eds., Academic Press: New York, 2000, 110.

Li, T.L.; King, J.M.; and Min, D.B. Quenching Mechanisms and Kinetics of Carotenoids in Riboflavin Photosensitized Singlet Oxygen Oxidation of Vitamin $D_2$. *J. Food Biochem.* **2000**, *24,* 477.

Macpherson, A.N.; Teflwer, A.; Barber, J.; and Truscott, T.G. Direct Detection of Singlet Oxygen from Isolated Photosystem II Reaction Centers. *Biochem. Biophys. Acta.* **1993**, *1143,* 301.

Min, D.B.; and Lee, E.C. Frontiers of Flavor. *Frontiers of Flavor,* Charalambous, G. Ed., Elsevier: Amsterdam, 1988 473-498.

Min, D.B.; and Lee, H.O. Chemistry of Lipid Oxidation. *Food Lipids and Health,* McDonald, R.E., and Min, D.B., Eds., Marcel Dekker: New York, 1996, 241.

Min, D.B.; and Boff, J.M. Chemistry and Reaction of Singlet Oxygen in Foods. *Comp. Rev. Food Sci. and Food Saf.* **2002**, *1*, 58.

Min, D.B.; Callison, A.L.; and Lee, H.O. Singlet Oxygen Oxidation for 2-Pentylfuran and 2-Pentenylfuran Formation in Soybean Oil. *J. Food Sci.* **2003**, *68*, 1175.

Nonell, S.; and Braslavsky, S.E. Time-Resolved Singlet Oxygen Detestion. *Methods in Enzymology,* Packer, L., Sies, H., Eds., Vol. 319, Academic Press: New York, 2000, 37-49.

Pecci, L.; Costa, M.; Antonucci, A.; Montefoschi, G.; and Cavallini, D. Methylene Blue Photosensitized Oxidation of Cysteine Sulfinic Acid and Other Sulfinates: The Involvement of Singlet Oxygen and the Azide Paradox. *Biochem. Biophys. Res. Com.* **2000**, *270*, 782.

Rawls, H.R.; and VanSanten, P.J. A Possible Role for Singlet Oxidation in the Initiation of Fatty Acid Autoxidation. *J. Am. Oil Chem. Soc.* **1970**, *47*, 121.

Schneider, T.; Gugliotti, M.; Politi, M.J.; and Baptista, M.S. Quantitative Determination of Singlet Oxygen by Laser Deflection Calorimetry. *An. Lett.* **2000**, *33*, 297.

Sharman, W.M.; Allen, C.M.; and E., v. L.J. Role of Activated Oxygen Species in Photodynamic Therapy., in *Methods in Enzymology,* Packer, L. and Sies, H., Eds., Academic Press: New York, 2000, 376.

Smagula, M.S.; Ho, C.T.; and Chang, S.S. The Synthesis of 2-(2-Pentenyl) Furans and Their Relationship to the Reversion Flavor of Soybean Oil. *J. Am. Oil Chem. Soc.* **1978**, *56*, 516.

Smouse, T.H.; and Chang, S.S. A Systematic Characterization of the Reversion Flavor of Soybean Oil. *J. Am. Oil Chem. Soc.* **1967**, *44*, 509.

Song, Y.Z.; An, J.Y.; and Jiang, L.J. ESR Evidence of the Photogeneration of Free Radicals (GDHB*, $O_2^*$) and Singlet Oxygen ($^1O_2$) by 15-Deacetyl-13-Glycine-Substituted Hypocrelin B. *Biochem. Biophys. Acta.* **1999**, *1472*, 307.

Stratton, S.P.; and Liebler, D.C. Determination of Singlet Oxygen-Specific versus Radical-Initiated Lipid Peroxidation in Photosensitized Oxidation of Lipid Bilayers: Effect of β-Carotene and α-Tocopherol. *Biochemistry* **1997**, *36*, 12911.

Takayama, F.; Egashira, T.; and Yamanaka, Y. Singlet Oxygen Generation from Phosphatidylcholine Hydroperoxide in the Presence of Copper. *Life Sci.* **2001**, *68*, 1807-1815.

Vaisey-Genser, M.; Malcomson, L.J.; Przybylski, R.; and Eskin, N.A.M. Consumer Acceptance of Stored Canola Oils in Canada. *The 12th Project Report Research on Canola, Seed, and Oil Meal, Canada,* Canola Council: Canada, 1999, 189.

Whang, K.; and Peng, I.C. Electron Paramagnetic Resonance Studies of the Effectiveness of Myoglobin and Its Derivatives as Photosensitizers in Singlet Oxygen Generation. *J. Food Sci.* **1988**, *53*, 1863.

Yang, W.T.; and Min, D.B. Chemistry of Singlet Oxygen Oxidation of Foods, in *Lipids in Food Flavors,* Ho, C.T., and Hartmand, T.G., Eds., American Chemical Society: Washington D.C., 1994, 15.

# 2

# Chemistry and Reactions of Reactive Oxygen Species in Lipid Oxidation

Eunok Choe
Department of Food and Nutrition, Inha University, 253 Yonghyundong, Namku, Incheon, Korea

## Introduction

The reactive oxygen species (ROS) include oxygen radicals, such as hydroxyl (HO·), alkoxyl (RO·), hydroperoxyl (HOO·), peroxyl (ROO·), and superoxide anion ($O_2^{-\cdot}$) radicals, as well as nonradical derivatives of oxygen, such as hydrogen peroxide ($H_2O_2$), ozone ($O_3$), and singlet oxygen ($^1O_2$). Molecular oxygen in air is normally in a triplet state, and its sequential univalent reduction produces more reactive superoxide anion, hydrogen peroxide, and hydroxyl radicals. ROS are interrelated to each other as shown in Fig. 2.1. ROS has a very short half-life; half-lives of hydrogen peroxide and organic hydroperoxides are in the range of minutes, peroxyl radical has half-life of seconds. Superoxide anion and alkoxyl radicals show about a μsec half-life, and hydroxyl radical has the shortest half-life of nsec (Kehrer, 2000).

ROS is mainly responsible for the initiation of oxidation reaction of foods and affects food quality and human health. Lipid oxidation by ROS produces undesirable volatile compounds, destroys essential fatty acids, and sometimes produces carcinogens. ROS also changes the functionality of lipids by forming oxidized dimers and trimers. The ROS in foods lowers the overall quality of foods during processing, storage, and marketing (Choe and Min, 2005).

Since chemistry and reactions of singlet oxygen is dealt in a previous chapter of this book, hydroxyl, alkoxyl, hydroperoxyl, peroxyl, and superoxide anion radicals, hydrogen peroxide, and ozone are among the ROS covered in this chapter.

## Hydroxyl Radicals

Hydroxyl radicals are formed by radiolysis of water in the presence of high-energy radiation. Water absorbs the energy and becomes ionized ($H_2O^+$) and excited ($H_2O^*$) within $10^{-16}$ s (Halliwell and Gutteridge, 2001). Excited water molecules undergo homolysis in $10^{-14}$ to $10^{-13}$ s and produce hydrogen atoms and hydroxyl radicals (Halliwell and Gutteridge, 2001). A reaction of ionized water with other water molecules also produces hydroxyl radicals (Jacobien et al., 1996).

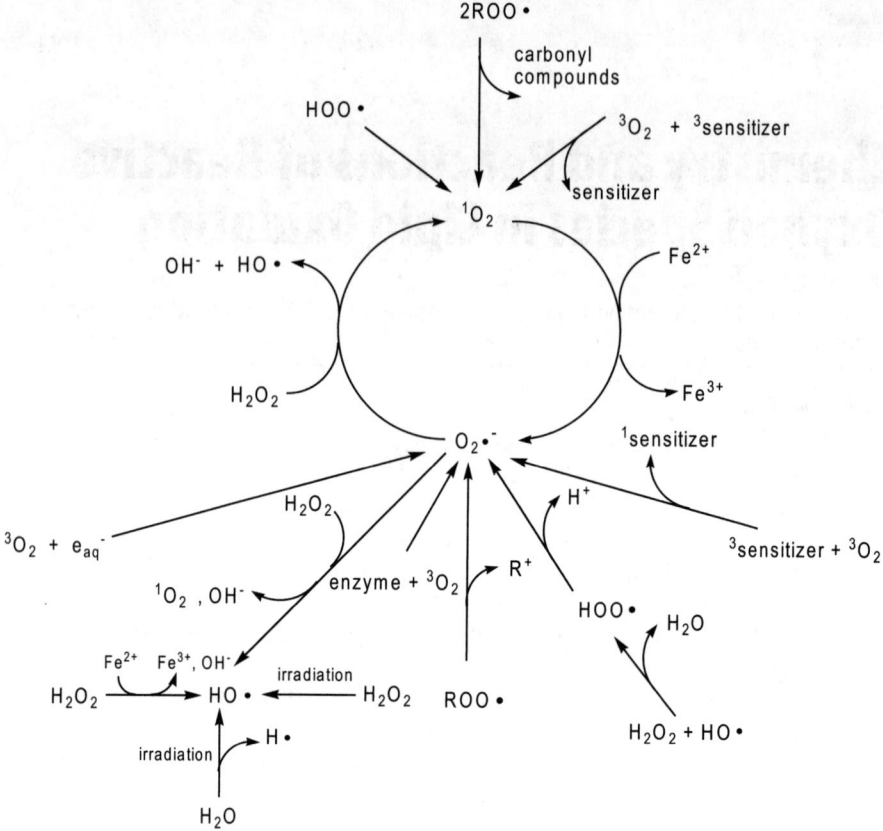

**Fig. 2.1.** Interrelationships among reactive oxygen species in foods.

$$2\ H_2O \xrightarrow{h\nu} H_2O^* + H_2O^+ + e_{aq}^-$$
$$H_2O^* \longrightarrow H\cdot + HO\cdot$$
$$H_2O^+ + H_2O \longrightarrow H_3O^+ + HO\cdot$$

Homolytic fission of the oxygen-oxygen bond in hydrogen peroxide upon exposure to UV light and Fenton reaction produce hydroxyl radicals. In Fenton reaction, hydrogen peroxide reacts with iron (Salem et al., 2000). At 25°C the rates to produce hydroxyl radical from the reaction of hydrogen peroxide with $Fe^{+2}$-ATP and $Fe^{+2}$-citrate complexes were $6.7 \times 10^3$ and $4.9 \times 10^3$ $M^{-1}s^{-1}$, respectively (Rush et al., 1990). Formate anion (HCOO·-) and superoxide anion radicals, hydroquinone, or cysteine promote hydroxyl radical production by reducing $Fe^{3+}$ to $Fe^{2+}$ (Watanabe et al., 2002). However, the reaction of hydrogen peroxide with human myoglobin containing iron did not produce hydroxyl radical (Witting et al., 2000).

Fenton reaction
$$H_2O_2 + Fe^{2+} \longrightarrow HO\cdot + OH^- + Fe^{3+}$$
$$Fe^{3+} + O_2^{\cdot-} \longrightarrow Fe^{2+} + O_2$$

Haber-Weiss reaction is another pathway for the production of hydroxyl radicals (Kehrer, 2000). In Haber-Weiss reactions, hydrogen peroxide and superoxide anion react with each other, but the reaction is very slow at 0.13 $M^{-1}s^{-1}$ (Weinstein and Bielski, 1979). Presence of transition metals increases the reaction rate (Hu and Jiang, 1996), and oxygen in a singlet state is also produced (Khan and Kasha, 1994). Ferrous ions are regenerated by the oxygen evolution reaction.

$$H_2O_2 + Fe^{2+} \longrightarrow HO\cdot + OH^- + Fe^{3+}$$
$$Fe^{3+} + O_2^{\cdot-} \longrightarrow Fe^{2+} + {}^1O_2$$
$$\overline{H_2O_2 + O_2^{\cdot-} \longrightarrow HO\cdot + OH^- + {}^1O_2}$$

Hydroxyl radical has a very high standard reduction potential (2.3 V) and is one of the most reactive species known (Choe and Min, 2005). It is an extremely strong oxidizing agent and powerful electrophilic radical. The electron accepting rate of hydroxyl radical is $10^9$ to $10^{10}$ $M^{-1}s^{-1}$ (Halliwell and Gutteridge, 2001).

Hydroxyl radical is mainly responsible for the initiation of lipid oxidation. It reacts with lipids unspecifically in a diffusion-limited mode (Kruk et al., 2005). Hydroxyl radical abstracts hydrogen, mostly allylic hydrogen of unsaturated lipids (RH) which have low bond dissociation energy. For example, the energy to break a bond between hydrogen and C11 of linoleic acid is 209 kJ/ mole. In contrast, the bond dissociation energy between hydrogen and C8, and that between hydrogen and C14 of linoleic acid is 314 kJ/ mole. More than 418 kJ/ mole are required to remove hydrogen at C17 or C18 (Min and Boff, 2002). It is easier to abstract hydrogen from a secondary or tertiary carbon than from a primary carbon in the lipid oxidation. In the oxidation of cholesterol by hydroxyl radical, the hydrogen at C7 is usually abstracted (Girotti, 1998). Abstraction of hydrogen from lipid (RH) results in lipid radical (R·), which can react with triplet oxygen for the free radical chain reaction as shown in Fig. 2.2. Triplet oxygen which has two unpaired electrons can react with lipid radicals, forming a lipid peroxyl radical, for example C9- and C13- peroxyl radicals in linoleic acid. The lipid peroxyl radical instantly propagates the chain reaction of lipid oxidation.

Hydroxyl radical can be added to the double bond of unsaturated lipids (Korycka-Dahl and Richardson, 1978; Lee et al., 2004) and form hydroxylated lipid radicals (Fig. 2.3). The hydroxylated lipid radical reacts with triplet oxygen and forms hydroxylated lipid peroxyl radicals. These radicals also propagate the chain reaction of lipid autoxidation.

β-carotene (CarH) decreases lipid oxidation by donating hydrogen to hydroxyl radicals and becomes a carotene radical (Car·). A carotene radical is a fairly stable spe-

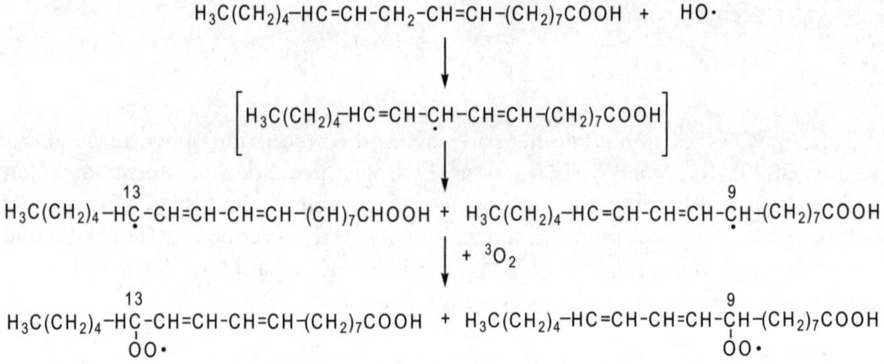

$H_3C(CH_2)_4-HC=CH-CH_2-CH=CH-(CH_2)_7COOH + HO\cdot$

$\downarrow$

$[H_3C(CH_2)_4-HC=CH-\overset{\cdot}{C}H-CH=CH-(CH_2)_7COOH]$

$\downarrow$

$\underset{13}{H_3C(CH_2)_4-\overset{\cdot}{C}H-CH=CH-CH=CH-(CH)_7CHOOH} + \underset{9}{H_3C(CH_2)_4-HC=CH-CH=CH-\overset{\cdot}{C}H-(CH_2)_7COOH}$

$\downarrow + {}^3O_2$

$\underset{13}{H_3C(CH_2)_4-\underset{OO\cdot}{\overset{|}{C}H}-CH=CH-CH=CH-(CH_2)_7COOH} + \underset{9}{H_3C(CH_2)_4-HC=CH-CH=CH-\underset{OO\cdot}{\overset{|}{C}H}-(CH_2)_7COOH}$

**Fig. 2.2.** Hydroxyl radical initiated oxidation of linoleic acid.

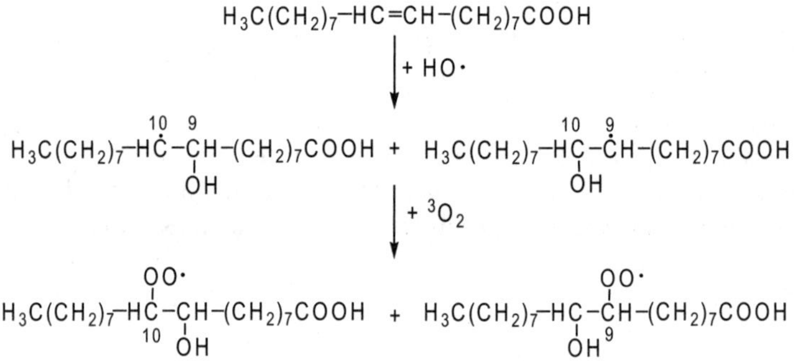

$H_3C(CH_2)_7-HC=CH-(CH_2)_7COOH$

$\downarrow + HO\cdot$

$\underset{OH}{\overset{10\quad 9}{H_3C(CH_2)_7-H\overset{\cdot}{C}-\overset{|}{C}H-(CH_2)_7COOH}} + \underset{OH}{\overset{10\quad 9}{H_3C(CH_2)_7-H\overset{|}{C}-\overset{\cdot}{C}H-(CH_2)_7COOH}}$

$\downarrow + {}^3O_2$

$\underset{10\ \ OH}{H_3C(CH_2)_7-H\overset{OO\cdot}{\overset{|}{C}}-\overset{|}{C}H-(CH_2)_7COOH} + \underset{OH\ ^9}{H_3C(CH_2)_7-H\overset{|}{C}-\overset{OO\cdot}{\overset{|}{C}}H-(CH_2)_7COOH}$

**Fig. 2.3.** Addition reaction of electrophilic hydroxyl radical to oleic acid.

cies due to delocalization of unpaired electron through its conjugated polyene system. When a carotene radical reacts with other radicals, such as peroxyl radicals, nonradical products (Car-OOR) are formed and the radical reactions stop.

CarH + ·OH $\longrightarrow$ Car· + $H_2O$
Car· + ROO· $\longrightarrow$ Car-OOR

Hydroxyl radicals efficiently oxidize α-tocopherol (TOH) through an indirect reaction, due to their low diffusibility. The hydroxyl radical first reacts with the solvent molecules (SH) and the solvent radicals (S·) will oxidize tocopherols afterward (Fukuzawa and Gebicki, 1983).

·OH + SH $\longrightarrow$ $H_2O$ + S·
S· + TOH $\longrightarrow$ SH + TO·

## Alkoxyl Radicals

Alkoxyl radicals are produced by homolysis of lipid hydroperoxides (ROOH). The energy required to cleave the oxygen-oxygen bond is 184 kJ/ mole (Hiatt et al., 1968), and heat, UV, or transition metals accelerate the homolysis (Heaton and Uri, 1961; Schaich, 1992; and Jadhav et al., 1996).

$$ROOH \xrightarrow{\text{heat, UV}} RO\cdot + \cdot OH$$

$$ROOH + Fe^{2+} \longrightarrow RO\cdot + Fe^{3+} + OH^-$$

Transition metal ion-mediated decomposition of lipid hydroperoxides is initiated by a one-electron reduction to an alkoxyl radical and is 10 times faster with $Fe^{2+}$ than with $Fe^{3+}$ (Kilic and Richards, 2003). Ascorbic acid induces hydroperoxide decomposition (Lee and Blair, 2000), and was more efficient at 25 µM to 2 mM than $Cu^{2+}$ or $Fe^{2+}$ in initiating the decomposition of 13-OOH of linoleic acid (Lee et al., 2001).

Alkoxyl radical undergoes homolytic β-scission of carbon-carbon bond and produces aldehydes, acids, alcohols, and short-chain hydrocarbons. Alkoxyl radical derived from C8-hydroperoxide of oleic acid can produce 1-decenol, decanal, 1-decene, 8-oxo-octanoic acid, 2-undecenal, 7-hydroxyheptanoic acid, and heptanoic acid by decomposition (Fig. 2.4). Pentane, hexanal, 2-heptanal, octanoic acid, nonanal, 2-decenal, and 2,4-decadienal are good examples of decomposition volatile compounds found in oxidized oils (Kanavouras et al., 2004), which give off-flavor. Alkoxyl radical can also abstract hydrogen from lipids to produce new lipid radicals since it has a standard reduction potential of approximately 1.6 V which is higher than that of unsaturated lipids (Choe and Min, 2005). The resulting lipid radicals continue the chain reaction of lipid oxidation. The reaction of alkoxyl radicals with lipids is at a rate constant of $10^6$-$10^7$ $M^{-1}s^{-1}$ (Simic et al., 1992). The lipid oxidation by alkoxyl radicals can be reduced by tocopherols. Tocopherols react with alkoxyl radicals and produce tocophroxyl radicals which are more stable than alkoxyl radicals due to their resonance structure (Fig. 2.5). The life-time of a tocophroxyl radical in low density lipoprotein was reported as 12.5 s (Ingold et al., 1993).

## Hydroperoxyl Radicals

A hydroperoxyl radical is a protonated form of a superoxide anion radical and is produced by reaction of hydrogen peroxide and a hydroxyl radical.

$$HO\cdot + H_2O_2 \longrightarrow HOO\cdot\ (H^+ + O_2\cdot^-) + H_2O$$

It is also produced by a reaction of triplet oxygen with hydrogen atoms which are produced from a radiolysis of water during pulsed electric field processing of food (Halliwell and Gutteridge, 2001).

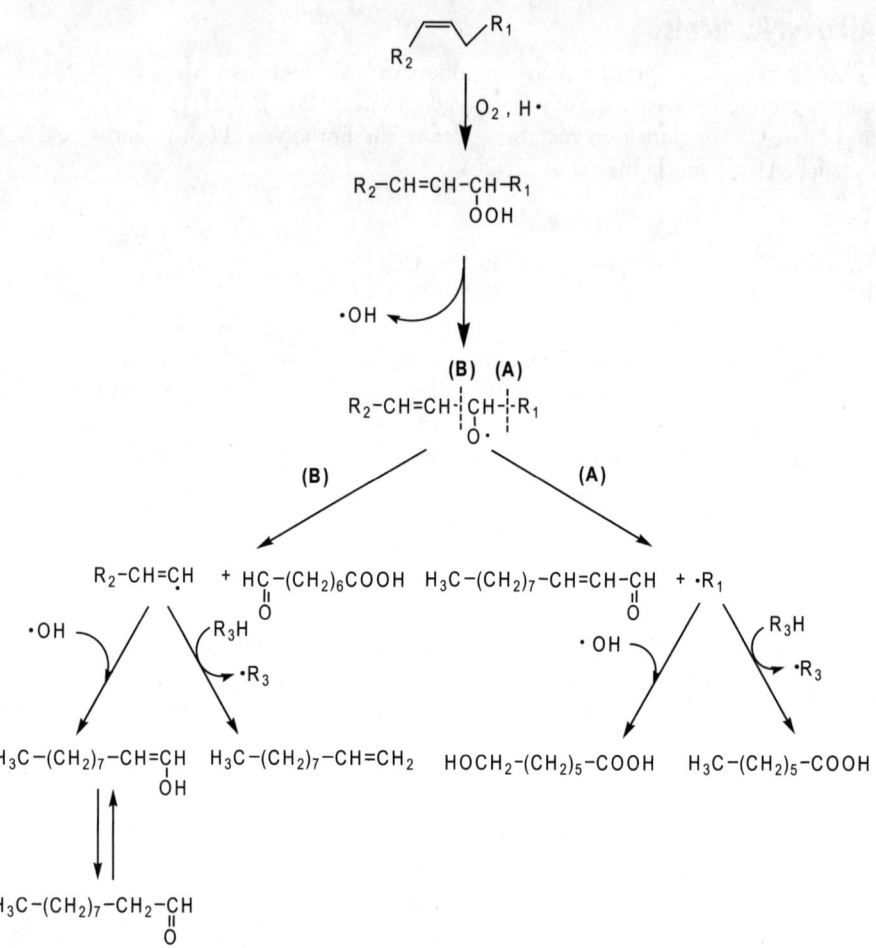

**Fig. 2.4.** Decomposition of hydroperoxides of oleic acid ($r_1$: -(ch2)$_6$cooh, $r_2$: -(ch$_2$)$_7$ch$_3$, $r_3$: alkyl group).

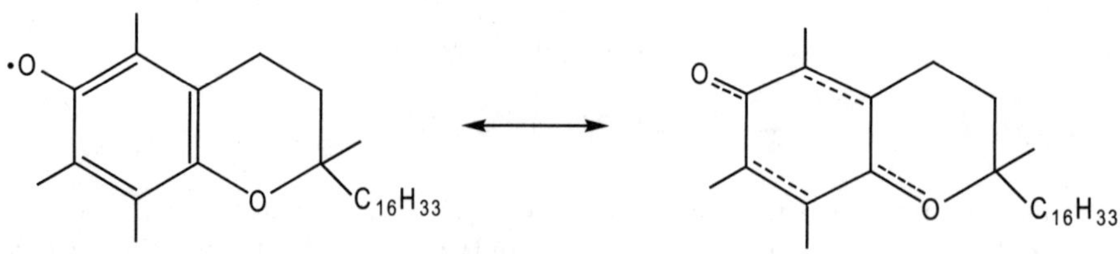

**Fig. 2.5.** Resonance form of α-tocophroxyl radical.

$$2\,H_2O \longrightarrow H_2O^+ + H_2O^* + e_{aq}^-$$
$$H_2O^* \longrightarrow H^\cdot + HO^\cdot$$
$$H^\cdot + O_2 \longrightarrow \mathbf{HOO^\cdot}$$

The hydroperoxyl radical has a reduction potential of 1.1-1.5 V (Choe and Min, 2005) and is thermodynamically capable of oxidizing unsaturated fatty acids whose reduction potential is approximately 0.6 V at neutral pH (Koppenol, 1990). It abstracts hydrogen, mostly from allylic carbons of unsaturated fatty acids (Bielski et al., 1983) and produces fatty acid radicals. The reaction rate of hydroperoxyl radicals increases as the unsaturation of lipids increases; $1.2 \times 10^3$, $1.7 \times 10^3$, and $3.1 \times 10^3\,M^{-1}s^{-1}$ for linoleic, linolenic, and arachidonic acids, respectively (Bielski et al., 1983; Aikens and Dix, 1991). Oleic acid did not react with hydroperoxyl radicals (Bielski et al., 1983). Hydroperoxyl radicals can also abstract hydrogen from lipid hydroperoxides and produces lipid peroxyl radicals (Aikens and Dix, 1991). Lipid radicals and peroxyl radicals can abstract hydrogen from other molecules, producing another radical, and the lipid oxidation propagates through a free radical chain reaction.

Lipid oxidation by hydroperoxyl radicals is slowed down by tocopherols. Tocopherol reacts with hydroperoxyl radicals at a rate of $2.0 \times 10^5\,M^{-1}s^{-1}$ (Arudi et al., 1983), which is a higher rate than that of a reaction between hydroperoxyl radical and lipid.

## Peroxyl Radicals

Triplet oxygen having two unpaired electrons reacts with lipid radicals and produces lipid peroxyl radicals. A lipid radical is formed by hydrogen donation from a lipid, which occurs mostly at allylic carbons, C8, C9, C10, or C11 for oleic acid, C9 or C13 for linoleic acid, and C9, C12, C13, or C16 for linolenic acid. Triplet oxygen goes to one of these carbons, which have odd electrons, and produces peroxyl radicals. Figure 2.6 shows a formation of C9- and C13-peroxyl radicals from a reaction of linoleic acid with oxygen. The rates for the formation of peroxyl radicals depend on oxygen availability and temperature (Velasco et al., 2003).

Peroxyl radicals have 1.0 V of standard reduction potential (Choe and Min, 2005) and can easily abstract hydrogen from other lipid molecules. The peroxyl radical serves as a chain carrier in the free radical lipid oxidation.

$$ROO^\cdot + R'H \longrightarrow ROOH + R'^\cdot$$
$$R'^\cdot + O_2 \longrightarrow R'OO^\cdot$$

Reaction rates with polyunsaturated fatty acids are higher in peroxyl radicals than in hydroperoxyl radicals. The C13-peroxyl radical of methyl linoleate reacts with linoleic acid at a rate of $1.1 \times 10^6\,M^{-1}s^{-1}$ (Iulaino et al., 1995), and hydroperoxides are formed along with new linoleic acid radicals. The hydroperoxides are mostly stable at room temperature; however, in the presence of heat, UV, or transition metals they are decomposed to

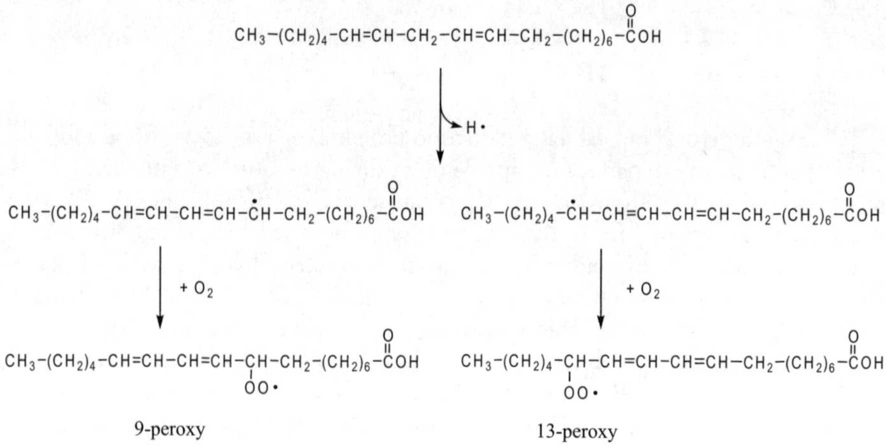

$$CH_3-(CH_2)_4-CH=CH-CH_2-CH=CH-CH_2-(CH_2)_6-\overset{O}{\overset{\|}{C}}OH$$

$$\downarrow -H\cdot$$

$CH_3-(CH_2)_4-CH=CH-CH=CH-\overset{\cdot}{C}H-CH_2-(CH_2)_6-\overset{O}{\overset{\|}{C}}OH$     $CH_3-(CH_2)_4-\overset{\cdot}{C}H-CH-CH=CH-CH=CH-CH_2-(CH_2)_6-\overset{O}{\overset{\|}{C}}OH$

$\downarrow +O_2$     $\downarrow +O_2$

$CH_3-(CH_2)_4-CH=CH-CH=CH-\underset{\underset{OO\cdot}{|}}{C}H-CH_2-(CH_2)_6-\overset{O}{\overset{\|}{C}}OH$     $CH_3-(CH_2)_4-\underset{\underset{OO\cdot}{|}}{C}H-CH=CH-CH=CH-CH_2-(CH_2)_6-\overset{O}{\overset{\|}{C}}OH$

9-peroxy    13-peroxy

**Fig. 2.6.** Formation of peroxyl radical from the reaction between linoleic acid and triplet oxygen.

lower molecular weight compounds via homolytic cleavage as described previously.

β-Carotene having 1.1 V of reduction potential (Edge et al., 2000) can compete with lipids to donate hydrogen to peroxyl radicals, which can decrease the reaction between lipid and peroxyl radicals. However, the actual occurrence is rarely observed. The bond dissociation energy of the most labile hydrogen in β-carotene is 309 kJ/mole, whereas the bond dissociation energy of O-H bond in hydroperoxides is about 370-380 kJ/mole (Luo and Holmes, 1994), which can thermodynamically transfer hydrogen from β-carotene to a peroxyl radical (Haila, 1999). Peroxyl radicals abstract hydrogen from β-carotene at low oxygen concentrations of less than 760 torr and produce a carotene radical. Carotene radicals undergo addition of oxygen and carotene and homolysis of the oxygen-oxygen bond repeatedly, and produce dicarbonyls of carotene (Beutner et al., 2001) as shown in Fig. 2.7.

Peroxyl radicals can be added to β-carotene, especially at higher than 150 torr of oxygen (Burton and Ingold, 1984) as shown in Fig. 2.8 (Decker, 2002). The peroxyl radical is added to the cyclic end group, or to the polyene chain of β-carotene especially at C15 and C15' position (Iannone et al., 1998), and produces 5,6-epoxides of carotene and peroxides of carotene radical (ROO-Car·), respectively. ROO-Car· reacts with another peroxyl radical and produces nonradical compounds, diperoxides of carotene (ROO-Car-OOR'). Cleavage between carbons having a peroxyl group in ROO-Car-OOR' and elimination of alkoxyl radicals produce aldehyde compounds containing cyclic groups. ROO-Car· can also react with molecular oxygen to produce peroxyl radicals of carotene peroxides (ROO-Car-OO·). These radicals react with lipid molecules and produce hydroperoxides of carotene and lipid radicals, which increase lipid oxidation (Tsuchihashi et al., 1995; Iannone et al., 1998). Burton and Ingold (1984) reported the prooxidant activity of β-carotene at a concentration of higher than $5 \times 10^{-4}$ M under the oxygen pressure of higher than 150 torr.

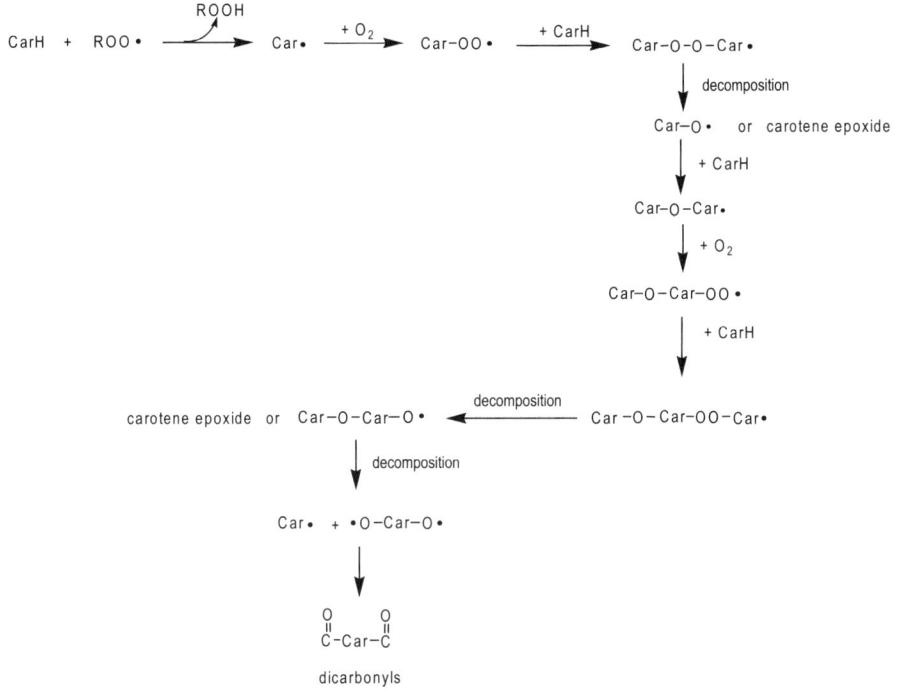

Fig. 2.7. Oxidation of β–carotene by peroxyl radical through carotene radical.

β-Carotene can donate an electron in its conjugated double bonds to peroxyl radicals and produce a β-carotene cation radical and a peroxyl anion (Liebler, 1993; Mortensen et al., 2001). β-Carotene cation radical is stable due to its resonance structure and rarely reacts with oxygen (Edge et al., 2000; Decker, 2002). The transfer of hydrogen or electrons from carotenoids to free radicals depends on the reduction potentials of the free radicals and the chemical structures of the carotenoids, especially the presence of oxygen-containing functional groups (Edge et al., 1997). The rates of electron transfer increase with the number of coplanar conjugated double bonds and is decreased by the presence of hydroxyl and keto groups (Mortensen et al., 2001).

Tocopherols have 0.5 V of standard reduction potential (Buettner, 1993) and compete with unsaturated lipids for peroxyl radicals. Peroxyl radicals react with tocopherols at $10^6$-$10^9$ $M^{-1}s^{-1}$ which is faster than with lipid at 10-60 $M^{-1}s^{-1}$ or with β-carotene at $10^6$ $M^{-1}s^{-1}$ (Mortensen and Skibsted, 1998). Reactivity of peroxyl radicals with α-tocopherol is 32 times higher than with β-carotene (Tsuchihashi et al., 1995). The reaction rate of oleic acid peroxyl radical with α-tocopherol is 2.5 x $10^6$ $M^{-1}s^{-1}$ (Simic, 1980). One tocopherol molecule can protect about $10^3$-$10^8$ polyunsaturated fatty acid molecules at low peroxide value (Kamal-Eldin and Appelqvist, 1996). Tocopherol donates hydrogen at the 6-hydroxyl group on the chromanol ring to lipid peroxyl radicals and produces lipid hydroperoxides and tocophroxyl radicals having a resonance structure.

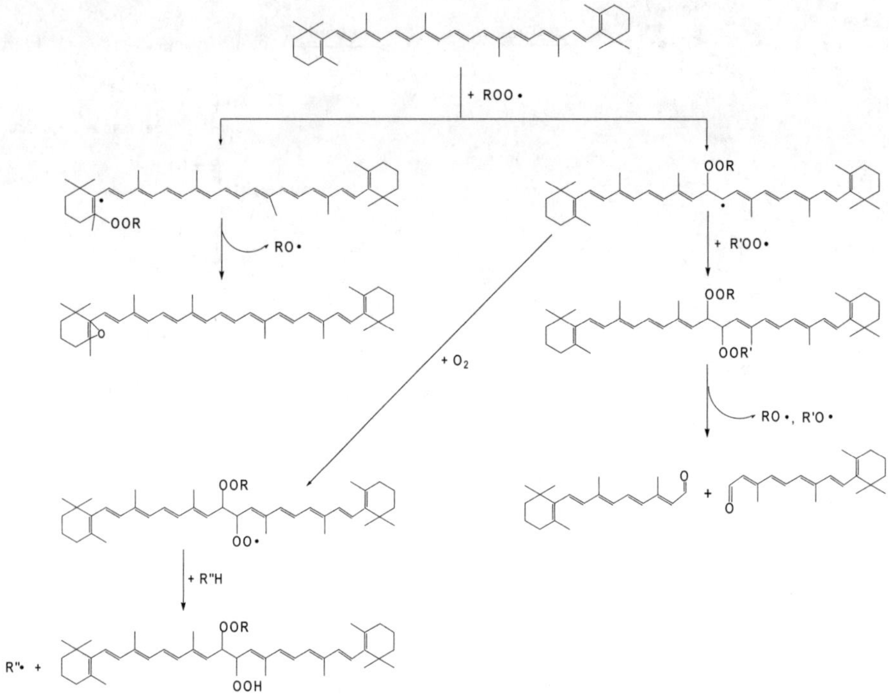

**Fig. 2.8.** Addition reaction of peroxyl radical to β-carotene.

The tocophroxyl radical can react with another peroxyl radical and produces tocopherol semiquinone, a potential antioxidant (Neuzil et al., 1997), or it reacts with another tocophroxyl radical and forms tocopherol dimer (Reische et al., 2002) as shown in Fig. 2.9.

Although tocopherols are generally decrease lipid oxidation by slowing down free radical reactions, they can increase lipid oxidation under certain conditions. At very low concentrations of peroxyl radicals, the tocophroxyl radical abstracts hydrogen from lipids at low rates to give tocopherol and lipid radicals, which promotes lipid oxidation. This is called tocopherol-mediated peroxidation (Bowry and Stocker, 1993; Yamamoto 2001). The rate of allylic hydrogen abstraction from ethyl oleate, ethyl linoleate, ethyl linolenate, and ethyl arachidonate by the tocophroxyl radical is $1.04 \times 10^{-5}$, $1.82 \times 10^{-2}$, $3.84 \times 10^{-2}$, and $4.83 \times 10^{-2}$ $M^{-1}s^{-1}$ at 25°C, respectively (Mukai and Okauchi, 1989). Linoleic acid in the tocopherol-mediated peroxidation favors the accumulation of 13-OOH over 9-OOH (Porter et al., 1980; Upston et al., 2002). Ascorbic acid prevents tocopherol-mediated peroxidation by reducing the tocophroxyl radical to tocopherol (Yamamoto, 2001).

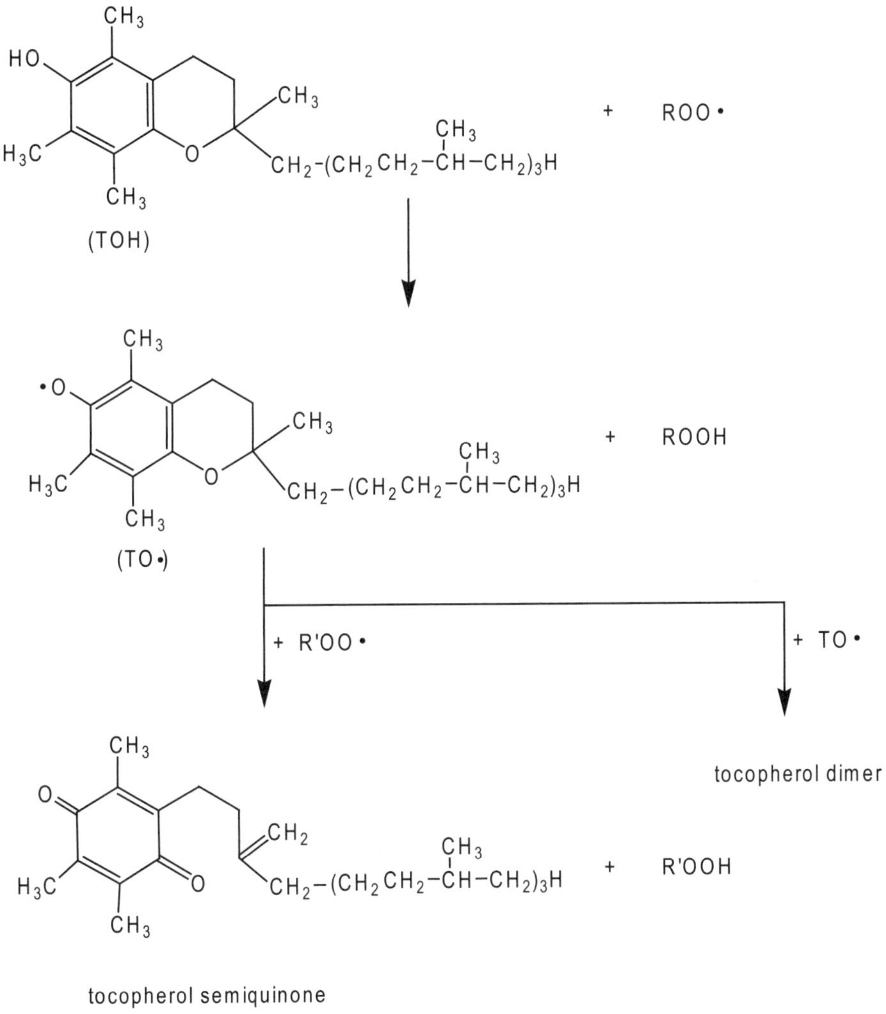

**Fig. 2.9.** Reaction of tocopherol with peroxyl radical.

## Superoxide Anion Radicals

Superoxide anion has one unpaired electron in the $2p\pi^*$ orbital as shown in Fig. 2.10 and is a radical. It is produced from triplet oxygen by the addition of one electron to one of the $2p\pi^*$ orbitals, which is reduction. Reaction of oxygen with xanthine or NADPH produces superoxide anion radicals and xanthine oxidase (Das and Das, 2002) and NADPH oxidase (Martinez and Moreno, 1996) catalyze the reaction, respectively.

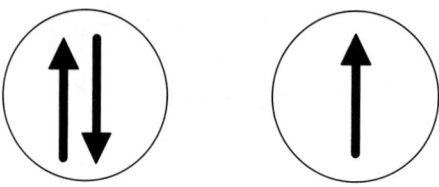

**Fig. 2.10.** Electronic configuration of $2p\pi^*$ orbital of superoxide anion radical.

$$\text{Hypoxanthine} + 2O_2 + H_2O \xrightarrow{\text{Xanthine oxidase}} \text{Xanthine} + \mathbf{2O_2^{\cdot-}} + 2H^+$$
$$\text{Xanthine} + 2O_2 + H_2O \longrightarrow \text{Uric acid} + \mathbf{2O_2^{\cdot-}} + 2H^+$$

$$\text{NADPH} + 2O_2 \xrightarrow{\text{NADPH oxidase}} \text{NADP}^+ + \mathbf{2O_2^{\cdot-}} + H^+$$

Reaction of aminoglycosides (AG) with oxygen produces superoxide anion (Ingold et al., 1997), and requires iron and fatty acids (FA) for the formation of FA-iron-AG complex (Lesniak et al., 2005). FA serves as an electron donor.

$$AG + Fe^{2+} + FA \longrightarrow [FA\text{-}Fe^{2+}\text{-}AG]$$
$$[FA\text{-}Fe^{2+}\text{-}AG] + O_2 \longrightarrow [FA\text{-}Fe^{2+}\text{-}O_2\text{-}AG]$$
$$[FA\text{-}Fe^{2+}\text{-}O_2\text{-}AG] \longleftrightarrow [FA\text{-}Fe^{3+}\text{-}O_2^{\cdot-}\text{-}AG]$$
$$[FA\text{-}Fe^{3+}\text{-}O_2^{\cdot-}\text{-}AG] \longleftrightarrow [FA\text{-}Fe^{3+}\text{-}AG] + \mathbf{O_2^{\cdot-}}$$

Photoactivation of sensitizers such as riboflavin (RF) produces superoxide anion (Buettner and Oberley, 1980). Light energy at a specific wavelength converts the RF to the excited RF ($^1RF^*$). The $^1RF^*$ becomes an excited triplet RF ($^3RF^*$) via intersystem crossing by emitting some of its energy. $^3RF^*$ can react with triplet oxygen and produce superoxide anion and a RF radical (Haseloff and Ebert, 1989). The reaction usually needs ethanol as a solvent, and the reaction rate depends on the concentration of ethanol (Athar et al., 1988; Haseloff and Ebert, 1989).

$$^1RF \xrightarrow{h\nu} {}^1RF^*$$
$$^1RF^* \longrightarrow {}^3RF^*$$
$$^3RF^* + O_2 \longrightarrow RF^{\cdot+} + \mathbf{O_2^{\cdot-}}$$

Superoxide anion is formed by radiolysis of a dilute aqueous solution, in which most of the energy is absorbed by water and produces electrons along with ionized water ($H_2O^+$) and excited water ($H_2O^*$) in a short time (Halliwell and Gutteridge, 2001). The electrons are surrounded by water molecules within $10^{-12}$-$10^{-11}$ s. The hydrated electrons ($e_{aq}^-$) have a standard reduction potential of -2.8 V and reduce oxygen to superoxide anion in the presence of enough oxygen.

$$2 H_2O \longrightarrow H_2O^+ + H_2O^* + e_{aq}^-$$
$$^3O_2 + e_{aq}^- \longrightarrow \mathbf{O_2^{\cdot-}}$$

Superoxide anion radical is relatively stable under anhydrous conditions, but not in aqueous media. It rarely oxidizes the lipids directly because its reduction potential (0.9 V) is not strong enough to abstract hydrogen from lipids (Bielski et al., 1983). Superoxide anion contributes indirectly to the lipid oxidation via formation of hydroxyl radical, a strong oxidant by Haber-Weiss reaction. Superoxide anion slowly and irreversibly oxidizes tocopherols in organic solvent and produces relatively stable tocophroxyl radical (Csallany and Ha, 1992), which can decrease lipid oxidation. But under aqueous conditions this reaction rarely occurs (Arudi et al., 1983; Halliwell and Gutteridge, 2001). Tocopherols react with hydroxyl radicals and produce tocopherol dimer, tocopherol dihydroxy dimer, or tocopheryl quinone, under aqueous conditions (Csallany and Ha, 1992).

## Hydrogen Peroxide

Hydrogen peroxide, the protonated form of peroxide ion, is formed by the reaction of two molecules of hydroxyl radicals (5 x $10^9$ $M^{-1}s^{-1}$; Halliwell and Gutteridge, 2001) and the dismutation of superoxide anion (Bielski et al., 1983).

$$O_2^{\cdot-} + O_2^{\cdot-} + 2H^+ \xrightarrow{\text{Dismutation}} \mathbf{H_2O_2} + O_2$$

Superoxide dismutase greatly accelerates the dismutation reaction at 1.6 x $10^9$ $M^{-1}s^{-1}$ (Halliwell and Gutteridge, 2001). Uncatalyzed dismutation of superoxide anion depends strongly on the pH of the solution. Hydrogen peroxides are formed faster at low pH; 1 x $10^2$ and 5 x $10^5$ $M^{-1}s^{-1}$ at pH 11 and pH 7, respectively (Halliwell and

Gutteridge, 2001). This is because superoxide anion radical with a pKa of 4.88 is present in its protonated form, hydroperoxyl radical, at low pH. Reactions among hydroperoxyl radicals produce hydrogen peroxide (Buettner and Hall, 1987). Hydrogen peroxide is not directly involved in the initiation of lipid oxidation due to its low reduction potential (0.3 V), however, it contributes indirectly to the lipid oxidation by generating hydroxyl radicals in the presence of light or iron (Choe and Min, 2005).

## Ozone

Ozone is produced by the photodissociation of molecular triplet oxygen into oxygen atoms, which then react with oxygen molecules (Halliwell and Gutteridge, 2001).

$$O_2 \xrightarrow{h\nu} O\cdot + O\cdot$$
$$O\cdot + O_2 \longrightarrow O_3$$

Ozone has a standard reduction potential of 2.1 V (Eren, 2006) and is a powerful oxidizing agent. Treatment of sardine (*Sardina pilchardus*) with ozone increased significantly ($p < 0.05$) the formation of peroxides and thiobarbituric acid reactive substances (Losada et al., 2004). Ozone readily reacts with unsaturated lipids and the reaction is faster in lipids with higher unsaturation, however, it hardly reacts with saturated lipids (Fredrick and Heath, 1970). Exposure of ground soybeans to 1.5 ppm ozone induced 1.7-5.4% destruction of linoleic acid and 10-22% destruction of linolenic acid (Brooks and Csallany, 1978), but there was no reaction of stearic acid with ozone (Pryor et al., 1991).

Ozone adds directly to double bonds in unsaturated lipids and produces primary ozonide (molozonide; 1,2,3-trioxolane), which decomposes to aldehydes and carbonyl oxide (Fig. 2.11). The carbonyl oxide can subsequently react with an aldehyde to give a secondary ozonide, Criegee ozonide (Squadrito et al., 1992). Ozonation of methyl oleate in hexane produced an 89% yield of the Criegee ozonides (Pryor and Wu, 1992). The secondary ozonides are more stable than the primary ozonides; however, they can decompose to produce aldehydes and carboxylic acid. Nonanedioic acid and nonanoic acid are good examples of oxidation products of oleic acid by ozone (Sparks et al., 2006).

Carbonyl oxides can react with water to produce metastable hydroxyl hydroperoxide which hydrolyzes to hydrogen peroxide and aldehydes (Pryor et al., 1991; Santrock et al., 1992) as shown in Fig. 2.12. Although hydrogen peroxide is a principal compound in the oxidation of unsaturated lipid emulsions by ozone, hydrogen peroxide was not found in the reaction of ozone with unsaturated lipids in carbon tetrachloride having no water (Pryor et al., 1991). Ozone produced stable ethoxyhydroperoxides by oxidizing linoleic acid (Heath and Tappel, 1976), phosphatidylcholine (Tagiri-Endo et al., 2002), and cholesterol (Endo et al., 2004). In aqueous solutions, ozone decomposes and forms hydroxyl radical at a low rate (Halliwell and Gutteridge, 2001).

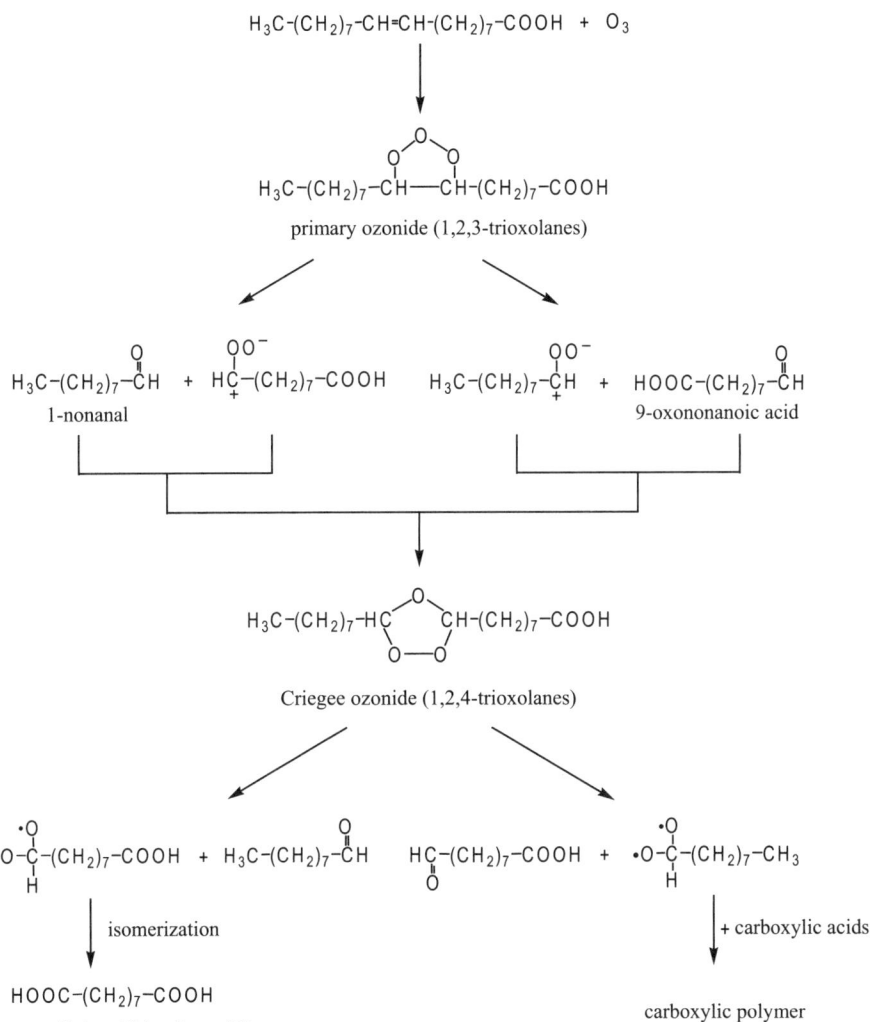

Fig. 2.11. Reaction of ozone with oleic acid.

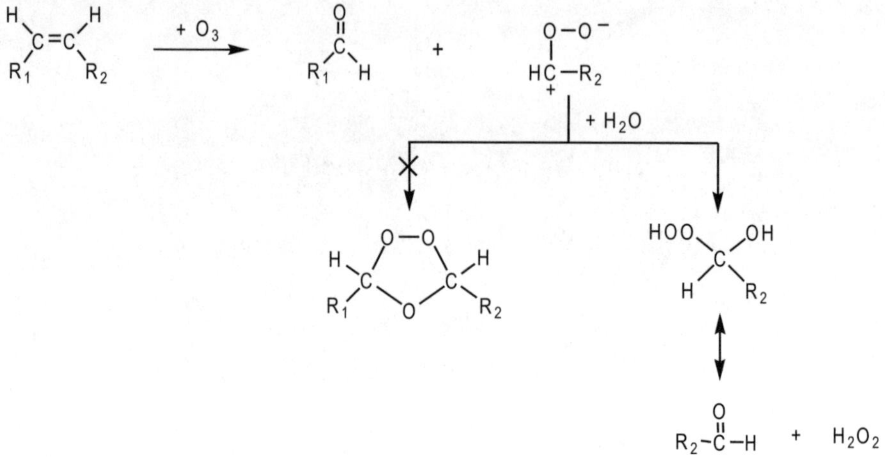

**Fig. 2.12.** Production of hydrogen peroxide and aldehydes from the reaction of lipids with ozone in the presence of water.

## References

Aikens, J.; and J.A. Dix. Perhydroxyl Radical (HOO·) Initiated Lipid Peroxidation, *J. Biol. Chem.* **1991**, *266*, 15091.

Arudi, R.L.; M.W. Sutherland; and B.H.J. Bielski. Oxy Radicals and Their Scavenger Systems. *Molecular Aspects*, Cohen, G., and R.A. Greenwald, Eds.; Elsevier: Amsterdam, 1983; Vol. 1, pp. 26-31.

Athar, M.; H. Mukhtar; C.A. Elmets; . Tarif Zaim; J.R. Lloyd; and D.R. Bickers. In situ Evidence for the Involvement of Superoxide Anions in Cutaneous Porphyrin Photosensitization, *Biochem. Biophys. Res. Commun.* **1988**, *151*, 1054-1059.

Beutner, S.; B. Bloedorn; S. Frixel; I.H. Blanco; T. Hoffmann; H-D. Martin; B. Mayer; P. Noack; C. Ruck; M. Schmidt; et al. Quantitative Assessment of Antioxidant Properties of Natural Colorants and Phytochemicals: Carotenoids, Flavonoids, Phenols, and Indigoids. The Role of β-Carotene in Antioxidant Functions. *J. Sci. Food Agr.* **2001**, *81*, 559-568.

Bielski, B.J.H.; R.L. Arudi; and M.W. Sutherland. A Study of The Reactivity of $HO_2/O_2^-$ with Unsaturated Fatty Acids. *J. Biol. Chem.* **1983**, *258*, 4759-4761.

Bowry, V.W.; and R. Stocker. Tocopherol-Mediated Peroxidation. The Prooxidant Effect of Vitamin E on the Radical-Initiated Oxidation on Human Low-Density Lipoprotein. *J. Am. Chem. Soc.* **1993**, *115*, 6029-6044.

Brooks, R.I.; and A.S. Csallany. Effects of Air, Ozone, and Nitrogen Dioxide Exposure on the Oxidation of Corn and Soybean Lipids. *J. Agric. Food Chem.* **1978**, *26*, 1203-1209.

Buettner, G.R. The Pecking Order of Free Radicals and Antioxidants: Lipid Peroxidation, α-Tocopherol, and Ascorbate. *Arch. Biochem. Biophys.* **1993**, *300*, 535–543.

Buettner, G.R.; and T.D. Hall. Superoxide, Hydrogen Peroxide, and Singlet Oxygen in Hematoporphyrin Derivative-Cysteine, -NADH and -Light Systems. *Biochim. Biophys. Acta.* **1987**, *923*, 501-507.

Buettner, G.R.; and L.W. Oberley. The Apparent Production of Superoxide and Hydroxyl

Radicals by Hematoporphyrin and Light as Seen by Spin-Trapping. *FEBS Lett.* **1980**, *121*, 161-164.

Burton, G.W.; and K.U. Ingold. β-Carotene: An Unusual Type of Lipid Antioxidant. *Science* **1984**, *224*, 569-573.

Choe, E.; and D.B. Min. Chemistry and Reactions of Reactive Oxygen Species in Foods. *J. Food Sci.* **2005**, *70*, R142-R159.

Csallany, A.S.; and Y.L. Ha. α-Tocopherol Oxidation Mediated by Superoxide Anion ($O_2^-$). I. Reactions in Aprotic and Protic Conditions. *Lipids* **1992**, *27*, 195-200.

Das, K.C.; and C.K. Das. Curcumin (diferuloylmethane), a Singlet Oxygen ($^1O_2$) Quencher. *Biochem. Biophys. Res. Commun.* **2002**, *295*, 62-66.

Decker, E.A. Antioxidant Mechanisms. *Food Lipids*, 2nd edn.; Akoh, C.C., and Min, D.B., Eds.; Marcel Dekker Inc.: New York, 2002, pp. 517-542.

Edge, R.; E.J. Land; D.J. McGarvey; M. Burke; and T.G. Truscott. The Reduction Potential of the β-Carotene$^{-+}$/ β-Carotene Couple in an Aqueous Micro-Heterogeneous Environment. *FEBS Lett.* **2000**, *471*, 125-127.

Edge, R.; D.J. McGarvey; and T.G. Truscott. The Carotenoids as Anti-Oxidants - A Review. *J. Photochem. Photobiol. B-Biol.* **1997**, *41*, 189-200.

Endo, M.T.; K. Nakagawa; T. Sugawara; K. Ono; and T. Miyazawa. Ozonation of Cholesterol in the Presence of Ethanol: Identification of a Cytotoxic Ethoxyhydroperoxides Molecule. *Lipids* **2004**, *39*, 259-264.

Eren, H.A. Afterclearing by Ozonation: A Novel Approach for Disperse Dyeing of Polyester. *Color. Technol.* **2006**, *122*, 329-333.

Fredrick, P.E.; and R.L. Heath. Ozone-Induced Fatty Acid and Viability Changes in *Chlorella*. *Plant Physiol.* **1970**, *55*, 15-19.

Fukuzawa, K.; and J.M. Gebicki. Oxidation of α-Tocopherol in Micelles and Liposomes by the Hydroxyl, Perhydroxyl and Superoxide Free Radicals. *Arch. Biochem. Biophys.* **1983**, *226*, 242-251.

Girotti, A.W. Lipid Hydroperoxide Generation, Turnover, and Effector Action in Biological Systems. *J. Lipid Res.* **1998**, *39*, 1529-1542.

Haila, K. *Effects of Carotenoids and Carotenoid-Tocopherol Interaction on Lipid Oxidation in vivo*. Thesis; University of Helsinki: Viikki, Finland, 1999, p 22.

Halliwell, B.; and J.M.C. Gutteridge. *Free Radicals in Biology and Medicine*, 3rd edn., Oxford University Press Inc.: New York, 2001.

Haseloff, R.F.; B. and Ebert. Generation of Free Radicals by Photoexcitation of Pheophorbide a, Haematoporphyrin, and Protoporphyrin. *J. Photochem. Photobiol. Biology* **1989**, *3*, 593-602.

Heath, R.L.; and A.L. Tappel. A New Sensitive Assay for the Measurement of Hydroperoxides. *Anal. Biochem.* **1976**, *76*, 184-191.

Heaton, F.W.; N. and Uri. The Aerobic Oxidation of Unsaturated Fatty Acids and Their Esters. Cobalt Stearate-Catalyzed Oxidation of Linoleic Acid. *J. Lipid Res.* **1961**, *2*, 152-160.

Hiatt, R.; T. Mill; K.C. Irwin; T.R. Mayo; C.W. Gould; and J.K Castleman. Homolytic Decomposition of Hydroperoxides. *J. Org. Chem.* **1968**, *33*, 1416-1441.

Hu, Y-Z.; and L-J. Jiang. Generation of Semiquinone Radical Anion and Reactive Oxygen ($^1O_2$, $O_2^- \bullet$, and $\bullet OH$) During the Photosensitization of a Water-Soluble Perylenequinone Derivative. *J. Photochem. Photobiol.* **1996**, *33*, 51-59.

Iannone, A.; C. Rota; S. Bergamini; A. Tomasi; and L.M. Canfield. Antioxidant Activity of Carotenoids: An Electron-Spin Resonance Study on β-Carotene and Lutein Interaction with Free Radicals Generated in a Chemical System. *J. Biochem. Mol. Toxicol.* **1998**, *12*, 299-304.

Ingold, K.U.; V.W. Bowry; R. Stocker; and C. Walling. Autoxidation of Lipids and Antioxidation by α-Tocopherol and Ubiquinol in Homogeneous Solution and in Aqueous Dispersions of Lipids: Unrecognized Consequences of Lipid Particle Size as Exemplified by Oxidation of Human Low Density Lipoprotein. *Proc. Natl. Acad. Sci. USA* **1993**, *90*, 45-49.

Ingold, K.U.; T. Paul; M.J. Young; and L. Doiron. Invention of the First Azo Compound to Serve as a Superoxide Thermal Source under Physiological Conditions: Concept, Synthesis, and Chemical Properties. *J. Am. Chem. Soc.* **1997**, *119*, 12364-12365.

Iuliano, L.; J.Z. Pedersen; G. Rotilio; D. Ferro; and F. Violi. A Potent Chain-Breaking Antioxidant Activity of the Cardiovascular Drug Dipyridamole. *Free Radical Biol. Med.* **1995**, *18*, 239-47.

Jadhav, S.J.; S.S. Nimbalkar; A.D. Kulkarni; and D.L. Madhavi. Lipid Oxidation in Biological and Food Systems. *Food Antioxidants*, Madhavi, D.L., Deshpande, S.S., and Salunke, D.K., Eds.; Mercel Dekker Inc.: New York, 1996, pp. 5-63.

Kamal-Eldin, A.; and L.A. Appelqvist. The Chemistry and Antioxidant Properties of Tocopherols and Tocotrienols. *Lipids* **1996**, *31*, 671-701.

Kanavouras, A.; P. Hernandez-Munoz; F. Coutelieris; and S. Selke. Oxidation-Derived Flavor Compounds as Quality Indicators for Packaged Olive Oil. *J. Am. Oil Chem. Soc.* **2004**, *81*, 251-257.

Kehrer, J.P. The Haber-Weiss Reaction and Mechanisms of Toxicity. *Toxicology* **2000**, *149*, 43-50.

Khan, A.U.; and M. Kasha. Singlet Molecular Oxygen in the Haber-Weiss Reaction. *Proc. Natl. Acad. Sci. USA* **1994**, *91*, 12365-12367.

Kilic, B.; and M.P. Richards. Lipid Oxidation in Poultry Donor Kebap: Prooxidative and Antioxidative Factors. *J. Food Sci.* **2003**, *68*, 686-689.

Koppenol, W.H. Oxyradical Reactions: From Bond-Dissociation Energies to Reduction Potentials, *FEBS Lett.* **1990**, *264*, 165-167.

Korycka-Dahl, M.B.; and T. Richardson. Activated Oxygen Species and Oxidation of Food Components. *Crit. Rev. Food Sci. Nutr.* **1978**, *10*, 209-241.

Kruk, I.; H.Y. Aboul-Enein; T. Michalska; K. Lichszteld; and A. Kladna. Scavenging of Reactive Oxygen Species by the Plant Phenols Genistein and Oleuropein. *Luminescence* **2005**, *20*, 81-89.

Lee, J.; N. Koo, and D.B. Min. Reactive Oxygen Species, Aging, and Antioxidant Nutraceuticals. *Comprehensive Rev. Food Sci. Food Safety* **2004**, *3*, 21-33.

Lee, S.H.; and I.A. Blair. Characterization of 4-Oxo-2-nonenal as a Novel Product of Lipid Peroxidation. *Chem. Res. Toxicol.* **2000**, *13*, 698-702.

Lee, S.H.; T. Oe; and I.A. Blair. Vitamin C-Induced Decomposition of Lipid Hydroperoxides to Endogenous Genotoxins. *Science* **2001**, *292*, 2083-2086.

Lesniak, W.; V.L. Pecoraro; and J. Schacht. Ternary Complexes of Gentamicin with Iron and Lipid Catalyze Formation of Reactive Oxygen Species. *Chem. Res. Toxicol.* **2005**, *18*, 357-364.

Liebler, D.C. Antioxidant Reaction of Carotenoids. *Ann. N.Y. Acad. Sci.* **1993**, *691*, 20-31.

Losada, V.; J. Barros-Velazquez; J.M. Gallardo; and S.P. Aubourg. Effect of Advanced Chilling Methods on Lipid Damage during Sardine (*Sardina pilchardus*) Storage. *Eur. J. Lipid Sci. Technol.* **2004**, *106*, 844-850.

Luo, Y-R.; and J.L. Holmes. The Stabilization Energies of Polyenyl Radicals. *Chem. Phys. Lett.* **1994,** *228,* 329-332.

Martinez, J.; and J.J. Moreno. Influence of Superoxide Radical and Hydrogen Peroxide on Arachidonic Acid Mobilization. *Arch. Biochem. Biophys.* **1996,** *336,* 191-198.

Min, D.B.; and J.M. Boff. Lipid Oxidation of Edible Oil. *Food Lipids,* 2nd edn.; Akoh, C.C., and Min, D.B. Eds.; Marcel Dekker Inc.: New York, 2002, pp 335-364.

Mortensen, A.; and L.H. Skibsted. Reactivity of β-Carotene Towards Peroxyl Radicals Studied by Laser Flash and Steady-State Photolysis. *FEBS Letters* **1998,** *426,* 392-396.

Mortensen, A.; L.H. Skibsted; and T.G. Truscott. The Interaction of Dietary Carotenoids with Radical Species. *Arch. Biochem. Biophy.* **2001,** *385,* 13-19.

Mukai, K.; and Y. Okauchi. Kinetic Study of the Reaction Between Tocopheroxyl Radical and Unsaturated Fatty Acid Esters in Benzene. *Lipids* **1989,** *24,* 936-939.

Neuzil, J.; P.K. Witting; and R. Stocker. α-Tocopheryl Hydroquinone Is an Efficient Multifunctional Inhibitor of Radical-Initiated Oxidation of Low Density Lipoprotein Lipids. *Proc. Natl. Acad. Sci USA* **1997,** *94,* 7885-7890.

Porter, N.A.; B.A. Weber; and K.J. Smith. Autoxidation of Polyunsaturated Fatty Acids. Factors Controlling the Stereochemistry of Product Hydroperoxides. *J. Am. Chem. Soc.* **1980,** *102,* 5597-5601.

Pryor, W.A.; and M. Wu. Ozonation of Methyl Oleate in Hexane, in a Thin Film, in SDS Micelles, and in Distearoylphosphatidylcholine Liposomes: Yields and Properties of the Criegee Ozonide. *Chem. Res. Toxicol.* **1992,** *5,* 505-511.

Pryor, W.A.; B. Das; and D.F. Church. The Ozonation of Unsaturated Fatty Acids: Aldehydes and Hydrogen Peroxide as Products and Possible Mediators of Ozone Toxicity. *Chem. Res. Toxicol.* **1991,** *4,* 341-348.

Reische, D.W.; D.A. Lillard; and R.R. Eitenmiller. Antioxidants. *Food Lipids,* 2nd edn.; Akoh, C.C., and Min, D.B. Eds; Marcel Dekker Inc.: New York, 2002, pp. 489-516.

Rush, J.D.; Z. Maskos; and W.H. Koppenol. Reactions of Iron(II) Nucleotide Complexes with Hydrogen Peroxide. *FEBS Lett.* **1990,** *261,* 121-123.

Salem, I.A.; M. El-Maazawi; and A.B. Zaki. Kinetics and Mechanisms of Decomposition Reaction of Hydrogen Peroxide in Presence of Metal Complexes. *Int. J. Chem. Kinetics* **2000,** *32,* 643-666.

Santrock, J.; R.A. Gorski; and J.F. O'Gara. Products and Mechanism of the Reaction of Ozone with Phospholipids in Unilamella Phospholipids Vesicles. *Chem. Res. Toxicol.* **1992,** *5,* 134-141.

Schaich, K.M. Metals and Lipid Oxidation. Contemporary Issues. *Lipids* **1992,** *27,* 209-218.

Simic, M.C.; S.V. Jovanovic; and E. Niki. Mechanisms of Lipid Oxidative Processes and Their Inhibition. *Lipid Oxidation in Foods,* St. Angelo, A.J. Ed.; ACS Press: New York, 1992, pp. 14-32.

Simic, M.G. Kinetic and Mechanistic Studies of peroxyl, Vitamin E and Antioxidant Free Radicals by Pulse Radiolysis. Autoxidation in Foods and Biological Systems, Simic, M.G., and Karel, M., Eds.; Plenum: New York, 1980, pp.17–26.

Sparks, D.L.; R. Hernandez; T. French; H. Toghiani; R.K. Toghiani; E. Alley; and M.E. Zappi. Supercritical Fluid Oxidation of Oleic Acid. 2006 AIChE Annual Meeting, American Institute of Chemical Engineers. San Francisco, California, U.S.A. 2006. (http://aiche.confex.com/aiche/2006/preliminaryprogram/abstract_60342.htm)

Squadrito, G.L.; R.M. Uppu; R. Cueto; and W.A. Pryor. Production of the Criegee Ozonide during the Ozonation of 1-Palmitoyl-2-oleoyl-*sn*-glycero-3-phosphocholine Liposomes,

*Lipids* **1992**, *27*, 955-958.

Tagiri-Endo, M.; K. Ono; K. Nakagawa; M. Yotsu-Yamashita; and T. Miyazawa. Ozonation of PC in Ethanol: Separation and Identification of a Novel Ethoxyhydroperoxide. *Lipids* **2002**, *37*, 1007-1012.

Tsuchihashi, H.; M. Kigoshi; M. Iwatsuku; and E. Niki. Action of β-Carotene as an Antioxidant Against Lipid Peroxidation. *Arch. Biochem. Biophys.* **1995**, *323*, 137-147.

Upston, J.M.; A.C. Terentis; K. Morris; J.F. Keaney, Jr.; and R. Stocker. Oxidized Lipid Accumulates in the Presence of α-Tocopherol in Atherosclerosis. *Biochem. J.* **2002**, *363*, 753-760.

Velasco, J.; M.L. Andersen; and L.H. Skibsted. Evaluation of Oxidative Stability of Vegetable Oils by Monitoring the Tendency to Radical Formation. A Comparison of Electron Spin Resonance Spectroscopy with the Rancimat Method and Differential Scanning Calorimetry. *Food Chem.* **2003**, *77*, 623-632.

Von Frijtag Drabbe Kunzel, J.K.; J. Van der Zee; and A.P. Ijzerman. Radical Scavenging Properties of Adenosine and Derivatives in vitro. Drug Development Res. 1996 37, 48-54.

Watanabe, T.; H. Yeranishi; Y. Honda; and M. Kuwahara. A Selective Lignin-Degrading Fungus, *Ceriporiopsis subvermispora*, Produces Alkylitaconates that Inhibit the Production of a Cellulolytic Active Oxygen Species, Hydroxyl Radical in the Presence of Iron and $H_2O_2$. *Biochem. Biophys. Res. Comm.* **2002**, *297*, 918-923.

Weinstein, J.; and B.H.J. Bielski. Kinetics of the Interaction of Radicals with Hydrogen Peroxide: The Haber-Weiss Reaction. *J. Am. Chem. Soc.* **1979**, *101*, 58-62.

Witting, P.K.; D.J. Douglas; and A.G. Mauk. Reaction of Human Myoglobin and $H_2O_2$: Involvement of a Thiyl Radical Produced at Cysteine 110. *J. Biol. Chem.* **2000**, *275*, 20391-20398.

Yamamoto, Y. Role of Active Oxygen Species and Antioxidants in Photoaging. *J. Dermatol. Sci.* **2001**, *27*, S1-S4 Suppl, 1.

# 3

# Oxidation of Long-Chain Polyunsaturated Fatty Acids

Kazuo Miyashita
Faculty of Fisheries Sciences, Hokkaido University, Hakodate 041-8611, Japan

## Introduction

Lipids account for about 40% of total calories in most of the industrialized countries and play a major role in human nutrition and health. They provide a concentrated source of energy and reduce the bulk of the diet. They are indispensable parts of the human diet as they are major sources of essential fatty acids and provide fat-soluble substances like vitamins and carotenoids. The type, nature, and composition of lipids and fatty acids depend on the lipid source. Plants and animal sources account for most of the dietary lipids. Long-chain polyunsaturated fatty acids (PUFA) of the α-linolenic acid family (n-3 fatty acids) are typical of marine lipids, while PUFA belonging to linoleic acid (LA) family (n-6 fatty acids) are predominant in vegetable oils. However, some seed oils, such as flaxseed and perilla, do contain n-3 PUFA in considerable quantities; whereas, soybean, walnut, and rapeseed oils contain moderate amounts of n-3 fatty acids. Like vegetable oils, the fats derived from terrestrial animal sources largely consist of PUFA belonging to LA family. Marine lipids contain substantial amounts of long-chain PUFA. Metabolic effects of the long-chain PUFA, especially in human nutrition, have attracted a major interest among biochemists and nutritionists alike.

The importance of n-3 and n-6 PUFA on human health has been proven beyond any doubt through research across the globe. They have several beneficial health and physiological effects. The functions of each n-3 or n-6 PUFA have attracted consumer attention and are often used in functional foods and nutraceuticals. Eicosapentaenoic acid (EPA; 20:5n-3) and docosahexaenoic acid (DHA; 22:6n-3) are the two PUFA found in marine lipids. These two long-chain PUFA have been shown to cause significant biochemical and physiological changes in the body (Li et al., 2003; Lands, 2005; Sinclair, 2005; Shahidi, and Miraliakbari, 2006; Narayan et al., 2006a) that most often result in positive influence on human nutrition and health. However, lipid oxidation products cause undesirable flavors and lower the nutritional quality and safety of lipid-containing foods. Hence, oxidative deterioration of functional PUFA still remains as the biggest problem in utilizing PUFA-rich oil for food applications. Owing to this, lipid peroxidation has received considerable attention in all investigations on the dietary effects of these n-3 and n-6 PUFA because of its possible contribution to damage biological systems (Kaneda and Ishii, 1954; Benedetti et al., 1980; Piche et al., 1988; Garrido et al., 1989; Hu et al., 1989; Burns and Wagner, 1991; Fritsche and Johnston, 1988).

Lipid oxidation proceeds through a free radical chain reaction consisting of chain initiation, propagation, and termination processes (Fig. 3.1) (Frankel, 1980, 1982, 1985, 1998a; Min and Smouse, 1985, 1989; Porter et al., 1995; Kamal-Eldin et al., 2003). The rate-limiting step in the reaction is abstraction of hydrogen radical (H•) from substrate lipids (LH) to form lipid free radicals (L•). In the propagation stage of autoxidation, fatty alkyl radicals react with molecular oxygen to form peroxy radicals. The peroxy radical abstracts a hydrogen atom from another unsaturated fatty compound to form a hydroperoxide and an alkyl radical. The latter reacts with molecular oxygen in a repetition of the first propagation reaction. The initially formed mono-hydroperoxide (MHP) may decompose subsequently to yield free radicals such as alkoxy and hydroxy radicals. These radicals serve as initiators for these reactions. Therefore, the chain-breaking antioxidants are of considerable practical importance in preserving PUFA from oxidative deterioration. These antioxidants inhibit or retard lipid oxidation by interfering with either chain propagation or initiation by readily donating hydrogen atoms to lipid peroxy radicals (Fig. 3.1). Phenolic compounds with bulky alkyl substituents, such as BHA, BHT, TBHQ, and tocopherols, are effective chain-braking antioxidants as they produce stable and relatively unreactive antioxidant radicals (A•) (Fig. 3.1). Lipid peroxidation is initiated by minor compounds present in oils. Metal ions promote the initiation and decomposition of MHP. Pigments in foods can serve as photosensitizers by absorbing visible or near-UV light to become electronically excited. The photoxidation provides an important way to produce MHP that initiate lipid peroxidation and flavor reversion. Several types of compounds, such as metal inactivators, can inhibit lipid oxidation by mechanisms that do not involve deactivation of free radical chain reactions.

Although mechanisms for free radical oxidation of PUFA and the inhibition of this reaction by antioxidants are well documented (Frankel, 1980, 1982, 1985, 1998a; Min and Smouse, 1985, 1989; Porter et al. 1995; Kamal-Eldin et al., 2003), autoxidation of long-chain PUFA, such as EPA and DHA, has received much attention owing to the problems associated with oxidation of these essential PUFA which makes their utilization in food systems a difficult proposition. Added to this, high intake of these PUFA may increase the oxidative stress on biological systems due to oxidation-related problems. The aim of this article is to highlight effect of oxidation of these PUFA on their stability and usage.

## Oxidation of EPA and DHA

The oxidative stability of each PUFA is inversely proportional to the number of bisallylic positions in the molecule or the degree of unsaturation of the PUFA; therefore, when the relative oxidative stability of typical PUFA are compared in air (bulk phase) or in organic solvents, DHA is most rapidly oxidized, followed by EPA, arachidonic

## Lipid oxidation

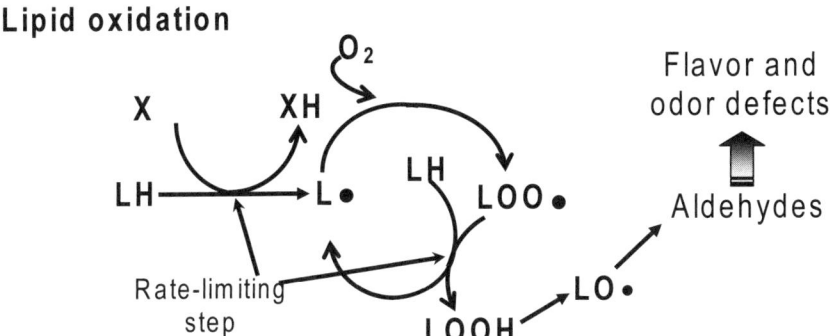

## Lipid antioxidant action

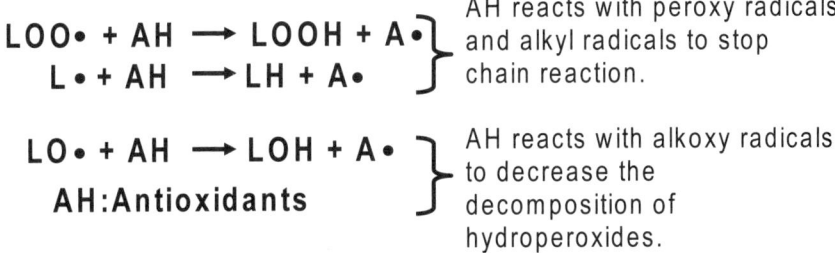

AH reacts with peroxy radicals and alkyl radicals to stop chain reaction.

AH reacts with alkoxy radicals to decrease the decomposition of hydroperoxides.

**Fig. 3.1.** Lipid oxidation and antioxidant mechanism.

acid (AA, 20:4n-6), α-linolenic acid (18:3n-3), and LA (18:2n-6), respectively (Gunstone and Hilditch, 1945; Holman and Elmer, 1947; Miyashita and Takagi, 1986; Cosgrove et al., 1987; Cho et al., 1987a, 1987b; Miyashita et al., 1990). In the free radical oxidation of EPA, hydrogen abstraction occurs at the C7, C10, C13, C16 position, which result in production of a pentadienyl radical between C5 and C9, C8 and C12, C11 and C15, or C14 and C18, respectively (Fig. 3.2). The radical then reacts at either end with oxygen to produce a mixture of 5-MHP and 9-MHP, 8-MHP and 12-MHP, 11-MHP and 15-MHP, or 14-MHP and 18-MHP, respectively (Yamauchi et al., 1983). Since DHA has five bis-allylic methylene groups (Fig. 3.2), there are five possible positions for hydrogen abstraction viz., carbons 6, 9, 12, 15, or 18. Thus, oxygen can attack carbons at either end of the pentadienyl radical, and the resulting MHP isomers are those with hydroperoxide substitution on carbons 4, 8, 7, 11, 10, 14, 13, 17, 16, 20 for DHA (VanRollins and Murphy, 1984).

Oxidation of PUFA containing more than two double bonds, such as α-linolenic acid and AA, produces a significant amount of hydroperoxy epidioxides as the main oxidation products–apart from monohydroperoxides-at an early stage of oxidation (Chan et al., 1980; Coxon et al., 1981; Neff et al., 1981). Hydroperoxy epidioxides are formed in autoxidation of EPA (Yamauchi et al., 1985) and DHA. Hydroperox-

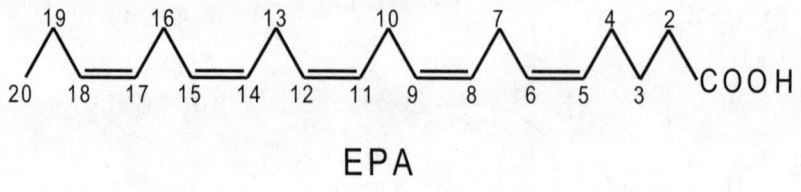

EPA

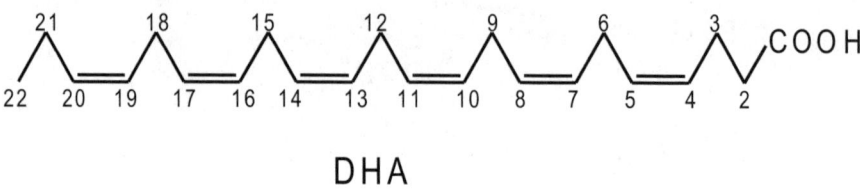

DHA

**Fig. 3.2.** Structures of EPA and DHA.

ides of EPA or DHA are more easily decomposed to form free radicals than those of LA, α-linolenic acid, or AA. This highly reactive EPA- or DHA-hydroperoxide may result in a different and complicated composition of oxidation products when compared with other PUFA.

When ethyl esters of α-linolenic acid, EPA and DHA were oxidized at 5°C in the dark (Cho et al., 1987a), the ratios of OOH-oxygen to total oxygen consumption in the oxidation of EPA or DHA were 50-70%, even in the early stages of oxidation. Remaining oxygen was used to form secondary oxidation products, mainly polymers. Gel permeation chromatography established that over 70% of the polar materials from both ethyl EPA and DHA to be dimers and polymers (Cho et al. 1987a). On the other hand, the ratio of OOH-oxygen in ethyl α-linolenate was slightly higher, 70-80%. In this case, most of the other oxygen was consumed in the formation of hydroperoxy epidioxides. Linoleate monohydroperoxides are much more stable than those of α-linolenate, arachidonate, EPA, and DHA. Therefore, the ratio of OOH-oxygen in the early stage of linoleate oxidation has been reported to be over 90% (Miyashita et al., 1982), although epoxyhydroxy, epoxy, dihydroxy, and trihydroxy compounds were identified from autoxidized methyl linoleate and further oxidation of linoleate MHP (Neff et al. 1978). Epoxy and dihydroxy compounds are also found in the autoxidation of methyl linolenate (Neff et al., 1981). However, the major secondary oxidation products in linolenate are hydroperoxy epidioxides (Neff et al., 1981). Although there are a limited number of reports on the oxidation products of EPA and DHA, the major secondary oxidation products will be polymers (Cho et al., 1987a, 1987b).

Although estimation of peroxide value (PV) is the most widely used method to assess lipid oxidation, it is not necessarily a good indication of oxidation in EPA and DHA because of the instability of hydroperoxides formed. The determination of secondary products-such as polymers, anisidine value, 2-thiobarbituric acid value, sensory analysis, and gas chromatographic (GC) methods for volatile compounds-are more important in evaluating the oxidative deterioration of EPA and DHA. Volatile compounds analysis is a better method. The GC method is capable of determining volatile oxidation products that are either directly responsible for or serve as markers of flavor development in oxidized EPA, DHA, and fish oil. Volatile compounds suitable for GC determinations are mainly aldehydes and hydrocarbons. Jacobsen (1999) conclusively showed that there is no correlation between PV and the taste panel response on fish oil enriched spreads. However, the data on volatile compounds obtained by headspace methods have been demonstrated to correlate well with sensory data (Frankel, 1998b). A large number of volatile compounds are found in fish oil oxidation (Karahadian and Lindsay, 1989; Hsieh et al., 1989; Aidos et al., 2002). Sixty different volatiles comprising alkenals, alkadienals, alkatrienals, and vinyl ketones have been identified in fish oil enriched milk (Venkateshwarlu et al., 2004a). 1-Penten-3-one, (Z)-4-heptenal, 1-octen-3-one, 1,5-octadien-3-one, (E,E)-2,4-heptadienal, and (E,Z)-2,6-nonadienal have been reported as the most potent odorants in fish oil (Venkateshwarlu et al., 2004a, 2004b). However, sensory impact of individual or combinations of volatile oxidation compounds in food systems containing fish oil is not yet established. Hence, further research is needed to use this method in evaluation of fish oil deterioration.

## Oxidative Stability of EPA and DHA in Phospholipids

Marine lipids contain a high percentage of long-chain PUFA, such as EPA and DHA. Due to their high degree of unsaturation, these PUFA are much more susceptible to oxidation compared to α-linolenic acid and LA found in vegetable oils. On the other hand, EPA and DHA are relatively stable to oxidation in marine biological systems. Furthermore, it has been reported that DHA ingestion does not increase lipid peroxides to the level one would expect from the 'peroxidizability' index of the tissue total lipids (Kubo et al., 1998). In particular, lipid peroxide levels in the brain and testis decreased when DHA was given to the animals. Ando et al. (2000) also examined the effects of fish oil on lipid peroxidation in rat organs and found that levels of phospholipid hydroperoxides and thiobarbituric acid reactive substances (TBARS) in rat organs fed a fish oil diet were similar to those of the safflower-oil diet group. Wander and Du (2000) measured plasma lipid peroxidation after supplementation with EPA and DHA from fish oil and tocopherol in postmenopausal women. They found that neither the concentration of plasma TBARS nor protein oxidation changed after fish oil supplementation. These results indicate a difference in the oxidative stability of highly unsaturated fatty acids, such as DHA and AA ,between biological systems and bulk phases.

Many foods and biological systems are complex, multi-component and heterogeneous systems, in which lipids are present along with various types of other components in aqueous medium. Furthermore, high levels of EPA and DHA in marine lipids imply the occurrence of strong antioxidant systems in marine animal tissues. Antioxidants, such as tocopherols, have been regarded to be most important to prevent marine lipid oxidation. Thus, complex systems with antioxidants will be important to prevent EPA and DHA oxidation.

In biological systems, a large amount of EPA and DHA are present as phospholipids (PL) in the membrane. PL has been shown to act synergistically with phenolic antioxidants, such as tocopherols. This is evidenced by the fact that PL containing EPA and DHA would be stable to oxidation in the presence of tocopherol (Hara et al., 1992; King et al., 1992; Ohshima et al., 1993; Segawa et al., 1994, 1995; Koga and Terao, 1995; Takeuchi et al., 1997; Saito and Ishihara, 1997; Bandarra et al. 1999). The mechanism responsible for the synergy of tocopherols and PL is not very well understood, but seems to be related to the involvement of the PL amino group in the enhancement of the antioxidant activity of tocopherols, regeneration of tocopherols, or chelation of prooxidant metal ions. The presence of EPA and DHA in a PL form may be important to prevent oxidation in biological systems (Ohshima et al., 1993; Nara et al., 2000).

Lipids from muscle and eye of squid are mainly composed of PL (>60%) and free cholesterol (30%). Figure 3.3 shows the comparison of oxidative stabilities of these lipids with those of other kinds of marine lipids. Marine lipids used contained high percentages of EPA and DHA, and showed higher average numbers (1.65-2.77) of bis-allylic positions per fatty acid molecule than those of vegetable oils, such as soybean oil (0.65) and safflower oil (0.75). The average number of bis-allylic positions in each lipid was 2.51, 2.77, 1.65, 2.19, and 1.92 for squid muscle, squid eye, tuna orbital, trout egg, and bonito oil, respectively. Judging from the average number of bis-allylic positions, squid eye total lipids (TL) and squid muscle TL were presumed to be oxidized rapidly, however oxidative stabilities of these lipids were higher than that of tuna orbital TL which showed the lowest number of bis-allylic positions. On the basis of PV determinations, tuna orbital TL oxidized rapidly without an induction period (Fig. 3.3). Bonito oil also oxidized rapidly, but it was less susceptible to oxidation than tuna orbital TL. The lower oxidative stabilities of these lipids were confirmed by determining the decrease in unoxidized PUFA (Fig. 3.4) (Nara et al., 2000). Antioxidants are the most important factors that influence oxidative stability of lipids. The difference in the oxidative stability between tuna orbital TL and bonito oil could be due to the higher content of tocopherols in bonito oil than that of tuna orbital TL (Table 3.1). However, it is difficult to explain the higher oxidative stability of squid tissue lipids and trout egg TL only by tocopherol content. As shown in Table 3.2, three kinds of squid tissue-TL and trout egg TL contained PL, suggesting the important role of PL on higher oxidative stabilities of these TL than those of tuna orbital TL and bonito oil.

# Oxidation of Long-Chain Polyunsaturated Fatty Acids

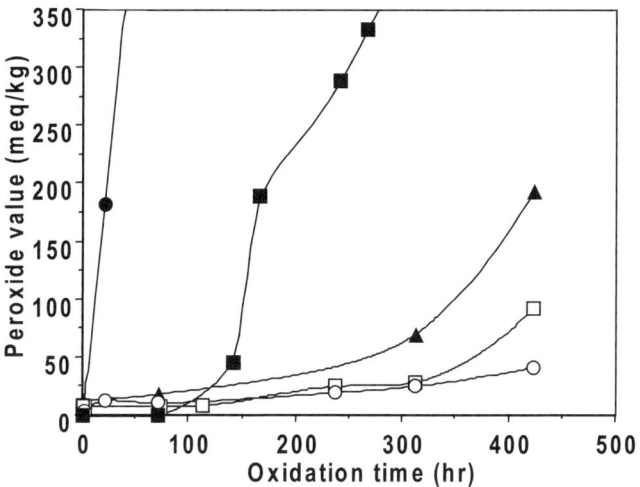

**Fig. 3.3.** Changes in Peroxide Value (PV) of Lipids from marine organisms during autoxidation at 37°C. Squid muscle total lipids (TL) (open circle); squid eye TL (open square); tuna orbital TL (solid circle); trout egg TL (solid triangle); bonito oil (solid square).

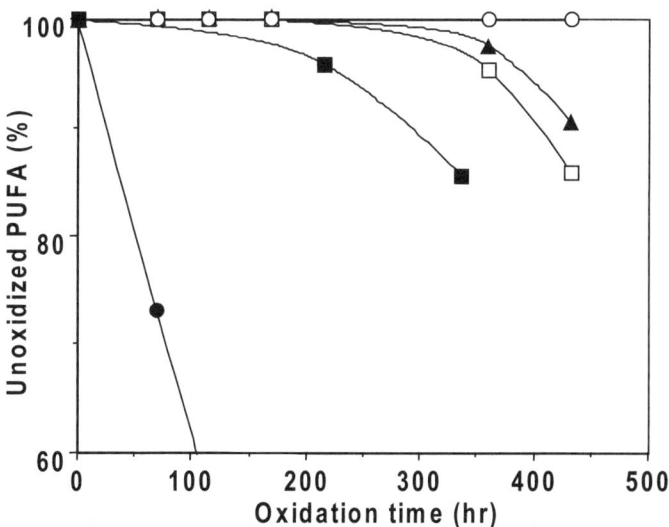

**Fig. 3.4.** Changes in unoxidized polyunsaturated fatty acids (PUFA) in lipids from marine organisms during autoxidation at 37°C. Squid muscle total lipids (TL) (open circle); squid eye TL (open square); tuna orbital TL (solid circle); trout egg TL (solid triangle); bonito oil (solid square).

**Table 3.1.** Tocopherol Contents of Lipids from Marine Organisms

| Tocopherol (μg/g Lipid) | Squid Muscle TL | Squid Eye TL | Tuna Orbital TL | Trout Egg TL | Bonito Oil |
|---|---|---|---|---|---|
| α-Tocopherol | 649.8 | 1198.8 | 541.3 | 215.5 | 253.4 |
| β-Tocopherol | ND | ND | ND | ND | 193.3 |
| γ-Tocopherol | ND | ND | ND | ND | 703.6 |
| δ-Tocopherol | ND | 9.2 | ND | 9.2 | 496.3 |
| Total | 649.8 | 1208.0 | 541.3 | 215.5 | 1646.6 |

ND; not detected

**Table 3.2.** Lipid Profiles of Squid Tissue TL, Tuna Orbital TL, Trout Egg, and Bonito Oil

| Lipid | Squid Muscle TL | Squid Eye TL | Tuna Orbital TL | Trout Egg TL | Bonito Oil |
|---|---|---|---|---|---|
| Triacylglycerols | ND | ND | 99.3 | 76.8 | 99.6 |
| Free Fatty Acids | ND | ND | 0.4 | ND | 0.1 |
| Glycolipids | ND | 6.8 | ND | ND | ND |
| Sterols | 23.7 | 28.3 | ND | 2.2 | 0.3 |
| Phospholipids | 75.6 | 66.4 | 0.2 | 23.1 | ND |

ND; not detected

The relatively higher oxidative stability of marine lipids containing EPA and DHA in PL form in fish roe lipids compared to commercial fish oils has also been documented (Moriya et al., in press). Salmon roe and herring roe lipids contained 23.1 and 73.6% PL, respectively (Table 3.3), while the main lipid class of fish oils was triacylglycerol (TAG). GC analysis showed that the EPA and DHA content of fish roe lipids were higher than those of fish oils. PL contained two acyl chain moieties and one phosphoryl moiety (molecular weight: PC, 166.1; PE, 123.0; PS, 167.1), while TAG contained three acyl chain moieties. Furthermore, both fish roe lipids contained 6.3 and 9.7% cholesterol (MW: 382.7), respectively. Therefore, the EPA and DHA content of salmon and herring roe were relatively higher than those in whole lipid compared with fish oils, which were mostly composed of TAG. Calculation of EPA and DHA content of four kinds of lipids were fish-1, 256mg/g lipid; fish-2, 341mg/g lipid; salmon roe lipids, 351mg/g lipids; herring roe lipids, 291mg/g lipid. The total content of EPA and DHA in four kinds of marine lipids indicates that salmon roe lipids should be oxidized most rapidly followed by fish-2, herring roe lipids, and fish-1.

However, as shown in Figures 3.5 and 3.6, herring roe lipid was most oxidatively stable followed by salmon roe lipid, fish-2, and fish-1. The higher oxidative stability of fish-2 than fish-1 may be due to higher content of tocopherols of fish-2 (Table 3.4). However, it is difficult to explain the higher oxidative stability of fish roe lipids by tocopherol contents alone. The higher oxidative stability of both roe lipids compared with other fish oils would be due to the presence of PL in them. Kashima et al. (1984) reported the higher synergistic effect of PE than that of PC. As shown in Table 3.5, PE content in herring roe lipids was extremely high, but there was no PE in salmon roe. The higher oxidative stability of herring roe, therefore, might also be due to the antioxidant effect of herring roe PE.

Table 3.3. Lipid Class of Marine Lipids Used for Oxidation

| Lipid Class (% of TL) | Fish-1 | Fish-2 | Salmon Roe | Herring Roe |
|---|---|---|---|---|
| Triacylglycerols | 99.6 | 99.8 | 71.8 | 9.3 |
| Free Fatty Acids | 0.1 | 0.2 | ND | 3.8 |
| Phospholipids | ND | ND | 23.1 | 73.6 |
| Sterols + Monoacylglycerols | 0.3 | ND | 5.2 | 12.3 |

ND; not detected

Table 3.4. Tocopherol Contents of Marine Lipids

| Tocopherol (μg/g Lipid) | Fish-1 | Fish-2 | Salmon Roe | Herring Roe |
|---|---|---|---|---|
| α-Tocopherol | 253.4 | 60.2 | 19.6 | 22.9 |
| β-Tocopherol | 193.3 | 45.7 | 214.1 | 258.0 |
| γ-Tocopherol | 703.6 | 376.7 | 11.6 | 7.7 |
| δ-Tocopherol | 496.3 | 2670.9 | 11.3 | 11.5 |
| Total | 1472.6 | 3153.5 | 256.6 | 300.1 |

Table 3.5. PL Class Compositions of Herring Roe and Salmon Roe Lipids

| Lipid Class (% of TL) | Salmon Roe | Herring Roe |
|---|---|---|
| PC | 97.0 | 72.3 |
| PE | ND | 6.6 |
| PS | 2.6 | 8.7 |
| LysoPC | ND | 11.8 |

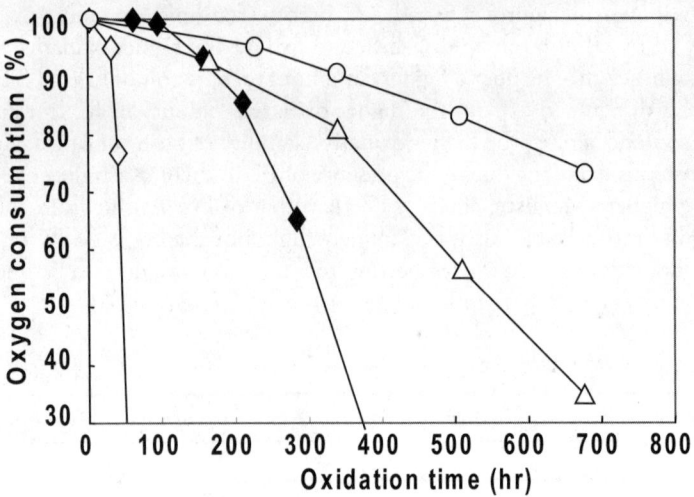

**Fig. 3.5.** Oxygen consumption during the oxidation of fish lipids at 37°C in the dark. Fish-1 (open diamond); fish-2 (solid diamond); salmon roe lipids (open triangle); herring roe lipids (open circle).

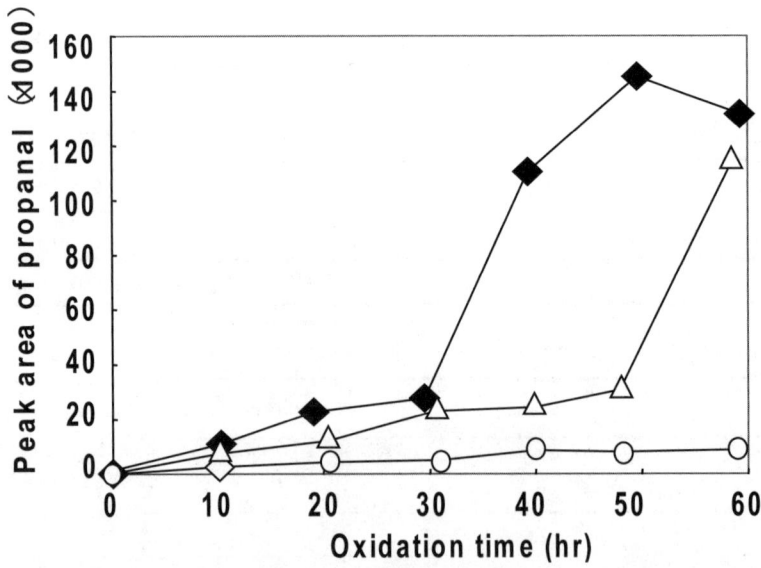

**Fig. 3.6.** Propanal formation during the oxidation of fish lipids at 37°C in the dark. Fish-2 (solid diamond); salmon roe lipids (open triangle); herring roe lipids (open circle).

## Oxidative Stability of DHA in Liposomes

The oxidative stability of PUFA in an aqueous system is dramatically different from that in the bulk phase (Miyashita et al., 1994a, 1995a, 1995b, 1997; Hirano et al., 1997 63; Miyashita, 2002). The oxidative stability in aqueous micelles rises with increasing number of bis-allylic positions in each molecule, and this relationship gets reversed in bulk phase or in organic solvent. A similar trend has also been observed in liposomes of phosphatidylcholine (PC) containing LA, AA, and DHA as a model of cell membrane (Miyashita et al., 1994b; Nara et al., 1995, 1997, 1998; Araseki et al., 2002). When three kinds of 1-palmitoyl (PA; 16:0)-2-PUFA-PC, 1-PA-2-linoleoyl (LA;18:2n-6)-PC (PA-LA); 1-PA-2-arachidonyl (AA; 20:4n-6)-PC (PA-AA); and 1-PA-2-docosahexaenoyl (DHA; 22:6n-3)-PC (PA-DHA), were oxidized in bulk phase (Araseki et al., 2002) or in t-BuOH solution (Kobayashi et al., 2003) (Fig. 3.7), PA-LA was most oxidatively stable followed by PA-AA and PA-DHA. However, the effect of the degree of unsaturation on the oxidative stability of PC in liposomes (Figures 3.7 and 3.8) was different from that in the bulk phase and in organic solutions. Analysis of oxidation products showed that PA-DHA was most oxidatively stable followed by PA-AA and OA-LA (Fig. 3.7). On the other hand, oxygen consumption analysis (Fig. 3.8 (A)) showed that the difference in the oxidative stability of three kinds of PC was relatively small.

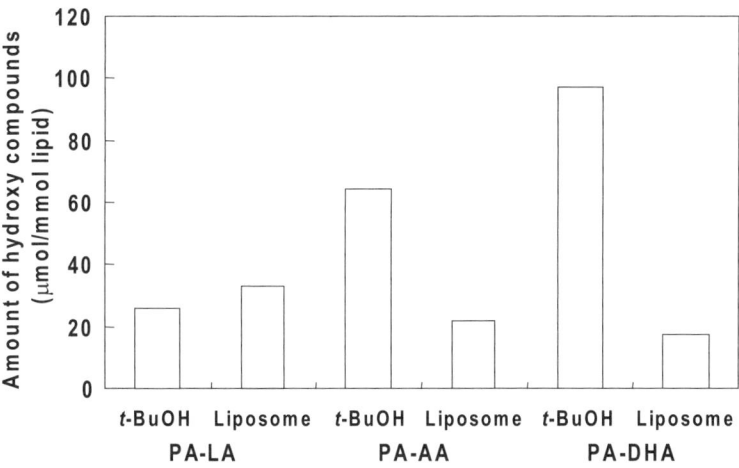

**Fig. 3.7.** Amounts of total hydroxy compounds derived from oxidation products of PUFA in PC, Formed in t-BuOH Solution and Liposomes. AAPH was used as oxidation inducer. The oxidation products were extracted with chloroform/methanol, then hydrogenated, transmethylated, trimethylsilylated (TMS), and then subjected to GC-MS analysis for the quantitative comparison of TMS-derivatives from mono-, di-, and tri-hydroxy compounds. These hydroxyl compounds are derived from monohydroperoxides, hydroperoxy epidioxides, and other oxidation products.

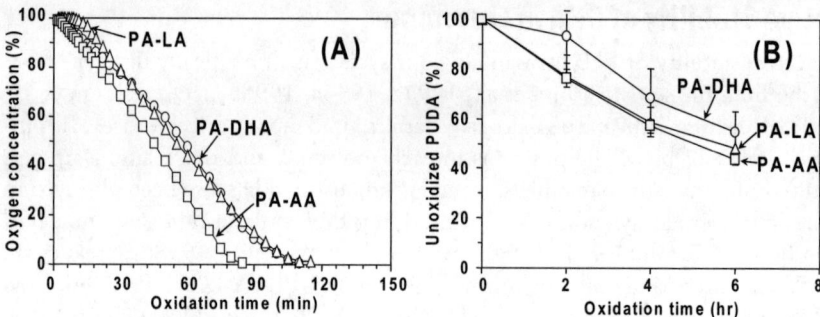

**Fig. 3.8.** Oxidative stability of PC in liposomes. PC (0.5 m*M*) was oxidized in liposomes in the presence of AAPH (3.0 m*M*). The oxidative stability was evaluated by the analysis of the decrease in oxygen concentration in the solution (A) or the decrease in unoxidized PUFA in the PC molecule (B).

Table 3.6 shows the proportion of MHP isomers formed in the oxidation of PC in different systems (Kobayashi et al., 2003). The uneven distribution of MHP in PC-AA liposomes indicates a preferential oxygen attack on the methyl-terminal side of the radical. However, in case of PC-DHA, the oxygen attack favored the carboxy terminal side of the pentadienyl radical. In PC-AA oxidation, amounts of 5-MHP + 9-MHP, 8-MHP + 12-MHP, and 11-MHP + 15-MHP were almost identical. This result indicates same rate of hydrogen abstraction at the three bis-allylic positions (C7, C10, and C13) and difficulty in the 1,3-cyclization of internal MHP. On the other hand, MHP distribution in PC-DHA revealed a favorable hydrogen abstraction at C18, C6, and C12.

PC in liposomes with highly unsaturated fatty acids, such as AA and DHA, are more permeable and show more flexibility in fatty acid chains than those formed from PC containing less unsaturated fatty acids such as LA. NMR analysis and molecular dynamics simulations of PC containing DHA in liposomes indicates the wide variety of DHA conformations (including back-bent, helical, and angle-iron conformations) occurring in liposome systems (Everts and Davis, 2000; Feller et al., 2001; Saiz and Klein, 2001; Huber et al., 2002). NMR analysis also showed the mobility of the hydrophobic part of the DHA molecule would higher than that of LA when forming liposomes. Liposomes of AA appear to have flexibility between those of DHA and LA. The flexibility of DHA chain conformation gives looser packing of the lipid chains (Huster et al., 1997; Olbrich et al., 2000). Looser packing of the membrane at the lipid-water interface brings about the high water permeability (Saiz and Klein, 2001). Molecular dynamics simulations also indicate remarkable overlap of water molecules with double bond regions of the DHA chain. The presence of water molecules near a DHA molecule will lower the density of the bis-allylic hydrogen and inhibits the hydrogen abstraction from bis-allylic positions of unoxidized fatty acid by peroxy radical of adjacent oxidized fatty acid in the propagation stage of free radical oxidation. The higher water permeability of DHA and its specific conformation may be a reason for the relatively higher oxidative stability of DHA in liposome.

Table 3.6. Isomeric Distribution of MHP Isomers Formed in the Oxidation of PUFA Esters

| PUFA | MHP Formation | | Positional Distribution | | |
|---|---|---|---|---|---|
| | Hydrogen Abstraction | Resulting MHP | In Chloroform[a] Solution | In PC[b] Liposome | In Cellular[c] PL |
| LA | C-13 | 9-MHP | 49.3±1.5 | 49.2±3.6 | 49.7±1.5 |
| | | 13-MHP | 50.7±1.5 | 50.8±3.6 | 50.3±1.5 |
| AA | C-7 | 5-MHP | 19.4±1.1 | 9.6±0.7 | 23.7±3.7 |
| | | 9-MHP | 15.3±0.4 | 23.8±2.0 | 6.6±1.3 |
| | C-10 | 8-MHP | 15.2±0.5 | 5.1±1.1 | 12.9±2.3 |
| | | 12-MHP | 15.0±1.1 | 26.3±2.4 | 23.2±3.9 |
| | C-13 | 11-MHP | 15.1±0.5 | 7.4±2.4 | 3.4±2.3 |
| | | 15-MHP | 19.9±1.3 | 27.9±2.1 | 30.2±3.4 |
| DHA | C-6 | 4-MHP | 17.7±1.3 | 17.9±2.4 | 25.7±2.2 |
| | | 8-MHP | 12.9±1.7 | 11.2±1.1 | 12.8±1.6 |
| | C-9 | 7-MHP | 10.5±0.7 | 6.2±1.1 | ND[d] |
| | | 11-MHP | 7.2±1.0 | 2.1±0.5 | ND[d] |
| | C-12 | 10-MHP | 4.2±0.5 | 13.0±1.1 | 10.2±1.3 |
| | | 14-MHP | 4.3±0.6 | 4.6±0.8 | 4.9±4.5 |
| | C-15 | 13-MHP | 3.5±1.0 | 5.5±1.0 | ND[d] |
| | | 17-MHP | 4.5±0.7 | 3.1±0.6 | ND[d] |
| | C-18 | 16-MHP | 8.7±1.4 | 21.3±4.0 | 24.6±1.5 |
| | | 20-MHP | 26.5±1.1 | 16.4±1.3 | 21.9±4.3 |

[a] The ethyl ester of each PUFA was oxidized in chloroform and in aqueous solution. The oxidation was induced by 2,2'-*azobis*(2,4-dimethyl-valeronitrile) (AMVN) or 2,2'-*azobis*(2-amidino-propane) dihydrochloride (AAPH) for chloroform and aqueous oxidation, respectively.
[b] 1-Palmitoyl-2-linoleoyl-phosphatidylcholine (LA-PC), 1-palmitoyl-2-arachidonoyl-phosphatidylcholine (AA-PC), and 1-palmitoyl-2-docosahexaenoyl-phosphatidylcholine (DHA-PC) were oxidized in liposomes. AAPH was used as oxidation inducer.
[c] Each PUFA was supplemented to the cell (HepG2). Supplementation with PUFA resulted in their incorporation into cellular lipids. Cellular oxidation was accelerated by the addition of $H_2O_2$.
[d] Not detected

## Oxidation of Long-chain PUFA in Cellular Lipids

The incorporation PUFA, such as LA, AA and DHA, into cellular lipids enhances the cellular lipid peroxidation (Igarashi and Miyazawa, 200; Araseki et al., 2005). Figure 3.9 shows the analysis of peroxidation level in the membrane PL of HepG2 cells. In this case, the average number of bis-allylic positions per mol was 0.57 for control cells, 0.64 for LA-supplemented cells, 0.77 for AA-supplemented cells, and 0.81 for DHA-supplemented cells. However, there was little difference in the lipid peroxidation level in the LA-, AA-, and DHA-supplemented cells.

Figure 3.9 also shows that the addition of H2O2 (0.5 mM) enhanced cellular lipid peroxidation levels in control, LA-, and AA-supplemented cells compared to those without H2O2. On the other hand, induction of lipid peroxidation by H2O2 was not observed in DHA-supplemented cells. Araseki et al. (2005) noticed the lower cellular oxidation level in DHA-supplemented cells after H2O2 addition during the analysis of total MHP content in the cellular PL (Table 3.7). Table 3.7 also shows that the main source for MHP was LA in all cases, and a significant amount of AA-MHP was observed only in AA-supplemented cells. A small amount of DHA-MHP was also observed in DHA-supplemented cells, but not in other cells. The order of lipid peroxidation levels of the cellular PL (Fig. 3.9; Table 3.7) was almost the same as that in the oxidative stability of PC in liposomes, suggesting that the characteristic cellular lipid peroxidation is also correlated with the PUFA conformation in the membrane PL.

With the oxidation of PC-AA and PC-DHA during the oxidation of DHA in the cellular PL, DHA-MHP was formed only by abstraction at C6, C12, and C18, but not at the C9 or C15 positions. This characteristic distribution of DHA-MHP isomers found in the cellular PL oxidation indicates uneven hydrogen abstraction from DHA molecules in the cell membrane lipid, which might be derived from the specific conformation of the DHA molecule and higher water permeability in the membrane lipids. Thus, further studies are necessary to reveal the relationship between the conformation and physicochemical properties of AA and DHA and their oxidative stability in biological membrane lipids.

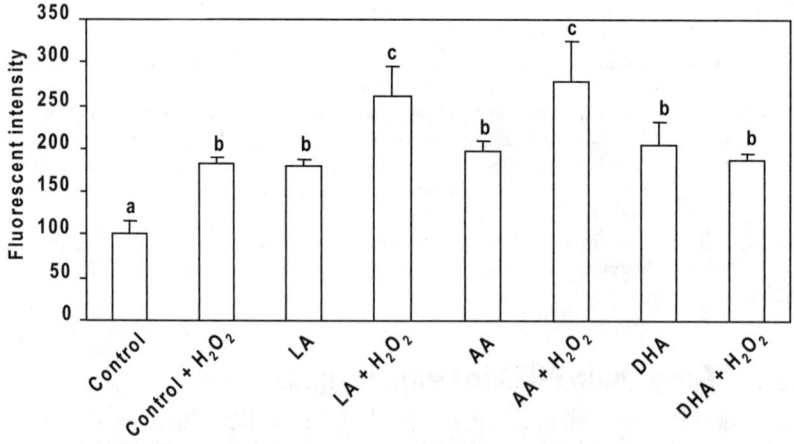

**Fig. 3.9.** Formation of MHP in HepG2 cells treated with or without $H_2O_2$. Values not sharing a common superscript are significantly different at $P < 0.01$. Values are mean $\pm$ S.D. (n = 3).

Table 3.7. Amount of MHP Formed in the Oxidation of Cellular PL with or without $H_2O_2$

| | Cell | Amount of MHP (mmol/mmol Lipid) | | | |
|---|---|---|---|---|---|
| | | LA-MHP | AA-MHP | DHA-MHP | Total MHP |
| Without $H_2O_2$ | Control | 1.80±0.43 | 0.15±0.05 | ND | 1.95±0.48 |
| | + LA | 3.06±0.40 | 0.17±0.08 | ND | 3.23±0.35 |
| | + AA | 2.17±0.35 | 1.23±0.11 | ND | 3.40±0.29 |
| | + DHA | 2.37±0.31 | 0.21±0.06 | 0.36±0.05 | 2.93±0.20 |
| With $H_2O_2$ | Control | 3.10±0.62 | 0.32±0.09 | ND | 3.42±0.71 |
| | + LA | 3.50±0.46 | 0.17±0.06 | ND | 3.67±0.41 |
| | + AA | 2.47±0.47 | 1.55±0.57 | ND | 4.02±1.03 |
| | + DHA | 2.27±0.47 | 0.20±0.09 | 0.33±0.10 | 2.79±0.65 |

Data re expressed as mean ± SD (*n* = 3)
ND; not detected

# Formation of Conjugated Trienoic Acids in Oils Containing Long-chain PUFA

Crude oil is refined by a series of processes to remove impurities that affect taste, smell, appearance, and stability of the oil. The refining processes involve degumming, alkali refining, bleaching and deodorization. PUFA, such as LA, α-linolenic acid, EPA, and DHA, found in vegetable and marine oils are more susceptible to oxidation during storage and refining of crude oils resulting in oxidation products comprising mainly of hydroperoxides. Hydroperoxides may also be formed by enzymatic oxidation of linoleate and linolenate by lipoxygenase during storage of oil seeds. Any oxidation is injurious to the flavor and oxidative stability of refined oils as it results in the formation and accumulation of oxidation products. The quality of a processed vegetable oil always depends on the quality of the crude oil used and the processing parameters selected. Indicators such as PV in the final products can be reduced by lowering storage temperature and minimizing exposure to prooxidants, like air, light, and heat, during processing. However, complete removal of all oxidation products from oils is impossible.

Yurawecz et al. (1993) found traces (up to 0.2%) of conjugated linolenic acid (CLN) in vegetable oils in their study of 27 oils for CLN content by UV measurement. The isomers were identified as α-eleostearic acid (9cis(c),11trans (t),13t-18:3), β-eleostearic acid (9t,11t,13t-18:3), and 8t,10t,12t-18:3. The possible mechanism for formation of these CLN isomers involves linoleate oxidation, reduction of hydroperoxide to hydroxide, and dehydration (Fig. 3.10) (Parr and Swoboda, 1976; Fishwick and Swoboda, 1977). CLN and other conjugated trienoic fatty acids are easily oxidized to induce lipid oxidation and flavor deterioration even if the content is very low (Suzuki et al., 2004).

HPLC analysis of purified soybean oil methyl esters indicates the presence of small amounts of conjugated linoleic acid (CLA) and CLN as shown by UV detection at 233 nm and 274 nm (Table 3.8) (Kinami et al., 2007). GC-MS after conversion of methyl esters to DMOX derivatives and by comparison with authentic CLN isomers on HPLC revealed the formation of 8,10,12- and 9,11,13-18:3 (c,t,t, or t,t,c, and t,t,t) in the purified soybean oil. HPLC chromatogram of crude soybean oil and processed soybean oil at different stages shows that a significant amount of CLN (8,10,12 or 9,11,13) was found in soybean oil after bleaching, although it could barely be detected in crude soybean oil or in the oil after degumming and alkali refining (Table 3.9). A slight decrease in CLN after deodorization may be due to the isomerization of the CLN to CLN with conjugated dienes (8,10,13-18:3). CLN contents in purified soybean oil from different companies in Japan are shown in Table 3.10. It varies from 387-1316 mg/kg oil, which corresponds to 0.039-0.13% (w/w). Similar values have also been reported by Yurawecz et al. (1993). CLN content is also affected by bleaching conditions (Kinami et al., 2007). Combinations of higher percentage of bleaching earth and lower bleaching temperature results in reduced CLN content. Similar effects of bleaching temperature and earth combinations have been reported by Van Den Bosch (1973a,1973b).

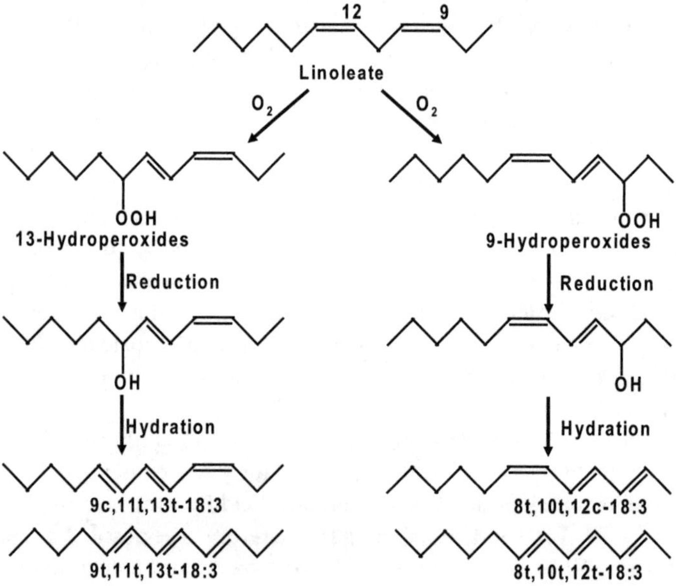

Fig. 3.10. A possible scheme for formation of CLN from linoleate hydroperoxides.

**Table 3.8.** Identification of *trans* Conjugated Fatty Acids in Commercial Soybean Oil

| Structure | Conjugation<br>*cis* (*c*), *trans* (*t*) Configuration |
|---|---|
| 8,10,13-18:3 (CLN) | Conjugated diene with *trans* configuration |
| 8,10,12-18:3 (CLN)<br>9,11,13-18:3 (CLN) | Conjugated triene;<br>c,t,t, or t,t,c |
| 7,9-18:2 (CLA)<br>10,12-18:2 (CLA) | Conjugated diene;<br>c,t or t,c |
| 8,10,12-18:3 (CLN)<br>9,11,13-18:3 (CLN) | Conjugated triene;<br>t,t,t |
| 7,9-18:2 (CLA)<br>9,11-18:2 (CLA | Conjugated diene;<br>t,t |

**Table 3.9.** Content of CLN (8,10,12 or 9,11,13) with Conjugated Triene at Different Stage of Soybean Oil Production

| Soybean Oil | CLN (mg/kg oil) | | |
|---|---|---|---|
| | c,t,t or t,t,c | t,t,t | Total |
| Crude | ND | ND | ND |
| Degumming | 10 | ND | 10 |
| Alkali Refining | 20 | ND | 20 |
| Bleaching | 250 | 358 | 608 |
| Deodorization | 217 | 315 | 532 |

ND; not detected

**Table 3.10.** CLN Content in Commercial Soybean Oil from Different Japanese Oil Companies

| Company | CLN (mg/kg oil) | | |
|---|---|---|---|
| | c,t,t or t,t,c | t,t,t | Total |
| A | 295 | 367 | 662 |
| B | 408 | 618 | 1026 |
| C | 198 | 230 | 427 |
| D | 753 | 559 | 1312 |
| E | 181 | 205 | 387 |
| F | 945 | 370 | 1316 |

Soybean oil tends to develop an undesirable flavor and odor known as reversion when PV is still as low as a few meq/kg. This reversion flavor is characterized as beany and grassy. Guth and Grosch (1991) reported the formation of a potent compound causing this flavor reversion as 3-methyl-2,4-nonanedione. The reversion flavor in soybean oil is mainly due to furan compounds such as 2-pentyl furan (Smouse and Chang, 1967) and 2-pentenyl furan (Ho et al., 1978; Smagula et al., 1979). Min et al. (2003) and Cho and Min (2006) clearly show that these furan compounds are formed from the photooxidation of LA and linolenic acid in soybean oil. Photooxidation is due to singlet oxygen that is produced from triplet oxygen in the presence of chlorophyll under light. On the other hand, conjugated fatty acids have also been considered as a possible source of furan fatty acids (Yurawecz et al., 1995). A trace amount of CLN plays a vital role in accelerating lipid oxidation. CLN may be one of the important precursors for off-flavor compounds since CLN is easily oxidized in bulk phase (Suzuki et al., 2004).

Table 3.11 indicates intensity of beany and grassy flavor of middle chain fatty acid (MCT) triacyglycerol (TAG) with or without bitter gourd TAG. The main fatty acids of bitter gourd TAG are 9c,11t,13t-18:3 (61.6%), 18:0 (25.7%), and 18:1n-9 (6.1%) (Suzuki et al. 2001; Narayan et al., 2006b). Little flavor was detected in MCT after heating at 180°C under light (7000 lux) for 16 hr. Beany and grassy flavors were detected in MCT by adding bitter gourd-TAG, although PV of the oil was less than 0.1. This result suggested that CLN (9c,11t,13t-18:3) of bitter gourd-TAG would be one of the potent compounds responsible for flavor reversion. The flavor significance of soybean oil after heating under the light is shown in Table 3.12. CLN content changed depending on bleaching earth content and temperature. In the purified soybean oil, the relationship between CLN content and the flavor intensity was not considerable. On the other hand, the addition of 0.5% activated carbon reduced beany and grassy flavor. Frankel (1998c) noted the importance of oxidative polymers on the flavor or oxidative deterioration of vegetable oils. The spontaneous decomposition of oxidative polymers has been referred to as "hidden oxidation" because it cannot be measured by PV. CLN is easily oxidized to produce polymers as main oxidation products (Suzuki et al., 2004), which may induce the oxidation of vegetable oils.

**Table 3.11.** Beany and Grassy Flavor Intensity of MCT with or without Bitter Gourd-TAG after Heating at 180°C under 700 Lux for 16 hr

| Bitter-Gourd TAG (%) | PV (meq/kg) | Flavor intensity |
|---|---|---|
| 0 | 0 | - |
| 0.01 | 0.1 | ++ |
| 0.05 | 0.1 | +++ |

**Table 3.12.** Beany and Grassy Flavor Intensity of Soybean Oil after Heating at 180°C under 7,000 Lux for 16 hr

| Bleaching Condition | | | CLN (mg/kg Oil) | | | PV (meq/ kg Oil) | Flavor |
|---|---|---|---|---|---|---|---|
| Activated Earth (%) | Temp. (°C) | Activated Carbon | c,t,t or t,t,c | t,t,t | Total | | |
| 1.5 | 110 | - | 173 | 284 | 457 | 1.3 | ++ |
| 1.5 | 110 | - | 153 | 255 | 408 | 1.2 | ++ |
| 0.5 | 110 | - | 122 | 166 | 288 | 2.3 | +++ |
| 1.5 | 50 | - | 117 | 186 | 303 | 1.4 | ++ |
| 0.5 | 50 | - | 68 | 81 | 149 | 2.1 | +++ |
| 0.5 | 50 | - | 65 | 75 | 140 | 1.9 | +++ |
| 0.5 | 50 | - | 106 | 127 | 233 | 1.1 | + |
| 0.5 | 50 | + | 67 | 84 | 151 | 0.9 | + |
| 0 | 50 | + | ND | 40 | 40 | 1.3 | + |

The presence of conjugated trienoic acids has also been reported in the purified fish oil from Japanese sardine and tuna by HPLC equipped with a photodiode-array detector (Fig. 3.11). GC-MS indicated that the main conjugated trienoic acids were those of 16:3, 20:5, and 22:6. Conjugated 16:3 trienoic acid was identified as cyclopropanoid acid with conjugated triene by NMR analysis. These conjugated trienoic acids were observed after deodorization, but not after degumming, alkali refining, or bleaching. The formation of conjugated EPA (CEPA) and conjugated DHA (CDHA) during deodorization can be due to the double bond migrations of EPA and DHA by heating. In the case of 16:4n-1, double bond migration and cyclization occur to produce corresponding cyclopropanoid 16:3 acid. These reactions were confirmed by the formation of conjugated trienoic acids from the heat treatment (200°C for 2hr) of purified methyl esters of 16:4n-1, EPA, and DHA.

The oxidation rate of conjugated trienoic acids of linolenate were much higher than those of α-linolenic acid esters and CLA esters (Suzuki et al., 2004), suggesting the higher susceptibility of conjugated trienoic fatty acid to oxidation than conjugated dienoic acids and fatty acid having the same number of double bonds. Since oxidation rate increases with increasing double bonds, 16:4n-1, EPA and DHA are more easily oxidized than α-linolenic acid. Therefore conjugated trienoic fatty acids originating from 16:4n-1, EPA and DHA in fish oil would be more susceptible to oxidation. They may have an important role in the oxidative deterioration of fish oil in spite of these conjugated fatty acids being present in trace amounts.

## Conclusions

It is generally accepted that lipid oxidation of DHA and EPA is a major problem in

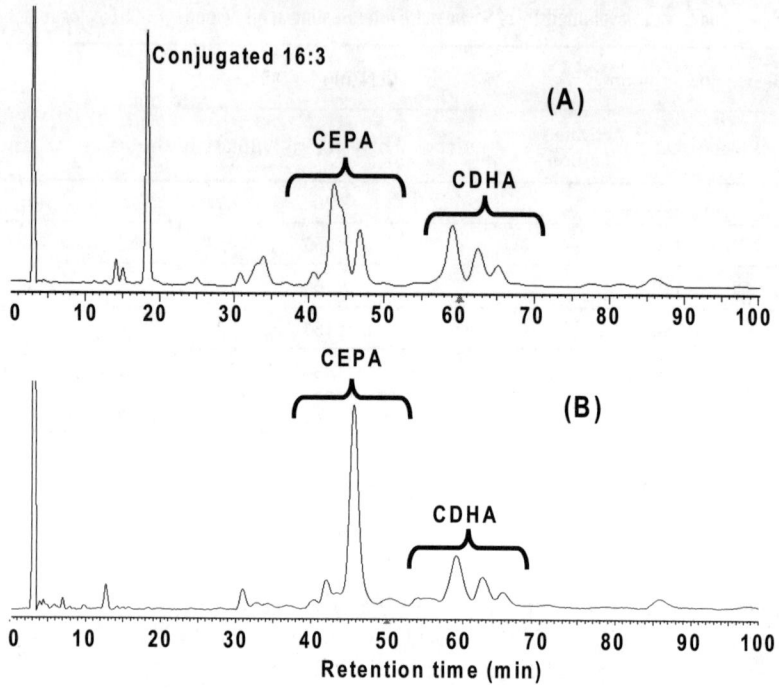

**Fig. 3.11.** HPLC of methyl esters from purified sardine oil (A) and purified tuna oil (B). HPLC was carried out with an analytical reversed-phase column (C30). A mixture of methanol and water (85:15, vol/vol) at a flow rate of 1.0 mL/min (analytical scale) was used as a mobile phase. The HPLC instrument housed a photodiode-array spectrophotometric detector.

using fish oil to make food materials, and a high intake of these PUFA may increase the oxidative stress of biological systems. Susceptibility of PUFA to oxidation depends on the availability of bis-allylic hydrogens. Oxidative stability of each PUFA is inversely proportional on the number of bis-allylic positions in the molecule or the degree of unsaturation of the PUFA. Due to the higher degree of unsaturation, MHP of EPA and DHA are easily decomposed, resulting in complex oxidation products when compared with those of other PUFA, such as LA and α-linolenic acid. A large amount of polar oxidation products, mainly polymers, form even at an early stage of autoxidation of EPA and DHA. Peroxide value is not necessarily a good indication of oxidation in EPA and DHA because of the instability of their hydroperoxides.

On the other hand, oxidative stability of EPA and DHA in biological systems is different from those in bulk phase or in organic solution. In biological systems lipids are present with various types of other components in aqueous medium. DHA is relatively stable in biological systems due to the complex, multi-component, and heterogeneous nature of biological systems. PL is synergistic in combination with phenolic antioxidants such as tocopherols. In biological systems large amounts of

EPA and DHA are present as PL in the membrane. Furthermore, the high levels of DHA in membrane lipids imply the occurrence of a strong antioxidant system in animal tissues.

In addition, the oxidative stability of PUFA in a liposome system is different from that in bulk phase. The oxidative stability of DHA is slightly higher than AA and LA in liposomes. NMR analysis and molecular dynamic simulations of PC containing DHA in liposomes indicates that DHA chain conformation gives looser packing of the lipid chains in liposome. The looser packing of the membrane at the lipid-water interface brings about the high water permeability. The presence of water molecules near a DHA molecule will lower the density of the bis-allylic hydrogen and reduce the chain-carrying reaction of lipid peroxidation. The higher water permeability of DHA and its specific conformation may be a reason for the relatively higher oxidative stability of DHA in liposomes. The difference in the oxidative stability of PC in liposomes and in the bulk phase or an organic solvent is due to the specific conformation of PC bilayers in the liposomes. The order of lipid peroxidation levels of the cellular PL was almost the same as that in the oxidative stability of PC in liposomes, suggesting that the characteristic cellular lipid peroxidation found in the present study is also correlated with the PUFA conformation in the membrane PL.

Finally, the impact of conjugated trienoic fatty acids, such as CLN, CEPA, and CDHA, on oxidative deterioration and flavor reversion is also an important factor in lipid oxidation. It can be said that mechanisms to prevent PUFA oxidation have been well understood when applied to lipid-containing foods. However, In conclusion, it is still difficult to completely prevent the PUFA oxidation, especially flavor deterioration. The formation of conjugated triene compounds and their strong impact on oxidative deterioration show the importance of the original quality and refining process of oils.

## References

Aidos, I.; C. Jacobsen; B. Jensen; J.B. Luten; A. van der Padt; and R.M. Boom. Volatile Oxidation Products Formed in Crude Herring Oil under Accelerated Oxidative Conditions. *Eur. J. Lipid Sci. Technol.* **2002**, *104,* 808-818.

Ando, K.; K. Nagata; R. Yoshida; K. Kikugawa; and M. Suzuki. Effect of n-3 Polyunsaturated Fatty Acid Supplementation on Lipid Peroxidation of Rat Organs. *Lipids* **2000**, *35,* 401-407.

Araseki, M.; K. Yamamoto; and K. Miyashita. Oxidative Stability of Polyunsaturated Fatty Acid in Phosphatidylcholine Liposomes. *Biosci. Biotechnol. Biochem.* **2002**, *66,* 2573-2577.

Araseki, M.; H. Kobayashi; M. Hosokawa; and K. Miyashita. Lipid Peroxidation of a Human Hepatoma Cell Line (HepG2) after Incorporation of Linoleic Acid, Arachidonic Acid, and Docosahexaenoic Acid. *Biosci. Biotechnol. Biochem.* **2005**, *69,* 483-490.

Bandarra, N.M.; R.M. Campos; I. Batista; M.L. Nunes; and J.M. Empis. Antioxidant Synergy of α-Tocopherol and Phospholipids. *J. Am. Oil Chem. Soc.* **1999**, *76,* 905-913.

Benedetti, A.; M. Comporti; and H. Esterbauer. Identification of 4-Hydroxynonenal as a Cytotoxic Product Originating from the peroxidation of liver microsomal lipids, Biochim.

Biophys. Acta **1980**, *620*, 281-296.
Burns, C.P.; and B.A. Wagner. Heightened Susceptibility of Fish Oil Polyunsaturated-Enriched Neoplastic Cells to Ethane Generation during Lipid Oxidation. *J. Lipid Res.* **1991**, *32*, 79-87.
Chan, H.W.-S.; J. Matthew; and D.T. Coxon. A Hydroperoxy-Epidioxide from the Autoxidation of a Hydroperoxide of Methyl Linolenate. *J. C. S. Chem. Comm.* **1980**, 235-236.
Cho, E.; and D.B. Min. Chemistry and Reactions of Reactive Oxygen Species in Foods. *Critical Rev., Food Sci. Nutr.* **2006**, *46*, 1-22.
Cho, S.-Y.; K. Miyashita; T. Miyazawa; K. Fujimoto; and T. Kaneda. Autoxidation of Ethyl Eicosapentaenoate and Docosahexaenoate. *J. Am. Oil Chem. Soc.* **1987a**, *64*, 876-879.
Cho, S.-Y.; K. Miyashita; T. Miyazawa; K. Fujimoto; and T. Kaneda. Autoxidation of Ethyl Eicosapentaenoate and Docosahexaenoate under Light Irradiation. *Nippon Suisan Gakkaishi* **1987b**, *53*, 813-817.
Cosgrove, J.P.; D.F. Church; and W.A. Pryor. The Kinetics of the Autoxidation of Polyunsaturated Fatty Acids. *Lipids* **1987**, *22*, 299-304.
Coxon, D.T.; K.R. Price; and H.W.-S. Chan. Formation, Isolation and Structure Determination of Methyl Linolenate Diperoxides. *Chem. Phys. Lipids* **1981**, *28*, 365-378.
Everts, S.; and J.H. Davis. $^1$H and $^{13}$C NMR of Multilamellar Dispersions of Polyunsaturated (22:6) Phospholipids. *Biophys. J.* **2000**, *79*, 885-897.
Feller, S.E.; K. Garwrish; and A.D. MacKerell Jr. Polyunsaturated Fatty Acids in Lipid Bilayers Intrinsic and Environmental Contributions to Their Unique Physical Properties. *J. Am. Chem. Soc.* **2001**, *124*, 318-326.
Fishwick, M.J.; and P.A.T. Swoboda. Measurement of Oxidation of Polyunsaturated Fatty Acids by Spectrophotometric Assay of Conjugated Derivatives. *J. Sci. Food Agric.* **1977**, *28*, 387-393.
Frankel, E.N. Lipid Oxidation. *Prog. Lipid Res.* **1980**, *19*, 1-22.
Frankel, E.N. Volatile Lipid Oxidation Products. *Prog. Lipid Res.* **1982**, *22*, 1-33.
Frankel, E.N. Chemistry of Free Radical and Singlet Oxidation of Lipids. *Prog. Lipid Res.* **1985**, *23*, 197-221.
Frankel, E.N. Chemistry of autoxidation: mechanism, products and flavor significance, in *Flavor Chemistry of Lipid Foods* Min, D., and Smouse, T.H., eds, AOCS Press, Champaign, Illinois, 1985.
Frankel, E.N. *Lipid Oxidation.* The Oily Press: Dundee, Scotland, 1998a, pp. 13-77.
Frankel, E.N. *Lipid Oxidation,* The Oily Press: Dundee, Scotland, 1998b, pp. 99-114.
Frankel, E.N. *Lipid Oxidation,* The Oily Press: Dundee, Scotland, 1998c, pp. 115-127.
Fritsche, K.L.; and P.V. Johnston. Rapid Autoxidation of Fish Oil in Diets without Added Antioxidants. *J. Nutr.* **1988**, *118*, 425-426.
Garrido, A.; F. Garrido; R. Guerra; and A. Valenzyela. Ingestion of High Doses of Fish Oil Increases the Susceptibility of Cellular Membranes to the Induction of Oxidation Stress. *Lipids* **1989**, *24*, 833-835.
Gunstone, F.D.; and T.P. Hilditch. The Union Gaseous Oxygen with Methyl Oleate, Linoleate, and Linolenate. *J. Chem. Soc.* **1945**, 836-841.
Guth, H.; and W. Grosch. Detection of Furanoid Fatty Acids in Soya-Bean Oil-Cause for the Light-Induced Off-Flavor. *Fat Sci. Technol.* **1991**, *93*, 249-255.
Hara, S.; N. Okada; H. Hibino; and Y. Totani. Antioxidative Behavior of Phospholipids for Polyunsaturated Fatty Acids of Fish Oil. *J. Jpn. Oil Chem. Soc.* **1992**, *41*, 130-135.
Hirano, S.; K. Miyashita; and T. Ota; M. Nishikawa; K. Maruyama; and S. Nakayama. Aque-

ous Oxidation of Ethyl Linoleate, Ethyl Linolenate, and Ethyl Docosahexaenoate. *Biosci. Biotechnol. Biochem.* **1997,** *161,* 281-285.

Ho, C.T.; M.S. Smagula; and S.S. Chang. The Synthesis of 2-(1-Pentenyl) Furan and Its Relationship to the Reversion Flavor of Soybean Oil. *J. Am. Oil Chem. Soc.* **1978,** *55,* 233-237.

Holman, R.T. and Elmer, O.C. The rates of oxidation of unsaturated fatty acids and esters, *J. Am. Oil Chem. Soc.*, **1947,** *24,* 127-129.

Hsieh, T.C.Y.; S.S. Williams; W. Vejaphan; and S.P. Meyers. Characterization of Volatile Components of Menhaden Fish (*Brevoortia tyrannus*) Oil. *J. Am. Oil Chem. Soc.* **1989,** *66,* 114-117.

Hu, M.-L.; E.N. Frankel; B.E. Leibovitz; and A.L. Tappel. Effect of Dietary Lipids and Vitamin E on in vitro Lipid Peroxidation in Rat Liver and Kidney Homogenates. *J. Nutr.* **1989,** *119,* 1574-1582.

Huber, T.; K. Rajamoorthi; V.F. Kurze; K. Beyer; and M.F. Brown. Structure of Docosahexaenoic Acid-Containing Phospholipid Bilayers as Studied by $^2$H NMR and Molecular Dynamics Simulations. *J. Am. Chem. Soc.* **2002,** *124,* 298-309.

Huster, D.; A.J. Jin; K. Arnold; and K. Gawrisch. Water Permeability of Polyunsaturated Lipid Membranes Measured by $^{17}$O NMR. *Biophys. J.* **1997,** *73,* 855-864.

Igarashi, M.; and T. Miyazawa. The Growth Inhibitory Effect of Conjugated Linoleic Acid on a Human Hepatoma Cell Line, HepG2, Is Induced by a Change in Fatty Acid Metabolism, But Not the Facilitation of Lipid Peroxidation in the Cells. *Biochim. Biophys. Acta* **2001,** *1530,* 162-171.

Jacobsen, C. Sensory Impact of Lipid Oxidation in Complex Food Systems. *Fett/Lipid* **1999** *101,* 484-492.

Kamal-Eldin, A., Mäkinen, M., Lampi, A.-M. The challenging contribution of hydroperoxides to the lipid oxidation mechanism, in *Lipid Oxidation Pathways*, Kamal-Eldin ed, pp. 1-36, AOCS Press, Champaign, Illinois, 2003.

Kaneda, T.; and S. Ishii. Nutritive Value or Toxicity of Highly Unsaturated Fatty Acids. I. *J. Biochem.* **1954,** *41,* 327-335.

Karahadian, C.; and R.C. Lindsay. Evaluation of Compounds Contributing Characterizing Fishy Flavors in Fish Oils. *J. Am. Oil Chem. Soc.* **1989,** *66,* 953-960.

Kashima, M.; G.-S. Cha; Y. Isoda; J. Hirano; and T. Miyazawa. The Antioxidant Effects of Phospholipids on Perilla Oil. *J. Am. Oil Chem. Soc.* **1984,** *61,* 950-954.

Kinami, T.; N. Horii; B. Narayan; S. Arato; M. Hosokawa; K. Miyashita; H. Negishi; J. Ikuina; R. Noda; and S. Shirasawa. Occurrence of Conjugated Linolenic Acids in Purified Soybean Oil. *J. Am. Oil Chem. Soc.* **2007,** *84,* 23-29.

King, M.F.; L.C. Boyd; and B.W. Sheldon. Effects of Phospholipids on Lipid Oxidation of a Salmon Oil Model System. *J. Am. Oil Chem. Soc.* **1992,** *69,* 237-242.

Kobayashi, H.; M. Yoshida; and K. Miyashita. Comparative Study of the Product Components of Lipid Oxidation in Aqueous and Organic Systems. *Chem. Phys. Lipids* **2003,** *126,* 111-120.

Koga, T.; and J. Terao. Phospholipids Increase Radical-Scavenging Activity of Vitamin E in a Bulk Oil Model System. *J. Agric. Food Chem.* **1995,** *43,* 1450-1454.

Kubo, K.; M. Saito; T. Tadokoro; and A. Maekawa. Dietary Docosahexaenoic Acid Does Not Promote Lipid Peroxidation in Rat Tissue to the Extent Expected from Peroxidizability Index of the Lipids. *Biosci. Biotechnol. Biochem.* **1998,** *62,* 1698-1706.

Lands, W.E.M. *Fish, Omega-3 and Human Health, 2nd Edn.*; AOCS Press: Champaign, Illinois, 2005.

Li, D.; O. Bode; H. Drummond; and A.J. Sinclair. Omega-3 (n-3) Fatty Acids. In *Lipids for Functional Foods and Nutraceuticals,* Gunstone, F.D. Ed., The Oily Press; Bridgwater, England, 2003 pp. 225-262.

Min, D.; and T.H. Smouse. *Flavor Chemistry of Fats and Oils,* AOCS Press: Champaign, Illinois, 1985.

Min, D.B., Lee, S.H., Lee, E.C. Singlet oxygen oxidation of vegetable oils, in *Flavor Chemistry of Lipid Foods,* Min, D., and Smouse, T.H., eds, pp. 57-97, AOCS Press, Champaign, Illinois, (1989.

Min, D.B.; A.L. Callison; and H.O. Lee. Singlet Oxygen Oxidation for 2-Pentylfuran and 2-Pentenylfuran Formation in Soybean Oil. *J. Food Sci.* **2003,** *68,* 1175-1178.

Miyashita, K. Polyunsaturated Lipids in Aqueous Systems Do Not Follow Our Preconceptions of Oxidative Stability. *Lipid Technol. News.* **2002,** *8,* 35-41.

Miyashita, K. Effects of Flexibility and Permeability of Polyunsaturated Fatty Acid Molecules on Their Oxidative Stability in Aqueous Systems. *Lipid Technol. News.* **2004,** *16,* 197-202.

Miyashita, K.; and T. Takagi. Study on the Oxidative Rate and Prooxidant Activity of Free Fatty Acids. *J. Am. Oil Chem. Soc.* **1986,** *63,* 1380-1384.

Miyashita, K.; K. Fujimoto; and T. Kaneda. Formation of Dimmers during the Initial Stage of Autoxidation in Methyl Linoleate. *Agric. Biol. Chem.* **1982,** *46,* 751-755.

Miyashita, K.; E.N. Frankel; W.E. Neff; and R.A. Awl. Autoxidation of Polyunsaturated Triacylglycerols. III. Syntheytic Triacylglycerols Containing Linoleate and Linolenate. *Lipids* **1990,** *25,* 48-53.

Miyashita, K.; N. Tateda; and T. Ota. Oxidative Stability of Free Fatty Acid Mixtures from Soybean, Linseed, and Sardine Oils in an Aqueous Solution. *Fisheries Sci.* **1994a,** *60,* 315-318.

Miyashita, K.; E. Nara; and T. Ota. Comparative Study on the Oxidative Stability of Phosphatidylcholines from Salmon Egg and Soybean in an Aqueous Solution. *Biosci. Biotechnol. Biochem.* **1994b,** *58,* 1772-1775.

Miyashita, K.; M. Hirao; E. Nara; and T. Ota. Oxidative Stability of Triglycerides from Orbital Fat of Tuna and Soybean Oil in an Emulsion. *Fisheries Sci.* **1995a,** *61,* 273-275.

Miyashita, K.; G. Azuma; and T. Ota. Oxidative Stability of Geometric and Positional Isomers of Unsaturated Fatty Acids in Aqueous Solution. *J. Jpn. Oil Chem. Soc.* **1995b,** *44,* 425-430.

Miyashita, K; S. Hirano; Y. Itabashi; T. Ota; M. Nishikawa; and S. Nakayama. Oxidative Stability of Polyunsaturated Monoacylglycerol and Triacylglycerol in Aqueous Micelles. *J. Jpn. Oil Chem. Soc.* **1997,** *46,* 205-208.

Moriya, H.; T. Kuniminato; M. Hosokawa; K. Fukunaga; T. Nishiyama; and K. Miyashita. Oxidative Stability of Salmon and Herring Roe Lipids and Their Dietary Effect on Plasma Cholesterol Levels of Rats. *Fisheries Sci.,* in press.

Nara, E.; K. Miyashita; and T. Ota. Oxidative Stability of PC Containing Linoleate and Docosahexaenoate in an Aqueous Solution with or without Chicken Egg Albumin. *Biosci. Biotechnol. Biochem.* **1995,** *59,* 2319-2320.

Nara, E.; K. Miyashita; and T. Ota. Oxidative Stability of Liposomes Prepared from Soybean PC, Chicken Egg PC, and Salmon Egg PC. *Biosci. Biotechnol. Biochem.* **1997,** *61,* 1736-1738.

Nara, E.; K. Miyashita; T. Ota; and Y. Nadachi. The Oxidative Stabilities of Polyunsaturated

Fatty Acids in Salmon Egg Phosphatidylcholine Liposomes. *Fisheries Sci.* **1998**, *64,* 282-286.
Nara, E.; K. Yamamoto; A. Hirose; M. Kotake; and K. Miyashita. Antioxidant Systems in Squid Eyes: Lipid Profiles and the Presence of Water Soluble Antioxidants. *J. Jpn. Oil Chem. Soc.* **2000**, *49,* 53-58.
Narayan, B.; K. Miyashita; and M. Hosokawa. Physiological Effects of Eicosapentaenoic Acid (EPA) and Docosahexaenoic Acid (DHA)-A Review. *Food Rev. Inter.* **2006a**, *22,* 291-307.
Narayan, B.; M. Hosokawa; and K. Miyashita. Occurrence of Conjugated Fatty Acids in Aquatic and Terrestrial Plants and Their Physiological Effects. In *Nutraceutical and Specialty Lipids and Their Co-Products,* Shahidi, F. Ed., CRC Taylor and Francis: New York, 2006b, pp. 201-218.
Neff, W.E.; E.N. Frankel; C.R. Scholfield; and D. Weisleder. High Pressure Liquid Chromatography of Autoxidized Lipids: I. Methyl Oleate and Methyl Linoleate. *Lipids* **1978**, *13,* 415-421.
Neff, W.E.; E.N. Frankel; and D. Weisleder. High Pressure Liquid Chromatography of Autoxidized Lipids: II. Hydroperoxy-Cyclic Peroxides and Other Secondary Products from Methyl Linolenate. *Lipids* **1981**, *16,* 439-448.
Ohshima, T.; Y. Fujita; and C. Koizumi. Oxidative Stability of Sardine and Mackerel Lipids with Reference to Synergism between Phospholipids and α-Tocopherol. *J. Am. Oil Chem. Soc.* **1993**, *170,* 269-276.
Olbrich, K.; W. Rawicz; D. Needham; and E. Evans. Water Permeability and Mechanical Strength of Polyunsaturated Lipid Bilayers. *Biophys. J.* **2000**, *79,* 321-327.
Parr, L.J.; and P.A.T. Swoboda. The Assay of Conjugable Oxidation Products Applied to Lipid Deterioration in Stored Foods. *J. Food Technol.* **1976**, *11,* 1-12.
Piche, L.A.; H.H. Draper; and P.D. Cole. Malondialyhyde Excretion by Subjects Consuming Cod Liver Oil vs a Concentrate of n-3 Fatty Acids. *Lipids* **1988**, *23,* 370-371.
Porter, N.A.; S.E. Galdwell; and K.A. Mills. Mechanisms of Free Radical Oxidation of Unsaturated Lipids. *Lipids* **1995**, *30,* 277-290.
Privett, O.S. and Blank, M.L. The initial stages of autoxidation, *J. Am. Oil Chem. Soc.,* **1962**, *39,* 465-469.
Saito, H.; and K. Ishihara. Antioxidant Activity and Active Sites of Phospholipids as Antioxidants. *J. Am. Oil Chem. Soc.* **1997**, *74,* 1531-1536.
Saiz, L.; and M.L. Klein. Structural Properties of a Highly Polyunsaturated Lipid Bilayer from Molecular Dynamics Simulations. *Biophys. J.* **2001**, *81,* 204-216.
Segawa, T.; S. Hara; and Y. Totani. Antioxidative Behavior of Phospholipids for Polyunsaturated Fatty Acids of Fish Oil. II. Synergistic Effect of Phospholipids for Tocopherol. *J. Jpn. Oil Chem. Soc.* **1994**, *43,* 515-519.
Segawa, T.; M. Kamata; S. Hara; and Y. Totani. Antioxidative Behavior of Phospholipids for Polyunsaturated Fatty Acids of Fish Oil. III. Synergistic Mechanism of Nitrogen Including Phospholipids for Tocopherol. *J. Jpn. Oil Chem. Soc.* **1995**, *44,* 36-42.
Shahidi, F.; and M. Miraliakbari. Marine Oils: Compositional Characteristics and Health Effects. In *Nutraceutical and Specialty Lipids and Their Co-Products,* Shahidi, F. Ed., CRC Press: New York, 2006, pp. 227-250.
Sinclair, A.; J. Wallace; M. Martin; N. Attar-Bashi; R. Weisinger; and D. Li. The Effects of Eicosapentaenoic Acid in Various Clinical Conditions. In *Healthful Lipids,* Akoh, C.C.; and Lai, O-M. Eds., AOCS Press: Champaign, Illinois, 2005, pp. 361-394.

Smagula, M.S.; C. Ho; and S.S. Chang. The Synthesis of 2-(2-Pentenyl) Furans and Their Relationship to the Reversion Flavor of Soybean Oil. *J. Am. Oil Chem. Soc.* **1979**, *56*, 516-519.

Smouse, T.H.; and S.S, Chang. A Systematic Characterization of the Reversion Flavor of Soybean Oil. *J. Am. Oil Chem. Soc.* **1967**, *44*, 509-514.

Suzuki, R.; R. Noguchi; T. Ota; M. Abe; K. Miyashita; and T. Kawada. Cytotoxic Effect of Conjugated Trienoic Fatty Acids on Mouse Tumor and Human Monocytic Leukemia Cells. *Lipids* **2001**, *36*, 477-482.

Suzuki, R.; M. Abe; and K. Miyashita. Comparative Study of the Autoxidation of TAG Containing Conjugated and Nonconnugated $C_{18}$ PUFA. *J. Am. Oil Chem. Soc.* **2004**, *81*, 563-569.

Takeuchi, M.; S. Hara; Y. Totani; H. Hibino; and Y. Tanaka. Antioxidative Behavior of Polyunsaturated Phospholipids. I. Oxidative Stability of Marine Oil Containing Polyunsaturated Phospholipids. *J. Jpn. Oil Chem. Soc.* **1997**, *46*, 175-181.

Van Den Bosch, G. Bleaching of Vegetable Oils: I. Conversions in Soybean Oil, Triolein and Trilinolein. *J. Am. Oil Chem. Soc.* **1973a**, *50*, 421-423.

Van Den Bosch, G. Bleaching of Vegetable Oils: II. Conversions of Methyl Oleate and Linoleate. *J. Am. Oil Chem. Soc.* **1973b**, *50*, 487-493.

VanRollins, M.; and R.C. Murphy. Autoxidation of Docosahexaenoic Acid: Analysis of Ten Isomers of Hydroxydocosahexaenoate. *J. Lipid Res.* **1984**, *25*, 507-217.

Venkateshwarlu, G.; M.B. Let; A.S. Meyer; and C. Jacobsen. Chemical and Olfactometric Characterization of Volatile Flavor Compounds in a Fish Oil Enriched Milk Emulsion. *J. Agric. Food Chem.* **2004a**, *52*, 311-317.

Venkateshwarlu, G.; M.B. Let; A.S. Meyer; and C. Jacobsen. Modeling the Sensory Impact of Defined Combinations of Volatile Lipid Oxidation Products on Fishy and Metallic Off-Flavors. *J. Agric. Food Chem.* **2004b**, *52*, 1635-1641.

Wander, R.C.; and S.-H. Du. Oxidation of Plasma Proteins Is Not Increased after Supplementation with Eicosapentaenoic and Docosahexaenoic Acids. *Am. J. Clin. Nutr.* **2000**, *72*, 731-737.

Yamauchi, R.; T. Yamada; and Y. Ueno. Monohydroperoxides Formed by Autoxidation and Photosensitized Oxidation of Methyl Eicosapentaenoate. *Agric. Biol. Chem.* **1983**, *47*, 2897-2902.

Yamauchi, R.; T. Yamada; K. Kato; and Y. Ueno. Autoxidation and Photosensitized Oxidation of Methyl Eicosapentaenoate: Secondary Oxidation Products. *Agric. Biol. Chem.* **1985**, *49*, 2077-2082.

Yurawecz, M.P.; A.A. Molina; M. Mossoba; and Y. Ku. Estimation of Conjugated Octadecatrienes in Edible Fats and Oils. *J. Am. Oil Chem. Soc.* **1993**, *70*, 1093-1099.

Yurawecz, M.P.; J.K. Hood; M.M. Mossoba; J.A.G. Roach; and Y. Ku. Furan Fatty Acids Determined as Oxidation Products of Conjugated Octadecadienoic Acid. *Lipids* **1995**, *30*, 595-598.

# 4

# Oxidation of Conjugated Linoleic Acid

Taina I. Pajunen (née Hämäläinen) and Afaf Kamal-Eldin
Department of Chemistry, University of Helsinki, P.O. Box 55, FIN-00014 University of Helsinki, Finland, and Department of Food Science, Swedish University of Agricultural Sciences (SLU), Box 7051, 750 07 Uppsala, Sweden

## Introduction

Conjugated linoleic acid (CLA) is a generic name for a group of positional and geometric isomers of octadecadienoic acid in which the two double bonds are conjugated, that is single and double bonds alternate (1,3-diene structure). The most abundant CLA isomers in nature are $9Z,11E$-octadecadienoic acid ($9Z,11E$-CLA) and $10E,12Z$-octadecadienoic acid ($10E,12Z$-CLA).

Interest in CLA rose exponentially when an isomeric mixture of CLA was discovered to have anticancer activities (Ha et al., 1987). Today, numerous physiological properties are attributed to CLA, including effects on cancer, atherosclerosis, adiposity, and insulin resistance (Gnädig et al., 2001; Belury et al., 2003; Terpstra, 2004; Wahle et al., 2004). The physiological properties of CLA seem to be structure-specific. So far, only two CLA isomers, namely $9Z,11E$-CLA and $10E,12Z$-CLA, are known to possess biological activities, with $10E,12Z$-CLA being more active. Both isomers were claimed to inhibit carcinogenesis in animal models (Scimeca, 1999; Pariza et al., 2001) and $10E,12Z$-CLA was shown to affect lipid metabolism. The biochemical mechanisms of CLA action remain unclear. Potential carcinogenic reduction mechanisms (Belury, 2002) as well as obesity reduction (Evans et al., 2002) seem to include induction of fatty acid oxidation. In humans, CLA was found to increase free-radical-induced lipid peroxidation leading to increased levels of 8-iso-$PGF_{2\alpha}$ as well as to increase cyclooxygenase activity leading to increased levels of urinary 15-keto-dihydro-$PGF_{2\alpha}$ and C-reactive proteins (Basu et al., 2000; Risérus et al., 2002). These effects might explain the decrease in insulin sensitivity and glucose tolerance observed for CLA isomers, particularly for $10E,12Z$-CLA (Moloney et al., 2004; Risérus et al., 2004).

Despite the wide interest in CLA and the large number of scientific articles published recently (a regularly updated listing of the scientific literature on CLA is available at http://www.wisc.edu/fri/clarefs.htm), surprisingly little is known about the oxidation reactions of CLA. For example, no CLA primary autoxidation products were identified until 2001. Most studies use a complex mixture of CLA isomers (in the form of free acids, and methyl, ethyl, or glyceryl esters) and autoxidation of CLA is performed under widely different conditions (use of different temperatures, solvents, initiators, antioxidants etcetera; see Table 4.1). Thus, it is difficult to interpret the data and it is not possible to correlate the results with a single isomer or to compare the data from different studies directly. A recent study provided evidence that CLA isomerizes through [1,5]-sigmatropic rearrangement of hydrogen in heated oils (Destaillats and Angers, 2002) and therefore, the data produced at high temperatures should be interpreted cautiously. In fact, the question is whether oxidations performed at high temperatures should be considered as autoxidation reactions, because by definition autoxidation is a reaction performed at ambient pressure and temperature.

The aim of this chapter is to review the available information relevant to CLA oxidation for inspiration of new studies needed to discover the mechanisms behind the biological activity of CLA isomers. It is noted that future research needs to produce more data on the oxidation of single isomers in order to characterize the products of each of the isomers, and to understand and evaluate the oxidation mechanisms of CLA.

## Autoxidation of CLA

Available studies suggest that the autoxidation of conjugated fatty acids occurs through similar and different mechanisms and produces unique oxidation products compared to autoxidation of monounsaturated and nonconjugated polyunsaturated fatty acids (Table 4.1). Thus, interpretation of the data of CLA autoxidation and comparison of it with that of nonconjugated fatty acids should be done with sufficient care. Misinterpretations are easily made because the knowledge of the structures of CLA autoxidation products and of mechanisms is limited.

The identification of adequate method(s) to monitor CLA autoxidation is a challenging task. Banni et al. (1998) have pointed out that the controversial results concerning the involvement of CLA in oxidative stress are due mostly to the lack of suitable methodology. In addition, Van den Berg et al. (1995) and Suzuki et al. (2004) have shown that peroxide value (PV) measurements produce incorrect results in kinetic studies pertinent to CLA oxidation. PV measurement is an indirect method and cannot be used to distinguish between hydroperoxides (ROOH) and dialkyl peroxides (ROOR, for example, cyclic or oligomeric peroxides). The relative sensitivity of PV measurement to the peroxide structure and the access of its reagents to peroxide bonds that might be hidden inside global oligomers, however, are not known. Thus, using PV measurements to estimate the extent of autoxidation may lead to incorrect results if autoxidation primarily yields oligomeric products. The contribution of the dialkyl peroxide cross-links have been determined by reacting oxidized oil samples with $SF_4$ prior to PV measurement (Mallégol et al., 2000). Moreover, the hydroperoxides can be detected and distinguished from cyclic peroxides and from oligomeric peroxides using NMR spectroscopy. The hydroperoxide protons resonate clearly at different spectral regions from the characteristic signals of cyclic peroxides (Hämäläinen and Kamal-Eldin, 2005). Furthermore, integration of the proton signals can be used to differentiate between monohydroperoxides and oligomeric products with hydroperoxide groups.

Table 4.1. Examples of Oxidation Studies on Conjugated Linoleic Acid (CLA) as a Free Acid or as an Ester.

| Oxidizing Substrate | Conditions (other FA, antioxidant(s), initiator, solvent, T, etc.) | Analysis | Main results/Comments | Reference |
|---|---|---|---|---|
| 10,12-CLA | Oxidized in air at 37°C. | oxygen consumption | • Autoxidation of 10,12-CLA proceeds approximately at the same rate as LA at 37°C. | Holman and Elmer, 1947 |
| 10,12-CLA, Me 10,12-CLA, and mixture of CLA isomers | Oxidized in open test tubes while oxygen was bubbled through at 30, 65, and 90°C. | PV, hydrogen number, CD | • The rate and the amount of peroxides formed were dependent on the temperature; maximum was reached faster at elevated temperatures, but larger amount of peroxides was present at 30 than at 65 or 90°C. | Allen et al., 1949 |
| Me 10,12-CLA | Oxidized in a closed system after flushing with oxygen at 30°C. | PV, hydrogen number, CD | • No peroxide oxygen was formed until Me 10,12-CLA had been oxidized more than 100 hours. The reaction proceeded three times more slowly and in a different manner (less peroxides were formed) than the ML. | Allen et al., 1949 |
| Me 10,12-CLA | Oxidized in presence and absence of AIBN, di-t-butyl-peroxide or copper salt of α-ethyl-caproic acid at 50°C. | oxygen consumption | • Autoxidation is autocatalytical in character and the energy of activation is 17.5 kcal/mol.<br>• Use of an initiator or a catalyst results in the formation of polymeric peroxides. | Kern et al., 1955 |
| Me 9,11-CLA | Oxidized in absence of catalyst at 50 and 70°C and with peroxide/Cu²⁺ at 50°C. | oxygen consumption | • Me 9,11-CLA reacts in 1,2- and 1,4-positon with molecular oxygen to give relatively low molecular oxygen-co-polymers. | Kern et al., 1956 |
| Me 9,11-CLA | Oxidized in presence and absence of +0.1% NDGA at 30°C. | oxygen consumption, PV, UV, SnCl₂ and KI reduction, residual FA ester | • Autoxidation proceeds by an autocatalytic manner.<br>• 84% of the reaction products were polymers which were not fissioned by SnCl₂ and 16% monomeric or monomers produced by reaction with SnCl₂ when the oxidation extent was 17%.<br>• Virtually all the isolated E-unsaturation receded in the polymer fraction. | Privett, 1959 |

*Cont. on p. 80.*

Table 4.1., cont. Examples of Oxidation Studies on Conjugated Linoleic Acid (CLA) as a Free Acid or as an Ester.

| Oxidizing Substrate | Conditions (other FA, antioxidant(s), initiator, solvent, T, etc.) | Analysis | Main results/Comments | Reference |
|---|---|---|---|---|
| Mixture of Me E,Z-CLA isomers | Oxidized in presence and absence of 0.01% BHA, BHT, PG, sesamol, NDGA, α-Toc, L-thyroxine sodium salt, 4,4'-dihydroxy-3,5,3',5'-tetra-t-butyldiphenyl methane. | weighing, UV, IR, PV, MS | • BHA, BHT, PG, and sesamol lengthened the induction period seven to twelve times.<br>• The induction periods were shorter and the peroxide values lower with or without antioxidants for the conjugated dienoates than for the nonconjugated dienoates.<br>• After autoxidation to a weight gain of 10 mg per 1.5 g, the antioxidant containing samples had higher molecular weights and lower diene contents than the control samples. | Fukuzumi and Ikeda, 1970 |
| 9E,11E-CLA | Oxidized in 90% v/v aqueous acetic acid in presence and absence of copper, manganese, and cobalt acetate at 80°C. | UV | • The rate of autoxidation increases with the initial acid concentration.<br>• The autoxidation of 9E,11E-CLA is retarded by copper (reducible), manganese (oxidizable) salts.<br>• The cobalt salt retards the autoxidation at low concentrations and accelerates it at higher concentrations. | Pekkarinen, 1972 |
| Me 8E,10E-CLA | Photooxidized in MeOH for 5d using MB sensitizer. | IR, NMR, MS | • The photooxidation of Me 8E,10E-CLA produces 6-heptyl-3-(6-methoxycarbonylhexyl)-3,6-dihydro-1,2-dioxine. | Gunstone and Wijesundera, 1979 |
| Me 9E,11E-CLA | Photooxidized in CCl$_4$/MeOH (95/5, v/v) for 16h using MB sensitizer. | IR, $^1$H and $^{13}$C NMR, MS | • The photooxidation of Me 9E,11E-CLA produces 6-hexyl-3-(7-methoxycarbonylheptyl)-3,6-dihydro-1,2-dioxine in over 80% yield after purification by column chromatography. | Bascetta et al., 1984 |
| Mixture of nine CLA isomers | Oxidized in the presence of LA in phosphate buffer-water-ethanol mixture in presence and absence of ascorbic acid or α-Toc for 15d at 40°C. | PV | • CLA is much more resistant to oxidation than LA.<br>• CLA was more potent than α-Toc and almost as effective as BHT as an antioxidant. (Note: This result has later been proven not to be correct. See van den Berg et al. 1995, Chen et al. 1997, and Banni et al. 1998.) | Ha et al., 1990 |

**Table 4.1., cont.** Examples of Oxidation Studies on Conjugated Linoleic Acid (CLA) as a Free Acid or as an Ester.

| Oxidizing Substrate | Conditions (other FA, antioxidant(s), initiator, solvent, T, etc.) | Analysis | Main results/Comments | Reference |
| --- | --- | --- | --- | --- |
| Mixture of CLA isomers | Oxidized with PLPC membrane using initiator (AMVN, AAPH or $H_2O_2$ and $Fe^{2+}$) in presence and absence of vitamin E or BHT in buffer solutions. | GC/MS analysis of FA | • CLA does not act as an efficient radical scavenger in any way comparable to vitamin E or BHT under conditions of metal ion-dependent or independent oxidative stress. | van den Berg et al., 1995 |
| Mixture of CLA isomers | Oxidized as a thin film in air with and without LA/AA at rt. | CD, GC/MS analysis of FA | • The oxidative susceptibility of CLA was higher than that of LA and comparable to AA.<br>• Oxidation rates for individual CLA isomers were found to be virtually identical. | van den Berg et al., 1995 |
| Mixture of CLA isomers | Oxidized in methanol-water solution while flushing air through at 45, 40, and 48-69°C. | CG-MS, GC/FID | • Furan fatty acids identified as secondary oxidation products of CLA autoxidation. | Yurawecz et al., 1995 |
| Me 9E,11E-CLA | Oxidation as ~0.5 mm thick film in the presence of Co/Ca/Zr drier in air for 2d. | DCI-, FAB-, FD-, ESI-, and SI-MS | • Oxidative crosslinking produced oligomers that yielded signals consisting of groups of peaks 16 mass units apart, pointing to a series of oxygenated homologues.<br>• Conjugated FA crosslinks by radical addition to the double bond whereas nonconjugated FA by recombination of radicals. | Muizebelt and Nielen, 1996 |
| Mixture of CLA isomers (as free acids, Me esters or TAG) | Added to canola oil; Oxidized after flushing with air at 90°C. | oxygen uptake and changes in LA and α-LnA by GC/FID | • CLA and Me CLA accelerated lipid oxidation in canola oil dose-dependently. CLA-TAG had no influence on lipid oxidation in canola oil.<br>• CLA is not an antioxidant in fats and oils. | Chen et al., 1997 |

*Cont. on p. 82.*

Table 4.1., cont. Examples of Oxidation Studies on Conjugated Linoleic Acid (CLA) as a Free Acid or as an Ester.

| Oxidizing Substrate | Conditions (other FA, antioxidant(s), initiator, solvent, T, etc.) | Analysis | Main results/Comments | Reference |
|---|---|---|---|---|
| Mixture of CLA isomers | Mixed with LA (1/1, w/w); Oxidized after flushing with air at 90°C. | oxygen uptake and changes in LA by GC/FID | • CLA is remarkably less stable than LA in air. | Chen et al., 1997 |
| Mixture of CLA isomers (as free acids or TAG) | Mixed with LA, LnA, AA, DHA, and HA (equal amounts) or TAG containing these FA; Oxidized after flushing with air at 90°C. | FA analysis by GC/FID | • CLA as a free acid or TAG more susceptible to oxidation than LA, LnA and AA. The oxidative susceptibility of CLA was similar to that of DHA. | Zhang and Chen, 1997 |
| Mixture of CLA isomers | Oxidized in presence LA as a thin film in air at 0 and 37°C. | HPLC with DAD | • No significant antioxidant effect of CLA was detected. | Banni et al., 1998 |
| Mixture of Me CLA isomers | Photooxidized as pure Me CLA or mixed with ML in ethanol for 5d using MB sensitizer at 25°C. | PV, residual FA, residual MB | • Me CLA was lost to a lower extent and yielded less (hydro)peroxides but bleached MB at a higher rate than ML.<br>• Me CLA isomers were not equally lost and the photooxidation was accompanied by interisomerisation.<br>• Me CLA photooxidizes through different mechanisms than ML. | Jiang and Kamal-Eldin, 1998 |
| Me 9Z,11E-CLA | Epoxidized with m-CPBA and Oxone. | TLC, IR, $^1$H and $^{13}$C NMR | • A mixture of positional monoepoxides was generated using m-CPBA and diepoxides formed by using Oxone. | Lie Ken Jie and Pasha, 1998 |
| CLA isomers (as free acid or Me or Et esters) | Oxidized in the presence of LA in aqueous system (pH 7.5) with AAPH. | Oxygen uptake | • CLA is more stable than LA in aqueous system when AAPH was used as a free radical initiator.<br>• The oxidizability was in the order Et CLA>LA>Me CLA>ML>EL>CLA. | Seo et al., 1999 |

Oxidation of Conjugated Linoleic Acid    83

Table 4.1., cont. Examples of Oxidation Studies on Conjugated Linoleic Acid (CLA) as a Free Acid or as an Ester.

| Oxidizing Substrate | Conditions (other FA, antioxidant(s), initiator, solvent, T, etc.) | Analysis | Main results/Comments | Reference |
|---|---|---|---|---|
| CLA isomers (as free acid or Me or Et esters) | Oxidized in the presence LA in benzene with AMVN. | Oxygen uptake | • CLA is less stable than LA in benzene when AMVN was used as a free radical initiator.<br>• The oxidizability was in the order Et CLA>EL=ML>Me CLA>CLA>LA. | Seo et al., 1999 |
| 9Z,11E-CLA and 10E,12Z-CLA | Tested in the TOSC assay (Oxidation of KMBA by ABAP). | Oxidation of KMBA to ethylene | • 10E,12Z-CLA and 9Z,11E-CLA inhibited the oxidation of KMBA at 2-200 micromolar concentrations. At higher concentrations, 9Z,11E-CLA but not 10E,12Z-CLA enhanced the oxidation. | Leung and Liu, 2000 |
| Me 9Z,11E-CLA and Me 10E,12Z-CLA | Oxidative crosslinking in the presence of Co or Co/Ca/Zr drier and reactive diluents up to 4d. | SEC, SI- and ESI-MS, NMR | • Nonconjugated FA reacts by recombination of radicals forming mainly peroxy and ether cross-links and conjugated FA by radical addition to the double bond.<br>• Despite the differences in the mechanism the rates of both reactions were found to be similar.<br>• Reactive diluents are incorporated into the actual paint film. | Muizebelt et al., 2000 |
| Mixture of 12 CLA isomers | Oxidized in the presence of LA (1.5g/0.3g); Oxidized after flushing with air at 50°C. | GLC with FID and Ag-HPLC with UV | • CLA oxidized considerably faster than LA. The oxidation extent of CLA was > 80% after 110 h.<br>• The four Z,Z-CLA isomers were most unstable followed by four E,Z-CLA isomers. The four E,E-CLA isomers were relatively stable under the same experimental conditions. The different positional isomers of the same geometry oxidized at a similar rate. | Yang et al., 2000 |
| Mixture of 12 CLA isomers | CLA isomers added to canola oil at 10% level; Oxidized in air in presence and absence of BHT or GTC at 90°C. | oxygen consumption and FA analysis | • 200 ppm of GTC were more effective (and dose dependent) than 200 ppm BHT in protecting CLA from oxidation.<br>• CLA oxidized without the antioxidants considerably faster than LA: The oxidation extent of CLA after 35 h was 86%. | Yang et al., 2000 |
| Mixture of Me CLA isomers | Oxidized by heating at 100, 150, or 200°C for 3 h or illumination at 4000 lux for 14 d. | GC of residual FA, PV | • CLA oxidizes at a higher rate compared to LA. | Chen et al., 2001 |

*Cont. on p. 84.*

Table 4.1., cont. Examples of Oxidation Studies on Conjugated Linoleic Acid (CLA) as a Free Acid or as an Ester.

| Oxidizing Substrate | Conditions (other FA, antioxidant(s), initiator, solvent, T, etc.) | Analysis | Main results/Comments | Reference |
|---|---|---|---|---|
| Mixture of Me CLA isomers | Oxidized as neat oil in presence and absence of α-Toc in the dark at 40°C. | PV, GC-MS, HPLC, 1D and 2D NMR, residual CLA | • Hydroperoxides discovered as primary products in the CLA autoxidation. The NMR results revealed that CLA autoxidizes at least partly according to the Farmer's hydroperoxide theory. | Hämäläinen et al., 2001 |
| CLA, Et CLA | Oxidized in the bulk phase at 50°C. | oxygen consumption, PV, dimers/ polymers by HPSEC | • The oxidative stability of CLA was almost the same as that of LA, but ethyl CLA was oxidatively more stable than ethyl LA. | Suzuki et al., 2001 |
| Mixture of CLA isomers | Reacted with DPPH radical. | DPPH radical assayed by ESR | • CLA slightly diminished DPPH compared to BHT, α-tocopherol, and vitamin C. | Yu, 2001 |
| Me 9Z,11E-CLA | Oxidized as neat oil in presence of α-Toc in the dark at 40°C. | PV, GC-MS, HPLC, 1D and 2D NMR spectroscopy | • Structural characterization of seven hydroperoxides from autoxidation of Me 9Z,11E-CLA.<br>• The reaction was in favour of one geometric isomer and a new type of E,Z-conjugated diene hydroperoxide, where the Z-double bond was adjacent hydroperoxyl-bearing methine carbon was discovered.<br>• A mechanism for the autoxidation of Me CLA was proposed based on well-characterized primary oxidation products. | Hämäläinen et al., 2002 |

**Table 4.1., cont.** Examples of Oxidation Studies on Conjugated Linoleic Acid (CLA) as a Free Acid or as an Ester.

| Oxidizing Substrate | Conditions (other FA, antioxidant(s), initiator, solvent, T, etc.) | Analysis | Main results/Comments | Reference |
|---|---|---|---|---|
| CLA concentrate (=alkali isomerized safflower oil) | Oxidized as oil in presence of antioxidant ($\alpha$-Toc, BHA, BHT, GTE, PG, RE, and TBHQ) in the dark up to 44 d at 45°C. | PV | • The protective effect of 200 ppm antioxidant was in the order of TBHQ> BHA> PG> BHT> RE> $\alpha$-Toc> GTE. | Lee et al., 2003 |
| Me 9Z,11E-CLA | Epoxidized with m-CPBA, DMDO, MTO/$H_2O_2$, Oxone /tetrahydrothiopyran-4-one, and Novoenzyme 435/ $H_2O_2$. | IR, $^1$H and $^{13}$C NMR, EI-MS | • The reactions furnished mono-epoxides and a mixture of diastereomers of syn- and anti-diepoxy-stearate. | Lie Ken Jie et al., 2003 |
| 9Z,11E-CLA and 10E,12Z-CLA | Oxidized separately as thin films in the dark at 40 to 80°C. | Unoxidized CLA analyzed by GC | • 10E,12Z-CLA oxidized faster than 9Z,11E-CLA at all tested temperatures. Apparent activation energies determined as 65.4 and 72.1 kJ/mol, respectively. | Minemoto et al., 2003 |
| 9Z,11E-CLA | Oxidized with peroxygenase in an aqueous medium using t-butyl hydroperoxide as an oxidant. | HPLC with APCI-MS/ EI-MS and $^1$H and $^{13}$C NMR | • Main product (approx 90%) was up to 6 h 9,10E-epoxy-11E-octadecenoic acid.<br>• Acidic work up resulted in the formation of 1,2- and 1,4-diols. | Piazza et al., 2003 |

*Cont. on p. 86.*

Table 4.1., cont. Examples of Oxidation Studies on Conjugated Linoleic Acid (CLA) as a Free Acid or as an Ester.

| Oxidizing Substrate | Conditions (other FA, antioxidant(s), initiator, solvent, T, etc.) | Analysis | Main results/Comments | Reference |
|---|---|---|---|---|
| CLA-TAG (69.5% CLA) and Me esters prepared from CLA-TAG | Oxidized in the bulk phase in a sealed vial with and without 0.5% α-Toc or trolox in the dark at 50°C.<br>Note: soybean-TAG (56.1% LA), perilla-TAG (54.5% α-LnA), and bitter gourd-TAG (61.6% conjugated LnA) and Me esters prepared from these TAG was oxidized in the same conditions. | oxygen consumption, PV, size-exclusion HPLC | • Rates of oxygen consumption and polymer formation of CLA-TAG and bitter gourd-TAG were faster than those of soybean-TAG and perilla-TAG, respectively. (The same results were obtained in the oxidation of the Me esters.)<br>• The main oxidation products of CLA-TAG and bitter gourd-TAG were dimers and polymers whereas hydroperoxides were the main products in the oxidation of soybean-TAG and perilla-TAG.<br>• α-Toc and trolox inhibited the oxidation of the TAG. The inhibitory effect of these antioxidants was more effective against the oxidation of CLA-TAG and bitter gourd-TAG than that of soybean-TAG and perilla-TAG, respectively. | Suzuki et al., 2004 |
| 9Z,11E-CLA and 10E,12Z-CLA | Oxidized separately as thin films at 37°C. | Unoxidized substrates by GC, PV, PV, TBARS | • CLA oxidizes at a higher rate than LA. 9Z,11E-CLA was consumed at the same rate as LnA and was more stable than the 10E,12Z-CLA.<br>• The 9Z,11E-CLA and 10E,12Z-CLA gave low values of PV and TBARS compared to LA and LnA. | Tsuzuki et al., 2004 |
| CLA [isomer(s) not reported] | Encapsulated in WPC, GA, and a blend of WPC and maltodextrin 10 DE (1:1, w/w) matrixes at water activities from 0.108 to 0.892 at 35 and 45°C. | Static head-space GC | • The highest values of CLA degradation and lipid oxidation were observed in the range of water activities 0.103–0.429 for all matrices at 45°C, whereas the lowest CLA degradation and lipid oxidation were observed for WPC at a water activity of 0.743 and 35°C. | Jimenez et al., 2006 |

## Oxidation of Conjugated Linoleic Acid 87

**Table 4.1., cont.** Examples of Oxidation Studies on Conjugated Linoleic Acid (CLA) as a Free Acid or as an Ester.

| Oxidizing Substrate | Conditions (other FA, antioxidant(s), initiator, solvent, T, etc.) | Analysis | Main results/Comments | Reference |
|---|---|---|---|---|
| | | | | |

Abbreviations: AA = arachidonic acid, AAPH = 2,2′-azobis(2-amidinopropane) dihydrochloride, AIBN = 2,2′-azobis(buturonitrile), AMVN = 2,2′-azobis(2,4-dimethylvaleronitrile), BHA = butylated hydroxyanisole, BHT = butylated hydroxytoluene, CD = conjugated dienes, CLA = conjugated linoleic acid, m-CPBA = m-chloroperoxybenzoic acid, DAD = diode array detector, DCI = direct chemical ionization, DHA = docosahexaenoic acid, DMDO = dimethyl dioxirane, DPPH• = 2,2-diphenyl-1-picrylhydrazyl radical, ESI = electrospray ionization, ESR = electron spin resonance, Et = ethyl, EL = ethyl linoleate, FA = fatty acid, FAB = fast atom bombardment, FD = field desorption, GA = gum arabic, GLC = gas-liquid chromatography, GTC = jasmine green tea catechins, GTE = green tea extract, HA = heptadecanoic acid, KMBA = α-keto-α-methiolbutyric acid, LA = linoleic acid = 9Z,12Z-octadecadienoic acid, LnA = linolenic acid = 9Z,12Z,15Z-octadecatrienoic acid, MB = methylene blue, Me = methyl, MeOH = methanol, ML = methyl linoleate, MS = mass spectrometry, MTO = methyltrioxorhenium, NDGA = nordihydroguaiaretic acid, Oxone = potassium peroxymonosulfate, PG = propyl gallate, PLPC = 1-palmitoyl-2-linoleyl phosphatidylcholine, PV = peroxide value, RE = rosemary extract, rt = room temperature, SEC = size exclusion chromatography, SI = secondary ion, TAG = triacylglyceride, TBHQ = tert-butylhydroquinone, TLC = thin-layer chromatography, α-Toc = α-tocopherol, TOSC = total antioxidant scavenging capacity, WPC = whey protein concentrate.

## Rate and Routes of Autoxidation of CLA

A number of studies have shown that a mixture of CLA isomers oxidizes at a higher rate compared to linoleic acid (LA; 18:2, 9Z,12Z) by measuring residual unoxidized substrate or consumed oxygen (van den Berg et al., 1995; Zhang and Chen, 1997; Chen et al., 1997,2001; Yang et al., 2000). As a thin film 10E,12Z-CLA oxidizes faster than 9Z,11E-CLA at 40–80°C when the residual unoxidized substrate is monitored (Minemoto et al., 2003). This was confirmed in a study by Tsuzuki et al. (2004), where similar thin films of CLA were oxidized at 37°C and substrate consumption was followed. Moreover, 9Z,11E-CLA was consumed at the same rate as linolenic acid (LnA; 18:3, 9Z,12Z,15Z). However, when oxygen consumption was used to monitor oxidation, 9Z,11E-CLA was more similar to LA, the difference between the two CLA isomers was not that large, and LnA consumed oxygen at a comparable rate to α-eleostearic acid (ESA; 18:3, 9Z,11E,13E). Suzuki et al. (2004) showed that CLA-containing oil (69.5% CLA) oxidized faster than purified soybean oil (56.1% LA) and that bitter gourd oil (61.6% ESA) oxidized faster than perilla seed oil (54.5% LnA) when oxygen consumption was used to monitor oxidation at 50°C. In addition, they demonstrated the previous claim that CLA is oxidatively more stable than LA at room temperature came as a result of the use of PV measurement to monitor oxidation (Ha et al., 1990; Ip et al., 1991). Suzuki et al. (2004), Allen et al. (1949), and Holman (1954) showed that only small amount of peroxides accumulate during the oxidation of CLA, unlike the case of LA. The results of Tsuzuki et al. (2004) suggest that CLA and ESA absorb less oxygen per mole of oxidized substrate compared to their unsaturated counterparts, LA and LnA. It has already been established that the stability of CLA isomers decreases in the order $E,E>E,Z>Z,Z$ which shows a stabilizing effect of $E$-double bonds (Yang et al., 2000) as is seen for other fatty acids, for example elaidic and oleic acids (Lanser et al.,1986).

Allen and Kummerow (1951) and Kern et al. (1955) showed that the role of conjugated fatty acids in their own autocatalytic oxidation differ from that of their nonconjugated counterparts. For example the rate of the reaction is one-half order with respect to the catalytic product in the case of CLA while first order in the case of LA, that is

Rate of CLA autoxidation = k [CLA] $\sqrt{\text{[Oxidation product(s)]}}$
Rate of LA autoxidation = k [LA] [LOOH]

Further analysis of the Kern et al. (1955) data using empirical kinetics suggested that the oxidation of 9Z,11E-CLA is dominated by trimerization (up to 1% oxidation at 40°C, 2% oxidation at 50°C, and 15% oxidation at 70°C) followed by formation of monomers (Brimberg and Kamal-Eldin, 2003). These results suggest that the extent of oligomerization increases with increased temperature. Since the activation energies for both reactions leading to oligomers and monomers were almost identical, it was suggested that the monomers split from the oligomers. Comparing the oxidation kinetics of thin films of 9Z,11E-CLA and 10E,12Z-CLA at different temperatures,

Minemoto et al. (2003) found that the oxidation of $9Z,11E$-CLA followed an autocatalytic rate expression through the entire oxidation process while $10E,12Z$-CLA followed the autocatalytic rate expression only during the first half of the oxidation course, which was followed by first order kinetics. The apparent activation energies for $9Z,11E$-CLA and $10E,12Z$-CLA were approximately 15.5 and 17 kcal/mol, which were greater than those of LA (12 kcal/mol) and LnA (14.5 kcal/mol). The activation energy was the same during the overall oxidation regimen of the $10E,12Z$-CLA isomer. These authors found that the enthalpy-entropy compensation holds during autoxidation of $9Z,11E$-CLA, $10E,12Z$-CLA, and LA and suggested a "common" oxidation mechanism.

Seo et al. (1999) compared the oxidation of CLA and LA as equal mixtures of 9,11- and 10,12- isomers in the form of free acids and methyl and ethyl esters in the presence of methyl linoleate (ML). Both were examined in an aqueous system, where oxidation was induced by 2,2'-azobis(2-amidinopropane)dihydrochloride (AAPH) and in benzene, where oxidation was induced by 2,2'-azobis(2,4-dimethyl-valeronitrile) (AMVN) at 37°C. The oxidizability in aqueous system was much lower for CLA than for LA while it was much higher for CLA ethyl and methyl esters than for LA methyl and ethyl esters (Table 4.1). In benzene, CLA had a higher oxidizability than LA while the oxidizabilities of their esters were comparable. Indeed, the mixture of CLA in this study might have affected the results, especially in those where only small differences were found.

Important differences between CLA and LA oxidations relate to the effect of pro- and antioxidants on their oxidation rates. Jackson and Kummerow (1949) demonstrated that metallic naphthenates have less effect on the oxidation of CLA than LA. Moreover, Kern et al. (1955) found that copper(II)octanoate ($3 \times 10^{-4}$ mol Cu/mol ester) delayed the oxidation of methyl $10E,12Z$-CLA at 40°C without inhibiting the rate of oxidation afterwards. This result, which contradicts the enhancement of LA oxidation by metal ions, may be explained by inhibitory interactions of $\pi$-complexes with transition metal ions (Park et al., 2007). $\alpha$-Tocopherol was more potent in protecting CLA than purified soybean triacylglycerols (Suzuki et al., 2004). Moreover, while $\alpha$-tocopherol remains resistant to oxidation during the induction period of fatty acids with methylene-interrupted double bonds and then declines sharply, its degradation during the oxidation of CLA seems to follow a sigmoid curve (Suzuki et al., 2004). It was proposed that antioxidants retard the oxidation of CLA by two mechanisms, first by forming H-bonds with the $\pi$-system of the conjugated double bond and secondly by interfering with the propagation reactions (Fukuzumi and Ikeda 1970).

## Formation of Hydroperoxides During Autoxidation of CLA Methyl Ester

The identification of the primary autoxidation products is of crucial importance for understanding the subsequent secondary oxidation steps. The formation mechanism of the oligomeric products, which seem to dominate in CLA, remains more or less speculative without the knowledge of the primary species involved in their initiation.

It has been assumed that the primary autoxidation products of conjugated dienes are not similar to those of methylene-interrupted systems (Yurawecz et al., 1997). Contrary to this assumption, conjugated diene allylic monohydroperoxides were discovered as primary autoxidation products of CLA methyl esters (Hämäläinen et al., 2001). The conclusive evidence for hydroperoxide formation during autoxidation of CLA methyl esters with and without α-tocopherol was provided by NMR spectroscopy. The hydroperoxide protons of CLA hydroperoxides appeared as partly overlapping singlets at $\delta_H$ 7.98–7.81 in deuterochloroform and at $\delta_H$ 10.48–10.40 in deuteroacetone. These assignments were further confirmed by $D_2O$-test, and by comparison of the $^1H$ NMR spectrum of CLA hydroperoxides with that of ML hydroperoxides. In addition, based on proton signal integration and $^1H$-$^1H$ correlation experiment (COSY), it was evident that the main CLA hydroperoxides have a conjugated diene allylic monohydroperoxy structure.

Determination of an oxidation mechanism requires full characterization of the oxidation products. A good mechanism explains not only the formation of all the products but accounts for their relative proportions. Hämäläinen et al. (2002) produced hydroperoxides from 9Z,11E-CLA methyl ester oxidizing the fatty acid ester with 20% α-tocopherol under atmospheric oxygen at 40°C in the dark. In these conditions, as well as in the autoxidation of ML (Peers and Coxon, 1983), α-tocopherol does not exert its recognized antioxidant effect; it allows the autoxidation to proceed smoothly and quite rapidly to produce good yield of hydroperoxides. The CLA methyl ester hydroperoxides were isolated by flash column chromatography and subsequently reduced to corresponding hydroxy derivatives. By combining HPLC separation, UV, 1D and 2D NMR, and GC-MS techniques the structures of the individual hydroxy isomers were characterized. Detailed example of the structural characterization can be found elsewhere (Hämäläinen and Kamal-Eldin, 2005). The characterization of the primary products enabled the authors to propose a mechanism for the hydroperoxide formation during autoxidation of CLA methyl esters.

The autoxidation of 9Z,11E-CLA methyl ester through the hydroperoxide pathway is depicted in Scheme 4.1. The autoxidation yielded seven conjugated diene allylic monohydroperoxides as a pair of enantiomers: four with E,E-geometry and three with E,Z-geometry. Note that one of the E,Z-isomers was a new type of lipid hydroperoxide (11) where the Z-double bond is adjacent to the hydroperoxyl-bearing methine carbon atom. Furthermore, the autoxidation was diastereoselective in favor of one geometric isomer (9); namely methyl 13-(R,S)-hydroperoxy-9Z,11E-octadecadienoate (Me 13-OOH-9Z,11E).

The mechanism for the hydroperoxide pathway of CLA autoxidation shows similarities to both methyl oleate and ML autoxidation. The first step is an abstraction of one of the allylic hydrogen atoms from the starting material. Similarly to methyl oleate, there are four allylic hydrogen atoms available for abstraction. However, differing from methyl oleate the two allylic positions are not equal since one is allylic to an E-double bond and the other is allylic to a Z-double bond. Moreover, the H-atom abstraction leads to the formation of two pentadienyl radicals 2 and 3, where the lone

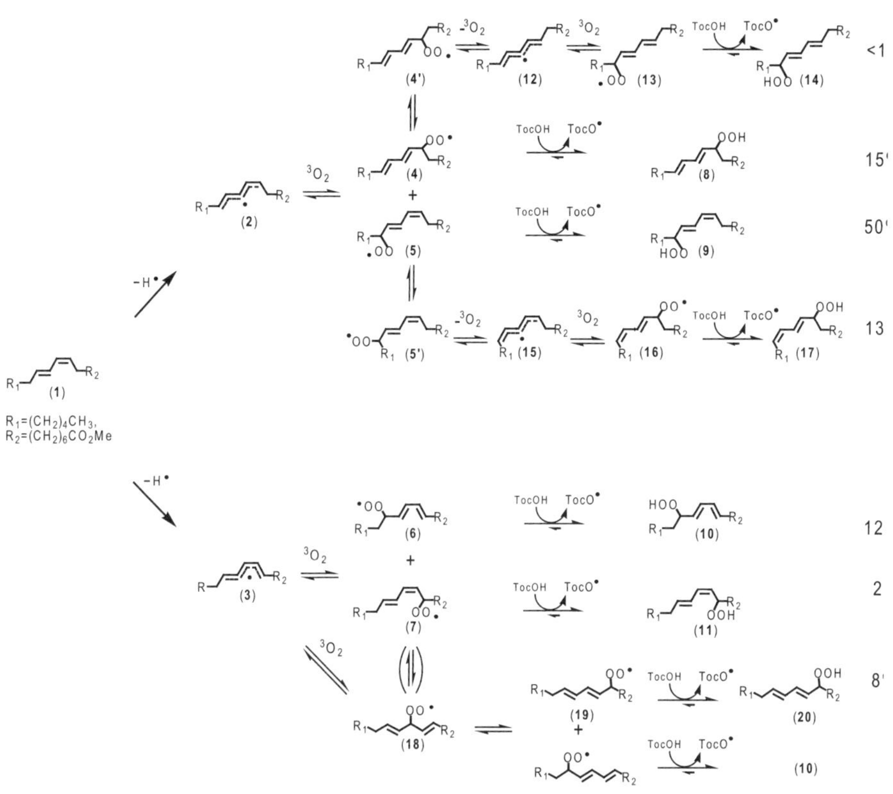

**Scheme 4.1.** Proposed Mechanism for the Hydroperoxide Pathway of CLA Methyl Ester Autoxidation (Source: Hämäläinen et al., 2002).

electron is delocalized not over three but five carbon atoms as in the autoxidation of ML. Subsequent peroxidation of the pentadienyl radicals, and H-atom abstraction yields four 'kinetic' hydroperoxides: Me 9-OOH-10$E$,12$E$ (8); Me 13-OOH-9$Z$,11$E$ (9); Me 12-OOH-8$E$,10$E$ (10); Me 8-OOH-9$Z$,11$E$ (11). The formation of Me 13-OOH-9$E$,11$E$ (14) and Me 9-OOH-10$E$,12$Z$ (17) may be explained by the β-fragmentation pathway in the same manner as the isomerization of the kinetic $E,Z$-hydroperoxides of ML autoxidation. The direct β-fragmentation pathway does not explain the formation of 8% of Me 8-OOH-9$E$,11$E$ (20). Because both termini of pentadienyl radical 3 have a partial $E$-double bond character, roughly the same amount of peroxyl radicals 6 and 7 is expected to be formed. Formation of significantly less Me 8-OOH-9$Z$,11$E$ (11) than Me 12-OOH-8$E$,10$E$ (10) suggested that hydroperoxide 20 might be formed trough through two successive [2,3]-allyl rearrangements of peroxyl radical 7 since the direct isomerization of pentadienyl radical 3 seems unlikely under normal oxygen pressure. Recently, Tallman et al. (2001, 2004) demonstrated that the nonconjugated diene bisallylic hydroperoxide is the main product in the autoxidation of ML in the presence of high amounts of α-tocopherol at oxidation extent of 2%. In light of this study, it seems likely that the nonconjugated peroxyl radical 18 is formed directly from pentadienyl radical 3 rather than from radical 7. This is supported by theoretical calculations (unpublished results by Pajunen and co-workers).

The origin of the diastereoselectivity in the autoxidation of 9Z,11E-CLA methyl ester lies in the selectivity of the initial H-atom abstraction and peroxidation steps; preferential formation of the more stable pentadienyl radical 2 over radical 3, and regioselectivity in the oxygen addition to the pentadienyl radical 2 in favor of peroxyl radical 5, and thereafter hydroperoxide 9. The selectivity in the H-atom abstraction can be explained by thermodynamic terms; invoking the Hammond postulate we can make an assumption that the geometry of the transition state resembles that of the pentadienyl radical. Since radical 2 is more stable than radical 3, the transition state leading to radical 2 should be more stable than that leading to radical 3. The 'energy barrier' $\Delta G^{\ddagger}$ is therefore smaller for the formation of radical 2 than it is for the formation of radical 3, and thus radical 2 is formed faster. The kinetic and ESR studies performed by Chan et al. (1978,1979) and Porter and Wujek (1984) using ML show that the peroxidation step is not only reversible but also regioselective in that the pentadienyl radical terminus having a partial $E$-double bond character reacts faster with triplet oxygen than that having a partial $Z$-double bond character.

The autoxidation of 9Z,11E-CLA methyl ester produces a complex mixture of hydroperoxides including the 9- and 13-positional isomers identical to those formed in the autoxidation of ML. This complexity results from the formation of two distinct pentadienyl radicals. Moreover, while ML autoxidation in the presence of α-tocopherol yields only two positional isomers both with $E,Z$-geometry, CLA autoxidation yields four positional isomers with three different geometries including a new type of $E,Z$-hydroperoxide with the $Z$-double bond adjacent to the hydroperoxyl-bearing carbon atom.

The proposed mechanism agrees with the kinetic evidence that the autoxidation is an autocatalytic free radical chain reaction (Kern et al., 1955; Minemoto et al., 2003). It is evident that CLA autoxidation at least partly follows Farmer's hydroperoxide theory, which was developed for nonconjugated olefinic compounds. Further confirmation for the mechanism has been provided by studying the autoxidation of the other biologically active CLA isomer; 10E,12Z-CLA methyl ester (Hämäläinen and Hopia, 2006). This mechanism correctly predicted the hydroperoxides formed and their isomeric distribution. The details of the mechanism are topics of future research.

It is necessary to explain why these primary hydroperoxides were not recognized by early investigators. Conjugated fatty acids are generally less studied that their nonconjugated counterparts because they are rare and have not been readily available until recently. The complexity of product mixtures might have been one of the obstacles against the discovery of hydroperoxides. Not only is there less hydroperoxide formation compared to ML, but the number of hydroperoxide isomers is greater. Since even today, autoxidation studies are often done with mixtures of CLA isomers, the resulting primary product mixture contains even greater numbers of different positional and geometric isomers of CLA hydroperoxides and is more difficult to analyze. In the absence of a good H-atom donor, a mechanism other than hydroperoxide formation prevails in the autoxidation of CLA in which radical addition to the double bonds occurs and leads to oligomeric products. However, knowledge of these hydroperoxide formation steps is crucial to understand the subsequent steps in the CLA autoxidation. Moreover, knowledge of the CLA hydroperoxide structures allows us to predict possible structures for the formed oligomers.

## Formation of Oligomers

The early studies report that CLA autoxidation (Table 4.1) yields, in an autocatalytic manner, relatively low molecular weight polymeric peroxides (Allen et al., 1949; Kern et al., 1955, 1956; Privett 1959). The data supports oxygen-carbon rather than carbon-carbon polymerization and that appreciable amounts of isolated $E$-unsaturation reside in the polymer fraction. Empirical kinetic reassessment of the early data by Brimberg and Kamal-Eldin (2003) suggests that oligomers having an average of three monomers would be kinetically favored at the beginning of the oxidation. This autoxidative polymerization can be envisioned to involve several basic chemical reactions in propagation and termination phases of the autoxidation.

### Polymerization Resulting from Propagation Reactions.
Mayo (1968) suggested that in many alkene oxidations, the abstraction mechanism (the abstraction of an H-atom from the oxidizing substrate by the peroxyl radical) and addition mechanism (the addition of the peroxyl radical to the double bond at the oxidizing substrate) occurs simultaneously and that alkenes having conjugated unsaturation are among those that preferably undergo oxidation by the latter mechanism. This reactivity in favor of addition is easy to rationalize since the addition of the peroxyl radical to the conjugated diene system leads to formation of resonance-stabilized allylic radical intermediates (Fig. 4.1). Differing from LA, CLA has been shown to interact with the radicals produced by decomposition of 2,2'-azobis(amidinopropane) (ABAP) (Leung and Liu, 2000) as well as 1,1-dipehnyl-2-picrylhydrazyl radical (DPPH•) (Yu, 2001). The propagative addition is apparently less favored in the oxidation of nonconjugated alkenes, such as polyunsaturated fatty acids, in which the resulting radical intermediates will not be resonance stabilized. Using the knowledge of the peroxyl radicals (CLA-OO•) preceding the CLA hydroperoxides, we can predict the structures for the dimeric and higher oligomers formed through addition mechanism as depicted in Fig. 4.1. It must be emphasized that the formation of these products remains to be confirmed by structural elucidation. However, in agreement with the early literature the addition mechanism would produce oligomers with oxygen-carbon rather than carbon-carbon links, and it would result in the disappearance of conjugated double bonds and in the appearance of isolated unsaturation. Importantly, these dimers would have 1,2- or 1,4-addition of oxygen in

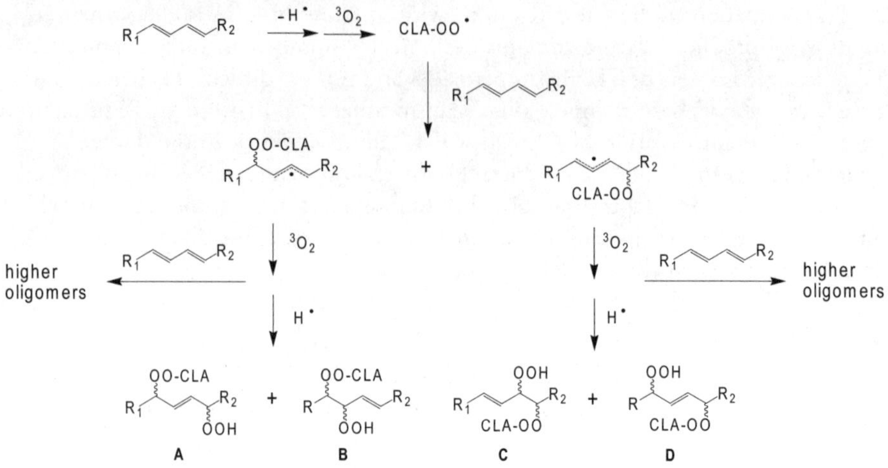

Fig. 4.1. Postulated autoxidative polymerization of CLA resulting from addition of peroxyl radicals to the double bonds of unoxidized CLA (Note: CLA-OO• refers to peroxyl radicals 4-7, 13, 16, and 19 in Scheme 4.1).

respect to one of the monomers agreeing with the data of Kern et al. (1955).

Evidence for the formation of dimers similar to those depicted in Fig. 4.1 with a hydroperoxyl or a hydroxyl group has been provided for ML autoxidation (Miyashita et al., 1982). The dimer fraction was complex because it was a mixture of positional and geometric isomers. In CLA autoxidation, the product mixture can be expected to be even more complex. Autoxidative polymerization of one pure CLA isomer would involve (at least) seven enantiomeric pairs, that is 14 isomers of peroxyl radicals (CLA-OO•), if the conditions from Hämäläinen et al. (2002) are used. Addition reaction of these peroxyl radicals with the starting material alone would lead to 56 dimeric allylic-radicals; subsequent peroxidation and H-atom abstraction yields 224 hydroperoxide dimers with a peroxide cross-link (represented by structures A to D) when the possibility of geometric isomerization in the last peroxidation step is excluded. Moreover, we can expect further complications. Whenever competition between addition and abstraction favors addition, the competition between peroxidation, which leads to dimeric or higher peroxides, and "unzipping" of the polyperoxides, which leads to epoxides and alkoxy radicals, becomes important (Mayo 1968). Thus formed alkoxy radicals would be highly reactive and could participate in various propagation and termination reactions. For example, the addition of the alkoxy radicals to the double bond system of CLA may lead to hydroperoxy dimers with ether cross-links, β-scission to aldehydes, H-atom abstraction to alcohols, and intramolecular cyclization to hydroperoxy epoxides (Fig. 4.2). We expect that the importance of these reactions grows in metal-catalyzed autoxidations because alkoxy radicals are also formed through homolytic cleavage of peroxides. Importantly, the unzipping ultimately increases the heterogeneity of the oligomeric fraction. It is also noted that unzipping of the trimeric peroxyl radical followed by subsequent H-atom abstraction by the dimeric alkoxy radical would lead to formation of hydroxy derivatives of the dimeric hydroperoxides A to D. In conclusion, since trimers (Fig. 4.3) seem to be kinetically favored, the complexity of the product mixture would be enormous as a result of the propagation reactions alone.

*Polymerization Resulting from Termination Reactions*
CLA is autoxidized partly according to Farmer's hydroperoxide theory producing hydroperoxides. Thus, we know that our autoxidation mixture contains pentadienyl and peroxyl radicals. It is easy to envision dimer formation through termination reactions of these radicals (radical-radical addition reactions), which by definition lead to non-radical products as depicted in equations 1 to 3 in Fig. 4.4. Since the peroxidation

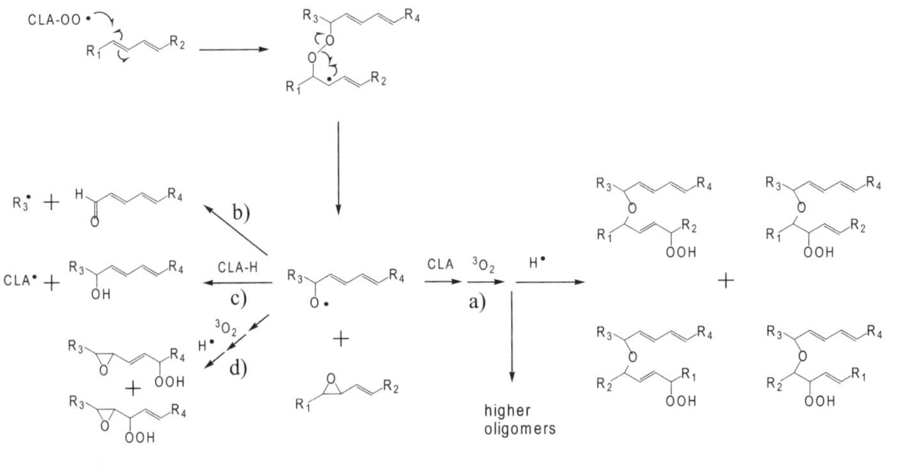

**Fig. 4.2.** Unzipping of the postulated dimeric radical leading to an epoxide and alkoxy radical, and subsequent propagation reactions of the alkoxy radical a) addition reaction, b) β-scission c) H-atom abstraction, and d) intramolecular cyclization.

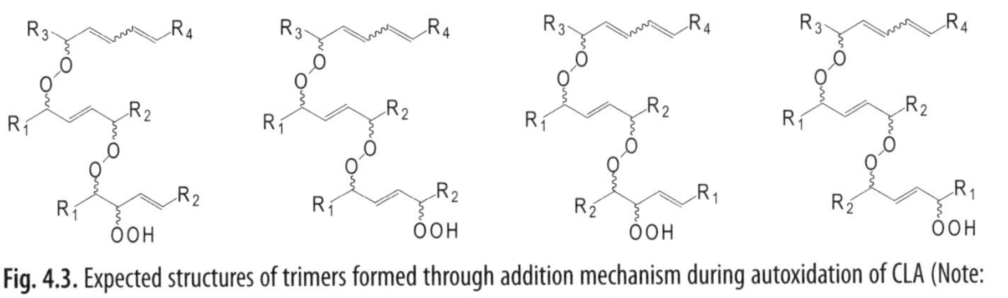

**Fig. 4.3.** Expected structures of trimers formed through addition mechanism during autoxidation of CLA (Note: These structures represent the trimers formed from dimer A in Fig. 4.1).

reaction of the pentadienyl radicals is extremely fast (occurring at diffusion-controlled rate) and the subsequent hydrogen atom abstraction step leading to the hydroperoxides is slow, it is safe to assume that most of the chain reaction-carrying radical present in the autoxidation mixture is CLA-OO• provided that oxygen is present at sufficient concentration. Hence, termination reactions involving pentadienyl radicals are unimportant and the only significant termination reaction under normal conditions is reaction 1. The self-reaction of CLA-OO• radicals yields a tetroxide, which decomposes by the Russell mechanism to stable end-products: secondary alcohol, ketone, and oxygen. The early studies do not support the dehydropolymerization reactions between two pentadienyl radicals (Eqn. 3) because they lead to carbon-carbon bonded dimers with the original amount of unsaturation. As discussed previously, unzipping of oligomeric peroxides and homolytic cleavage of peroxides in metal-catalyzed autoxidations produce alkoxy radicals. We expect that the concentration of alkoxy radicals is low due to high reactivity of the alkoxy radical. Hence, the radical-radical addition reactions 4 and 5, and the disproportionation reaction 6 would not be dominant termination reactions.

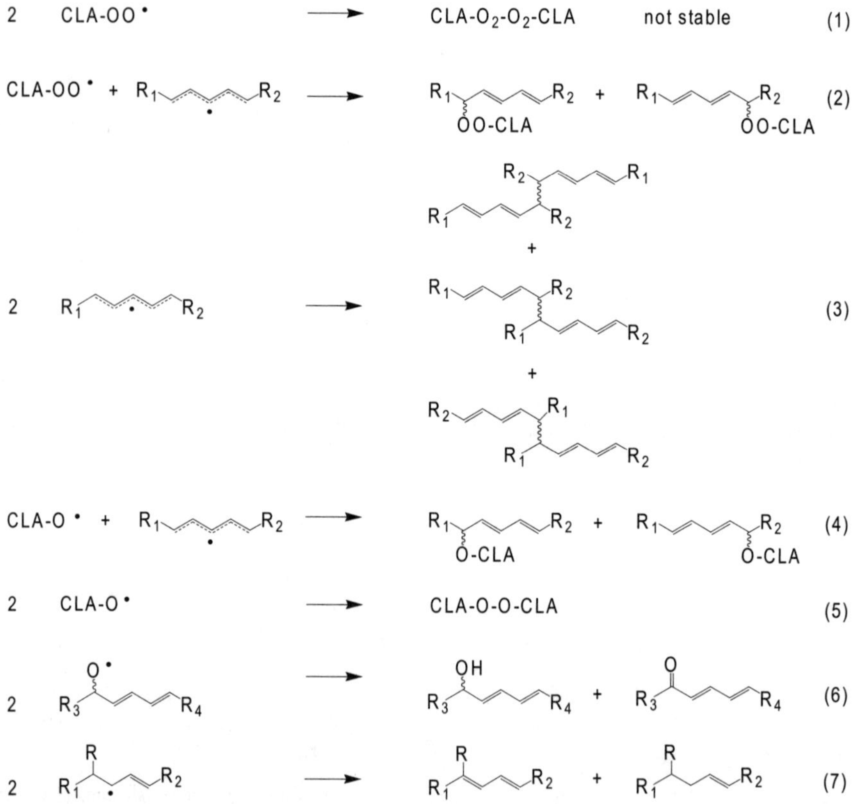

**Fig. 4.4.** Examples of termination reactions of peroxyl, pentadienyl, alkoxyl, and allylic radicals.

The balance between the different pathways of CLA autoxidation is known to be affected by the reaction conditions. The autoxidation of methyl 9$Z$,11$E$-CLA in the presence of 20% α-tocopherol at 40°C produces hydroperoxides as major products based on TLC analysis when the oxidation extent was 13% (Hämäläinen et al. 2002 and unpublished data). The reaction can also be shifted towards further polymerization by using metal catalysts as has been demonstrated by studies on oxidative crosslinking of unsaturated fatty acids in alkyd paints (van Gorkum, 2005). These studies are important to provide further insight to the autoxidative polymerization mechanism of nonconjugated and conjugated fatty acids.

Muizebelt and Nielen (1996) compared the oxidation products of methyl 9$E$,11$E$-CLA (methyl ricinoate, MR) and ethyl linoleate (EL) in the presence of a Co/Ca/Zr drier using mass spectrometry and showed that MR cross-linking yielded dimers to hexamers while EL yielded dimers to tetramers. Within each group of dimers to hexamers, oligomers varied in molecular weight suggesting structural heterogeneity (Fig. 4.5). This oxidative crosslinking under drier influence leads to formation of oligomers with ether, peroxy, and C-C cross-links. Some important mechanistic conclusions were drawn based on the mass spectra. In conjugated dienes, crosslinking occurs through radical addition to the double bond; in nonconjugated dienes it occurs through termination reactions. Furthermore, in the mass spectra the MR dimer peaks were all doubled with a mass difference of 2 Da as compared with EL dimer peaks. Thus, while dimerization in MR occurs through addition of radicals to the double bond with subsequent termination by disproportionation (Figure 4.4, eq. 7), EL produces dimers by recombination of radicals.

When oxidative cross-linking was performed in the presence of Co and/or Co/Ca/Zr catalysts and reactive diluents yet another difference was observed between the nonconjugated and conjugated fatty acid oligomers; the oxygen distribution in conjugated fatty acid oligomers was much narrower. This suggested that the carbon-centered radical formed from $E$,$Z$-CLA reacted rapidly with oxygen before adding to another fatty acid. Thus, the oligomers had a polyperoxide character (Muizebelt et al., 2000). Size exclusion chromatography indicated that in the oxidation of EL hydroperoxide formation is followed by formation of dimer, trimer, and higher oligomers whereas higher oligomers are formed form $E$,$Z$-CLA right from the beginning.

## Formation of Furan Fatty Acids and Other Secondary/Monomeric Oxidation Products

Four furan fatty acids (FFA), 8,11-epoxy-8,10-octadecadienoic acid ($F_{8,11}$), 9,12-epoxy-9,11-octadecadienoic acid ($F_{9,12}$), 10,13-epoxy-10,12-octadecadienoic acid ($F_{10,13}$), 11,14-epoxy-11,13-octadecadienoic acid ($F_{11,14}$), have been identified as secondary autoxidation products of a mixture of CLA isomers in water-methanol solution at 50°C by GC/MS and/or GC/FID detection (Yurawecz et al., 1995). The mechanism for the formation of these FFA and the precursor compounds are unknown (Fig. 4.6). However, FFA are suggested to arise from cyclic peroxides or possibly from dioxo fatty acids (Yurawecz et al., 1995; Eulitz et al., 1999). Bascetta et al. (1984) have shown that FFA can be produced from the major photooxidation

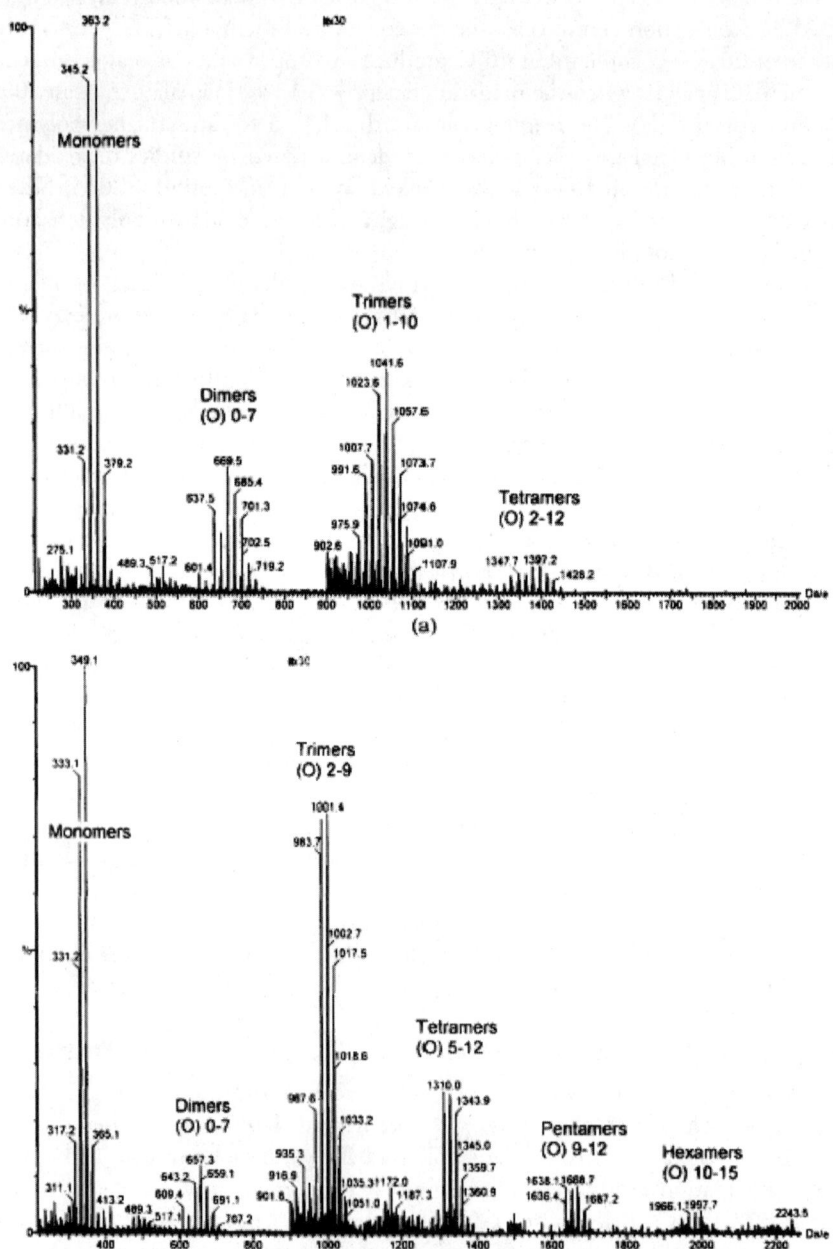

**Fig. 4.5.** ESI-MS of a) ethyl linoleate and b) methyl ricinoate after two days of reaction (Reproduced from Muizebelt and Nielen (1996) after permission from the publisher John Wiley and Sons).

product of CLA, 1,2-dioxine (See Singlet oxygen oxidation of CLA). Synthetically FFA can be produced from epoxyoxoene fatty acids (Hidalgo and Zamora, 1995) and oxoene fatty acids (Lie Ken Jie et al., 1998). More research is needed to illustrate the route for formation of FFA during autoxidation of CLA.

Besides FFA, autoxidation of CLA produced alkanals (hexanal, heptanal, and octanal), alkenals (2E-heptenal, 2E-octenal, 2E-nonenal, and 2E-decenal), alkanoates (heptanoate, octanoate, nonanoate, and decanoate), oxo-alkanoates (7-oxo-heptanoate, 8-oxo-octanoate, 9-oxo-nonanoate, 10-oxo-decanoate, and 11-oxo-undecanoate), and four α,β-unsaturated lactones (Yurawecz et al., 1995,1997; Sehat et al., 1998; Eulitz et al., 1999). The autoxidation of methyl $F_{9,12}$ as a neat oil at 50°C resulted in the formation of several secondary products, including 5-hexyl-2-furaldehyde, methyl 8-oxooctanate, methyl 13-oxo-9,12-epoxytrideca-9,11-dienoate, methyl 8-oxo-9,12-epoxy-9,11-octadecadienate, and methyl 13-oxo-9,12-epoxy-9,11-octadecadienate (Fig. 4.7). Since the autoxidation of CLA methyl esters in the presence of a good H-atom donor produces both identical and similar hydroperoxides as ML, we can expect in these conditions the formation of secondary products that are the same or homologues of the ones formed in the decomposition of ML hydroperoxides.

The autoxidation of CLA has also been suggested to yield products identical to singlet oxygen oxidation of CLA (Eulitz et al., 1999). The singlet oxygen oxidation

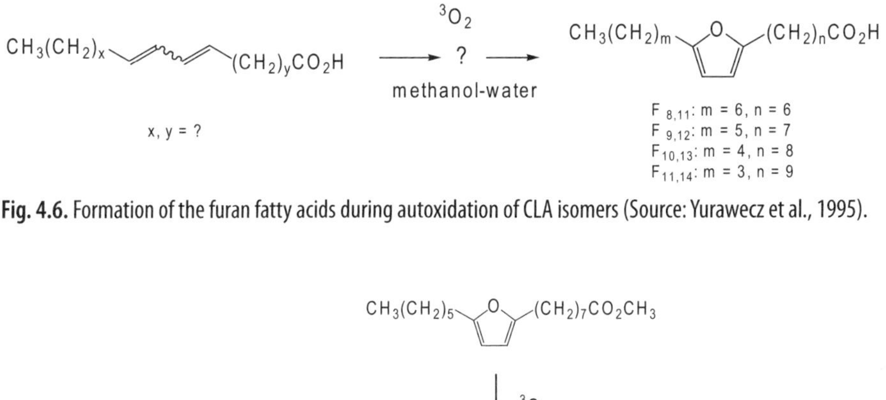

**Fig. 4.6.** Formation of the furan fatty acids during autoxidation of CLA isomers (Source: Yurawecz et al., 1995).

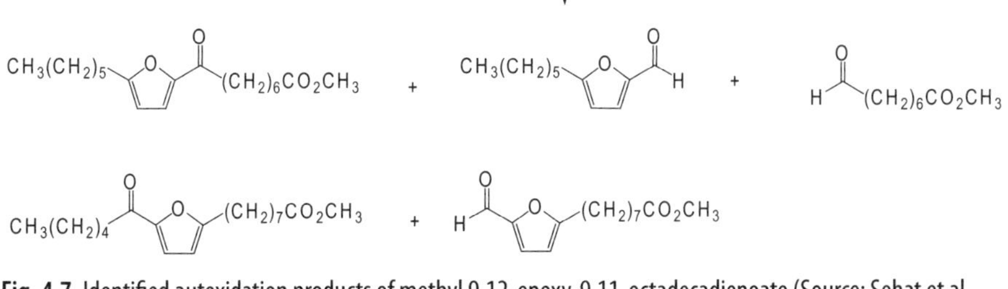

**Fig. 4.7.** Identified autoxidation products of methyl 9,12-epoxy-9,11-octadecadienoate (Source: Sehat et al., 1998).

was proposed to yield nonconjugated diene monohydroperoxides, 1,2-dioxine, and 1,2-dioxetanes as primary products (See structures in the next section). So far, none of these monomeric products have been isolated or characterized from CLA autoxidation. Formation of the nonconjugated diene hydroperoxides could be envisioned albeit not by a concerted ene-reaction as in photooxidation but through a free radical mechanism (Hämäläinen et al., 2001). The formation mechanism of the proposed cyclic peroxides as primary products during the autoxidation is difficult to understand. According to Wigner's spin-conservation rule the direct addition of triplet oxygen to the diene $\pi$ system gives a product in its triplet state. Since the formation of such products has a high energy barrier, it is not a plausible pathway for the formation of cyclic peroxides. Formation of these peroxides through splitting of oligomers seems more probable but remains to be confirmed. For example, decomposition of a dimeric alkoxy radical, formed by unzipping of a trimeric addition product, might yield a bis-epoxide and subsequent rearrangement a 1,2-dioxine as depicted in Fig. 4.8A. Alternatively, the 1,2-dioxines might be formed by intramolecular cyclization of the new type of $E,Z$-peroxyl radical where the double bond adjacent to the hydroperoxyl-bearing carbon atom is $Z$ and followed by a disproportionation reaction (Fig. 4.8B). The mechanism in Scheme 4.1 predicts that the autoxidation of the principal CLA isomers (9$Z$,11$E$/9$E$,11$Z$–CLA, 41%; 10$E$,12$Z$-CLA 44%) of the mixture used in the autoxidation study by Yurawecz et al. (1995) would produce those new types of $E,Z$-peroxyl radicals that would ultimately yield FFA, $F_{8,11}$, $F_{10,13}$, $F_{11,14}$. These FFA were indeed identified from the autoxidation mixture. However, formation of localized carbon radicals and loss of resonance stabilization (hyperconjugation of the alkylperoxy group) and of diene conjugation makes the cyclization step less desirable. If oxygen is present in sufficient concentration, we would also expect the formation of α-hydroperoxy-1,2-dioxines. In addition, the formation cyclic ethers (epoxides) is expected as the result of unzipping CLA oligomers. All these monomeric products remain to be isolated and characterized. An overview of the CLA autoxidation pathways is presented in Fig. 4.9.

## Singlet Oxygen Oxidation of CLA
### Formation of Primary Oxidation Products

The literature on CLA singlet oxygen oxidation is scarce. The major primary oxidation products of methylene blue sensitized oxidation of methyl 8$E$,10$E$-CLA (Gunstone and Wijesundera, 1979) in methanol and methyl 9$E$,11$E$-CLA (Bascetta et al., 1984) in tetrachloromethane/methanol (95/5, v/v) have been identified as 1,2-dioxines. These unsaturated cyclic peroxides are the result of a Diels-Alder type [2+4] addition of the $E,E$-fatty acid in s-*cis* conformation with singlet oxygen (Fig. 4.10).

It is evident that singlet oxygen reacts with conjugated double bond systems by an entirely different mechanism than the biradical triplet oxygen. In CLA, the 1,3-diene structure is not sterically hindered and therefore the [2+4] addition is the preferred reaction pathway. The photooxidation of methyl 9$E$,11$E$-CLA produces 6-hexyl-3-

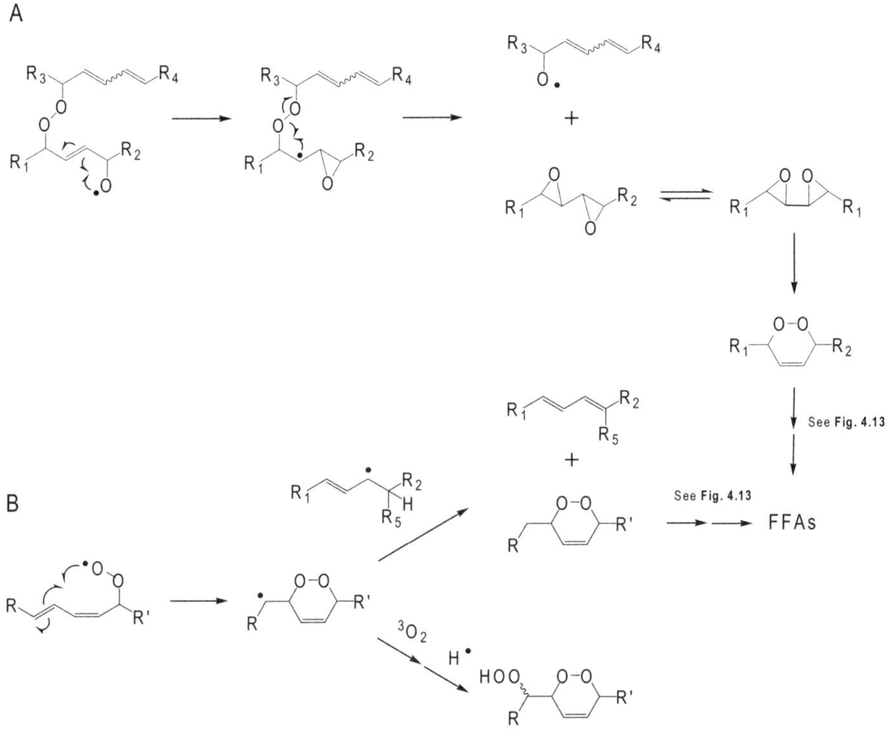

**Fig. 4.8.** Postulated formation of 1,2-dioxines from A) Dimeric alkoxy-radicals B) E,Z-Peroxyl radicals, where the double bond adjacent to the peroxyl-bearing methine carbon has Z-geometry.

(7-methoxycarbonylheptyl)-3,6-dihydro-1,2-dioxine in over 80% yield (Bascetta et al., 1984). Based on photooxidation studies on 1,3-dienes, the formation of Z-endoperoxides is favored (Manring and Foote, 1983; Clennan and L'Esperance, 1985a,b; O'Shea and Foote, 1988). Furthermore, these studies suggest that nonconjugated diene monohydroperoxides and dioxetanes are possible minor primary products of the photooxidation of CLA as depicted in Fig. 4.11. The influence of the ester group on the formation of the primary products, as in the autoxidation of ML (Tallman et al., 2004), can not be expected to be relevant. To the best of our knowledge, no hydroperoxides and no dioxetanes have been identified as primary products in the singlet oxygen oxidations of CLA isomers.

The formation of possible minor products can be envisioned to occur through concerted mechanisms: hydroperoxides through an ene-reaction and dioxetanes through [2+2] addition. The photooxidation studies on 1,3-dienes suggest a competing stepwise mechanism to form all of the primary products through a perepoxide intermediate (or transition state). This perepoxide is zwitterion/biradical in character and is affected by the polarity of the solvent (Fig. 4.12). Furthermore, this interme-

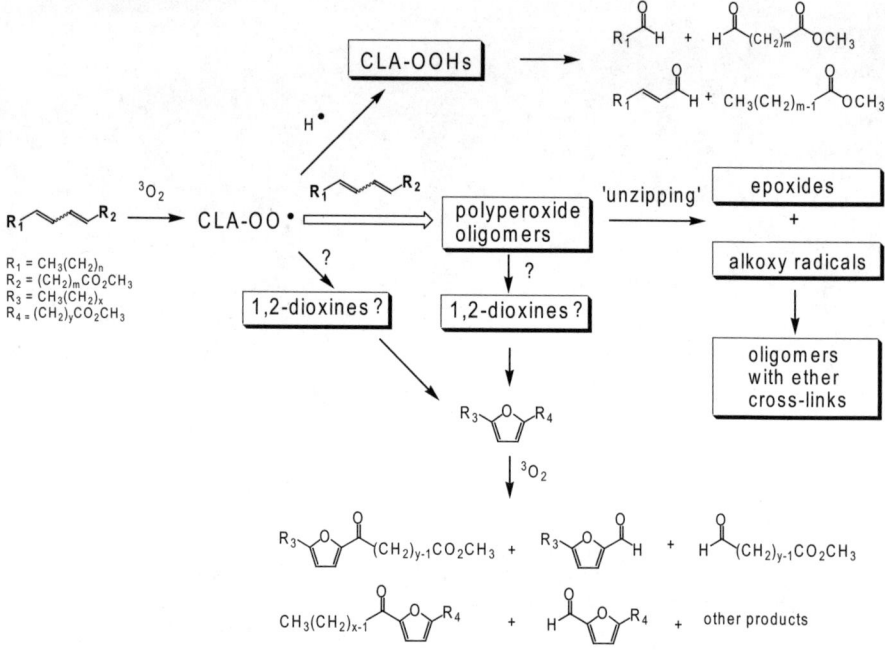

**Fig. 4.9.** Overview of CLA autoxidation.

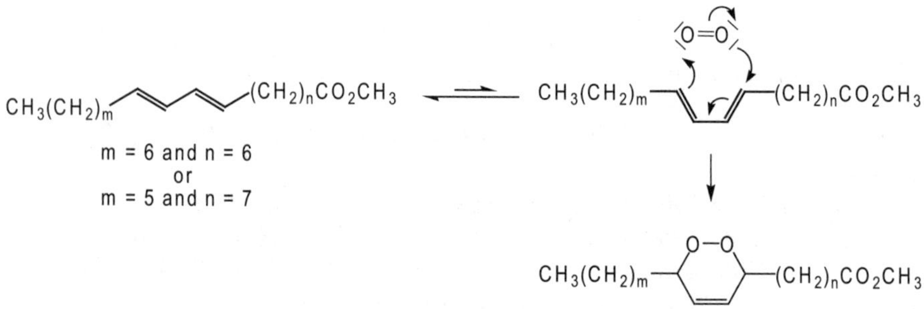

**Fig. 4.10.** Formation of the major primary singlet oxidation products of methyl 8E,10E-CLA and methyl 9E,11E-CLA through [2+4] addition.

diate is expected to be involved in nonreactive quenching of singlet oxygen and in geometric isomerization of 1,3-dienes (Manring et al., 1983; Manring and Foote, 1983; Clennan and L'Esperance, 1985b; O'Shea and Foote, 1988). Subsequently, the perepoxide intermediate collapses to cyclic peroxides or intramolecular removal of an α–hydrogen leads to the ene-reaction yielding a bisallylic hydroperoxide. Interestingly, this type of hydroperoxide is formed as a minor product in the autoxidation of ML (Haslbeck et al., 1983; Brash, 2000). In comparison, the reaction of ML with singlet oxygen proceeds through concerted ene-reactions and produces hydroperoxides.

# Oxidation of Conjugated Linoleic Acid   103

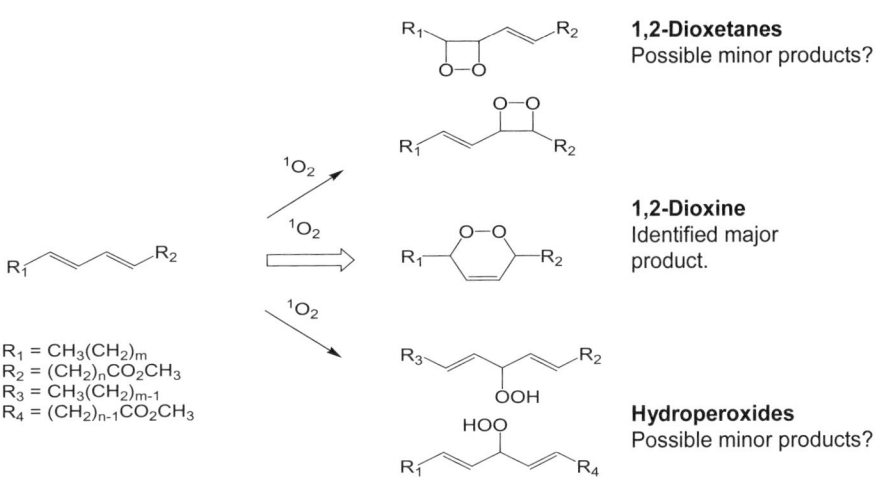

**Fig. 4.11.** The structures of singlet oxidation products of cla isomers based on 1,3-diene singlet oxygen oxidation studies.

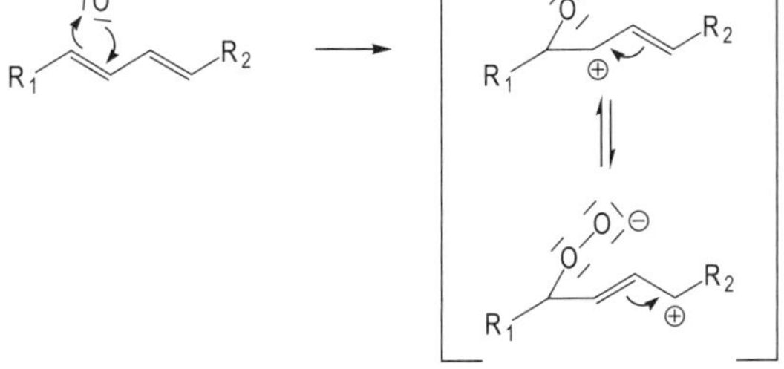

**Fig. 4.12.** Formation of a perepoxide intermediate in the singlet oxygen oxidation of 1,3-diene (Only the zwitterionic resonance structures are included).

These hydroperoxides are major products and have two conjugated or nonconjugated double bonds. In the latter, the hydroperoxide group is allylic with respect to one double bond and homoallylic with respect to the other (Frankel, 1998).

The photooxidation of a mixture of CLA methyl esters and ML has been compared using methylene blue as a sensitizer in ethanol solutions (Jiang and Kamal-Eldin, 1998). In this study, the PV measurements support the assumption that in CLA methyl ester photooxidation other primary oxidation products than (hydro)peroxides are formed and/or products that are decomposed at a rate similar to that of their formation. In addition, further differences were found between the singlet oxygen oxidation of the two fatty acids. Although CLA methyl esters were lost at a much lower rate and lower extent compared to ML during the course of oxidation, they bleached methylene blue at a much higher rate than ML. Moreover, different CLA methyl ester isomers were not lost at equal rates; only the level of *E,Z*-isomers decreased, the level of *Z,Z*-isomers remained constant, and the level of the *E,E*-isomers increased. In light of the photooxidation studies with 1,3-dienes, this increase and bleaching of methylene blue might be partly explained by isomerization and by nonreactive quenching of singlet oxygen.

## Formation of Secondary Oxidation Products

To the best of our knowledge, no secondary photooxidation products have been identified for CLA. An attempt to oxidize the major primary oxidation product of methyl 9*E*,11*E*-CLA, 1,2-dioxine, by further reaction with singlet oxygen gave the unchanged cyclic peroxide after 5 days. However, heating 1,2-dioxine in various solvents with high dielectric constant resulted in the formation of methyl $F_{9,12}$ and unidentified minor products. This thermal dehydration into the furanoid ester was also observed in the mass spectrum analysis of 1,2-dioxine. Treatment of 1,2-dioxine with ferrous ion in tetrahydrofuran produced methyl $F_{9,12}$ (Fig. 4.13) along with two minor products identified as diastereoisomeric 9,10:11,12-bis-epoxyoctadecanoates (Bascetta et al., 1984). These isomeric bis-epoxides can be produced from CLA synthetically by using various epoxidizing agents (Lie Ken Jie et al., 2003). Based on the

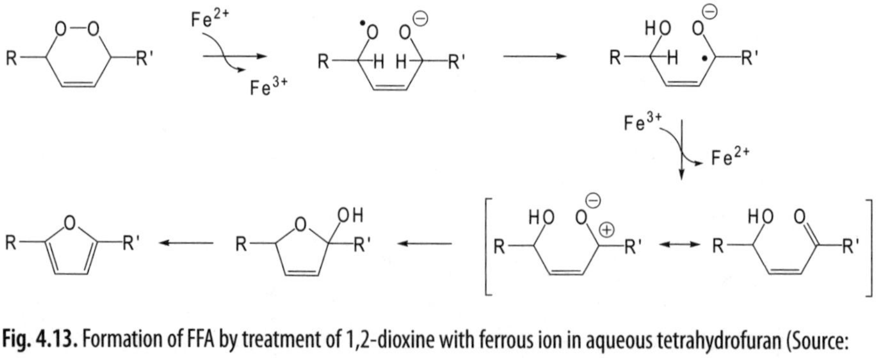

**Fig. 4.13.** Formation of FFA by treatment of 1,2-dioxine with ferrous ion in aqueous tetrahydrofuran (Source: Bascetta et al., 1984).

photooxidation studies on 2,4-hexadienes (O'Shea and Foote, 1988) we expect that the isolation of the postulated minor primary photooxidation products of CLA is challenging. However, the detection of aldehydes resulting from cleavage of the minor products should be possible.

## Conclusion

The literature review presented in this chapter clearly shows that the over-all process of CLA autoxidation is extremely complex and has many basic reactions occurring simultaneously. It has also been discussed that experimental conditions (such as temperature, film thickness, solvent, initiators, antioxidants, humidity, light, etcetera) differentially affect the course of the reaction and the composition of the end-products. In the presence of strong hydrogen atom donors, such as $\alpha$-tocopherol, hydroperoxides are formed from CLA to an appreciable extent and emphasizes their role at the initial stages of the reaction. In the absence of strong hydrogen atom donors, the main reaction pathway of CLA oxidation seems to be autoxidative polymerization in contrast to the hydroperoxide formation in the case of nonconjugated fatty acids. The heterogeneity of the polymeric products formed upon autoxidation of CLA needs extensive and thorough investigation.

Since CLA autoxidation is a cascade of a vast range of reactions occurring simultaneously and is significantly affected by reaction conditions, these should be carefully reported and taken into account when different studies are compared. Generalizations made on the basis of the results produced under simple reaction conditions with only few variables may be applied to more complicated system only to a certain extent. Reasoning by analogy, however, is seldom justified when generalizations are made on the basis of limited investigation(s) and/or on the basis of the results produced with very complex reaction mixtures. The CLA literature should be considered with a critical eye in order to estimate the validity of the conclusions drawn by the investigators. Particularly, conclusions drawn based on kinetic studies in which the oxidizing substrate is a mixture of a wide range of fatty acids in small amounts need to be justified. At this stage, studies pertinent to structural elucidation of products, and to understanding product formation and the kinetics of CLA autoxidation need to be performed with pure isomers of CLA.

The photooxidation of CLA occurs though an entirely different mechanism to CLA autoxidation and yields a 1,2-dioxine as the main product. The formation of minor products remains to be confirmed and the formation of the secondary photooxidation products of CLA isomers has not yet been studied.

For future research, careful consideration of the choice of method to follow both autoxidation and photoxidation reactions of CLA should be practiced, particularly when the aim is to compare these oxidation reactions with those that proceed entirely or partly by different mechanisms. In addition, more research with pure CLA isomers is required in order to elucidate the details of the oxidation mechanisms of CLA, and to develop a deeper understanding of the role of CLA oxidation in biological systems.

# References

Allen, R.R.; and F.A. Kummerrow. Factors Affecting the Stability of Highly Unsaturated Fatty Acids. III. The Autoxidation of Methyl Eleostearate. *J. Am. Oil Chem. Soc.* **1951**, *28*, 101–105.

Allen, R.R.; A. Jackson; and F.A. Kummerrow. Factors Which Affect the Stability of Highly Unsaturated Fatty Acids. I. Differences in the Oxidation of Conjugated and Nonconjugated Linoleic Acid. *J. Am. Oil Chem. Soc.* **1949**, *26*, 395–399.

Banni, S.; E. Angioni; M.S. Contini; G. Carta; V. Casu; G.A. Iengo; M.P. Melis; M. Deiana; M.A. Dessi; and F.P. Corongiu. Conjugated Linoleic Acid and Oxidative Stress. *J. Am. Oil Chem. Soc.* **1998**, *75*, 261–267.

Bascetta, E.; F.D. Gunstone; and C.M. Scrimgeour. Synthesis, Characterisation, and Transformations of a Lipid Cyclic Peroxide. *J. Chem. Soc. Perkin Trans.* **1984**, *1*, 2199–2205.

Basu, S.; U. Risérus; A. Turpeinen; and B. Vessby. Conjugated Linoleic Acid Induces Lipid Peroxidation in Men with Abdominal Obesity. *Clin. Sci.* **2000**, *99*, 511–516.

Belury, M.A. Inhibition of Carcinogenesis by Conjugated Linoleic Acid: Potential Mechanisms of Action. *J. Nutr.* **2002**, *132*, 2995–2998.

Belury, M.A.; A. Mahon; and S. Banni. The Conjugated Linoleic Acid (CLA) Isomer, $t10c12$-CLA, Is Inversely Associated with Changes in Body Weight and Serum Leptin in Subjects with Type 2 Diabetes Mellitus. *J. Nutr.* **2003**, *133*, 257–260.

Brash, A.R. Autoxidation of Methyl Linoleate: Identification of the Bis-Allylic 11-Hydroperoxide. *Lipids* **2000**, *35*, 947–952.

Brimberg, U.I.; and A. Kamal-Eldin. On the Kinetics of the Autoxidation of Fats: Substrates with Conjugated Double Bonds. *Eur. J. Lipid Sci. Technol.* **2003**, *105*, 17–23.

Chan, H.W.-S.; G. Levett; and J.A. Matthew. Thermal Isomerization of Methyl Linoleate Hydroperoxides. Evidence of Molecular Oxygen as a Leaving Group in a Radical Rearrangement. *J. Chem. Soc. Chem. Comm.*, **1978**, 756–757.

Chan, H.W.-S.; G. Levett; and J.A. Matthew. The Mechanism of the Rearrangement of Linoleate Hydroperoxides. *Chem. Phys. Lipids* **1979**, *24*, 245–256.

Chen, Z.Y.; P.T. Chan; K.Y. Kwan; and A. Zhang. Reassessment of the Antioxidant Activity of Conjugated Linoleic Acid. *J. Am. Oil Chem. Soc.* **1997**, *74*, 749–753.

Chen, J.F.; C.-Y. Tai; Y.C. Chen; and B.H. Chen. Effects of Conjugated Linoleic Acid on the Degradation and Oxidation Stability of Model Lipids During Heating and Illumination. *Food Chem.* **2001**, *72*, 199–206.

Clennan, E.L.; and R.P. L'Esperance. The Unusual Reactions of Singlet Oxygen with Isomeric 1,4-Di-tert-butoxy-1,3-butadienes. A 2S + 2a Cycloaddition. *J. Am. Chem. Soc.* **1985a**, *107*, 5178–5182.

Clennan, E.L.; and R.P. L'Esperance. Mechanism of Singlet Oxygen Addition to Conjugated Butadienes. Solvent Effects on the Formation of a 1,4-Diradical. The 1,4-Diradical/1,4-Zwitterion Dichotomy. *J. Org. Chem.* **1985b**, *50*, 5424–5426.

Destaillats, F.; and P. Angers. Evidence for [1,5] Sigmatropic Rearrangements of CLA in Heated Oils. *Lipids* **2002**, *37*, 435–438.

Eulitz, K.; M.P. Yurawecz; and Y. Ku. The Oxidation of Conjugated Linoleic Acid. In *Advances in Conjugated Linoleic Acid Research*, Yurawecz, M.P., Mossoba, M.M., Kramer, J.K.G., Pariza, M.W., and Nelson, G.J., Eds.; AOCS Press: Champaign, 1999, Vol. 1, pp. 55–63.

Evans, M.E.; J.M. Brown; and M.K. McIntosh. Isomer-Specific Effects of Conjugated Linoleic

Acid (CLA) on Adiposity and Lipid Metabolism. *J. Nutr. Biochem.* **2002**, *13,* 508–516.

Frankel, E.N. Lipid Oxidation, The Oily Press: Dundee, 1998, pp. 43-54.

Fukuzumi, K.; and N. Ikeda. The Effect of Antioxidants in the Autoxidation of Methyl Conjugated *cis*,*trans*-Octadecadienoates. *J. Am. Oil Chem. Soc.* **1970**, *47,* 369–370.

Gnädig, S.; R. Rickert; J.L. Sébedio; and H. Steinhart. Conjugated Linoleic Acid (CLA): Physiological Effects and Production. *Eur. J. Lipid Sci. Technol.* **2001**, *103,* 56–61.

Gunstone, F.D.; and R.C. Wijesundera. Fatty Acids, Part 54: Some Reactions of Long-Chain Oxygenated Acids with Special Reference to Those Furnishing Furanoid Acids. *Chem. Phys. Lipids* **1979**, *24,* 193–208.

Ha, Y.L.; N.K. Grimm; and M.W. Pariza. Anticarcinogens from Fried Ground Beef: Heat-Altered Derivatives of Linoleic Acid. *Carcinogenesis* **1987**, *8,* 1881–1887.

Ha, Y.L.; J. Storkson; and M.W. Pariza. Inhibition of Benzo(α)pyrene-Induced Mouse Forestomach Neoplasia by Conjugated Dienoic Derivatives of Linoleic Acid. *Cancer Res.* **1990**, *50,* 1097–1101.

Haslbeck, F.; W. Grosch; and J. Firl. Formation of Hydroperoxides with Unconjugated Diene Systems During Autoxidation and Enzymic Oxygenation of Linoleic Acid. *Biochim. Biophys. Acta* **1983**, *750,* 185–193.

Hämäläinen, T.I.; and A. Hopia. Characterization of the Primary Autoxidation Products of CLA Methyl Ester and the Mechanism of the Hydroperoxide Pathway. Abstract book: 97th AOCS Annual Meeting and Expo (April 30-May 3), St. Louis, MO, USA, AOCS Press: Champaign, 2006.

Hämäläinen, T.I.; and A. Kamal-Eldin. Analysis of Lipid Oxidation Products by NMR Spectroscopy. In *Analysis of Lipid Oxidation*, Kamal-Eldin, A., and Pokornÿ, J., Eds.; AOCS Press: Champaign, 2005, pp. 70–126.

Hämäläinen, T.I.; S. Sundberg; M. Mäkinen; T. Hase; S. Kaltia; and A. Hopia. Hydroperoxide Formation During Autoxidation of Conjugated Linoleic Acid Methyl Ester. *Eur. J. Lipid Sci. Technol.* **2001**, *103,* 588–593.

Hämäläinen, T.I.; S. Sundberg; T. Hase; and A. Hopia. Stereochemistry of the Hydroperoxides Formed During Autoxidation of CLA Methyl Ester in the Presence of α-Tocopherol. *Lipids* **2002**, *37,* 533–540.

Hidalgo, F.J.; and R. Zamora. Epoxyoxoene Fatty Esters: Key Intermediates for the Synthesis of Long-Chain Pyrrole and Furan Fatty Esters. *Chem. Phys. Lipids* **1995**, *77,* 1-11.

Holman, R.T. Autoxidation of Fats and Related Substances. In *Progress in the Chemistry of Fats and Other Lipids*, Holman, R.T., Lundberg, W.O., and Malkin, T., Eds.; Pergamon Press: London, 1954, pp. 51–98.

Holman, R.T.; and O.C. Elmer. The Rates of Oxidation of Unsaturated Fatty Acids and Esters. *J. Am. Oil Chem. Soc.* **1947**, *24,* 127–129.

Ip, C.; S.F. Chin; J.A. Scimeca; and M.W. Pariza. Mammary Cancer Prevention by Conjugated Dienoic Derivative of Linoleic Acid. *Cancer Res.* **1991**, *51,* 6118–6124.

Jackson, A.H.; and F.A. Kummerow. Factors Which Affect the Stability of Highly Unsaturated Fatty Acids. II. The Autoxidation of Linoleic and Alkali Conjugated Acid in the Presence of Metalic Naphthenates. *J. Am. Oil. Chem. Soc.* **1949**, *26,* 460–465.

Jiang, J.; and A. Kamal-Eldin. Comparing Methylene Blue-Photosensitized Oxidation of Methyl-Conjugated Linoleate and Methyl Linoleate. *J. Agric. Food Chem.* **1998**, *46,* 923–927.

Jimenez, M.; H.S. Garcia; and C.I. Beristain. Spray-Drying Microencapsulation and Oxidative Stability of Conjugated Linoleic Acid. *Eur. Food Res. Technol.* **2006**, *219,* 588–592.

Kern, W.; A.R. Heinz; and J. Stallman. Über die Autoxydation ungesättigter Verbindungen. IV. Die Autoxydation des 2,3-Dimethylbutadiens(1.3) und des 10,12-Octadecadiensäuremethylesters. *Macromol. Chem.* **1955**, *16*, 21–35.

Kern, W.; A.R. Heinz; and D. Höhr. Über die Autoxydation ungesättigter Verbindungen. VI. Die Autoxydation des 9,11-Octadecadiensäuremethylesters. *Macromol. Chem.* **1956**, *18/19*, 406–413.

Lanser, A.C.; E.A. Emken; and J.B. Ohlrogge. The Oxidation of Oleic and Elaidic Acids in Rat and Human-Heart Homogenates. *Biochim. Biophys. Acta* **1986**, *875*, 510–515.

Lee, J.; S.-M. Lee; I.-H. Kim; J.-H. Jeong; C. Rhee; and K.-W. Lee. Oxidative Instability of CLA Concentrate and Its Avoidance with Antioxidants. *J. Am. Oil Chem. Soc.* **2003**, *80*, 807-810.

Leung, Y.H.; and R.H. Liu. *trans*-10,*cis*-12-Conjugated Linoleic Acid Isomer Exhibits Stronger Oxyradical Scavenging Capacity than *cis*-9,*trans*-11-Conjugated Linoleic Acid Isomer. *J. Agric. Food Chem.* **2000**, *48*, 5469–5475.

Lie Ken Jie, M.S.F.; and M.K. Pasha. Epoxidation Reactions of Unsaturated Fatty Esters with Potassium Peroxomonosulfate. *Lipids* **1998**, *33*, 633-637.

Lie Ken Jie, M.S.F.; J. Mustafa; and K.M. Pasha. An Efficient Ultrasound-Assisted Zinc Reduction of Fatty Esters Containing Conjugated Enynol and Conjugated Enynone Systems. *Lipids* **1998**, *33*, 941–945.

Lie Ken Jie, M.S.F.; C.N.W. Lam; J.C.M. Ho; and M.M.L. Lau. Epoxidation of a Conjugated Linoleic Acid Isomer. *Eur. J. Lipid Sci. Tech.* **2003**, *105*, 391-396.

Mallégol, J.; J.L. Gardette; and J. Lemaire. Long-Term Behavior of Oil-Based Varnishes and Paints. Fate of Hydroperoxides in Drying Oils. *J. Am. Oil. Chem. Soc.* **2000**, *77*, 249-255.

Manring, L.E.; and C.S. Foote. Chemistry of Singlet Oxygen. 44. Mechanism of Photooxidations of 2,5-Dimethylhexa-2,4-diene and 2-Methyl-2-penten. *J. Am. Chem. Soc.* **1983**, *105*, 4710–4717.

Manring, L.E.; R.C. Kanner; and C.S. Foote. Chemistry of Singlet Oxygen. 43. Quenching by Conjugated Olefins. *J. Am. Chem. Soc.* **1983**, *105*, 4707–4710.

Mayo, F.R. Free-Radical Autoxidatons of Hydrocarbons. *Acc. Chem. Res.* **1968**, *1*, 193–201.

Minemoto, Y.; S. Adachi; Y. Shimada; T. Nagao; T. Iwata; Y. Yamauchi-Sato; T. Yamamoto; T. Kometani; and R. Matsuno. Oxidation Kinetics for *cis*-9,*trans*-11 and *trans*-10,*cis*-12 Isomers of CLA. *J. Am. Oil Chem. Soc.* **2003**, *80*, 675–678.

Miyashita, K.; K. Fujimoto; and T. Kaneda. Structures of Dimers Produced from Methyl Linoleate During Initial Stage of Autoxidation. *Agric. Biol. Chem.* **1982**, *46*, 2293–2297.

Moloney, F.; T.P. Yeow; A. Mullen; J.J. Nolan; and H.M. Roche. Conjugated Linoleic Acid Supplementation, Insulin Sensitivity, and Lipoprotein Metabolism in Patients with Type 2 Diabetes Mellitus. *Am. J. Clin. Nutr.* **2004**, *80*, 887–895.

Muizebelt, W.J.; and M.W.F. Nielen. Oxidative Crosslinking of Unsaturated Fatty Acids Studied with Mass Spectrometry. *J. Mass Spectrometry* **1996**, *31*, 545–554.

Muizebelt, W.J.; J.C. Hubert; M.W.F. Nielen; R.P. Klaasen; and K.H. Zabel. Crosslink Mechanisms of High-Solids Alkyd Resins in the Presence of Reactive Diluents. *Progress in Org. Coatings* **2000**, *40*, 121–130.

O'Shea, K.E.; and C.S. Foote. Chemistry of Singlet Oxygen. 51. Zwitterionic Intermediates from 2,4-Hexadienes. *J. Am. Chem. Soc.* **1988**, *110*, 7167–7170.

Park, Y.; Y.L. Ha; and M.W. Pariza. pi-Complex Formation of Conjugated Linoleic Acid with Iron. *Food Chem.* **2007**, *100*, 972–976.

Pariza, M.W.; Y. Park; and M.E. Cook. The Biologically Active Isomers of Conjugated Linoleic Acid. *Progr. Lipid Res.* **2001,** *40,* 283–298.

Peers, K.E.; and D.T. Coxon. Controlled Synthesis of Monohydroperoxides by α-Tocopherol Inhibited Autoxidation of Polyunsaturated Lipids. *Chem. Phys. Lipids* **1983,** *32,* 49–56.

Pekkarinen, L. The Effect of Copper, Manganese and Cobalt Acetates on the Autoxidation of *trans*-9,*trans*-11-Octadecadienoic Acid in 90% v/v Aqueous Acetic Acid. *J. Am. Oil Chem. Soc.* **1972,** *46,* 354–356.

Piazza, G.J.; A. Nuñez; and T.A. Foglia. Isolation of Unsaturated Diols After Oxidation of Conjugated Linoleic Acid with Peroxygenase. *Lipids* **2003,** *38,* 255–261.

Privett, O.S. Autoxidation and Autoxidative Polymerization. *J. Am. Oil Chem. Soc.* **1959,** *36,* 507–512.

Porter, N.A.; and D.G. Wujek. Autoxidation of Polyunsaturated Fatty Acids, an Expanded Mechanistic Study. *J. Am. Chem. Soc.* **1984,** *106,* 2626–2629.

Risérus, U.; S. Basu; S. Jovinge; G. Fredrikson; J. Ärnlöv; and B. Vessby. Supplementation with Conjugated Linoleic Acid Causes Isomer-Dependent Oxidative Stress and Elevated C-Reactive Protein: A Potential Link to Fatty Acid-Induced Insulin Resistance. *Circulation* **2002,** *106,* 1925–1929.

Risérus, U.; B. Vessby; J. Ärnlöv; and S. Basu. Effects of *cis*-9,*trans*-11 Conjugated Linoleic Acid Supplementation on Insulin Sensitivity, Lipid Peroxidation, and Proinflammatory Markers in Obese Men. *Am. J. Clin. Nutr.* **2004,** *80,* 279–283.

Scimeca, J.A. Cancer Inhibition in Animals. In *Advances in Conjugated Linoleic Acid Research,* Yurawecz, M.P., Mossoba, M.M., Kramer, J.K.G., Pariza, M.W., and Nelson, G.J., Eds.; AOCS Press: Champaign, 1999, Vol. 1, pp. 420–443.

Sehat, N.; M.P. Yurawecz; J.A.G. Roach; M.M. Mossoba; K. Eulitz; E.P. Mazzola; and Y. Ku. Autoxidation of Furan Fatty Acid Ester, Methyl 9,12-Epoxyoctadeca-9,11-dienoate. *J. Am. Oil Chem. Soc.* **1998,** *75,* 1313–1319.

Seo, H.-S.; Y. Endo; and K. Fujimoto. Kinetics for Autoxidation of Conjugated Linoleic Acid. *Biosci. Biotechnol. Biochem.* **1999,** *63,* 2009–2010.

Suzuki, R.; K. Nakao; M. Kobayashi; and K. Miyashita. Oxidative Stability of Conjugated Polyunsaturated Fatty Acids and Their Esters in Bulk Phase. *J. Oleo Sci.* **2001,** *50,* 491–495.

Suzuki, R.; M. Abe; and K. Miyashita. Comparative Study of the Autoxidation of TAG Containing Conjugated and Nonconjugated $C_{18}$ PUFA. *J. Am. Oil Chem. Soc.* **2004,** *81,* 563–569.

Tallman, K.A.; D.A. Pratt; and N.A. Porter. Kinetic Products of Linoleate Peroxidation: Rapid β-Fragmentation of Nonconjugated Peroxyls. *J. Am. Chem. Soc.* **2001,** *123,* 11827-11828.

Tallman, K.A.; B. Roschek, Jr.; and N.A. Porter. Factors Influencing the Autoxidation of Fatty Acids: Effect of Olefin Geometry of the Nonconjugated Diene. *J. Am. Chem. Soc.* **2004,** *126,* 9240–9247.

Terpstra, A.H. Effect of Conjugated Linoleic Acid on Body Composition and Plasma Lipids in Humans: An Overview of the Literature. *Am. J. Clin. Nutr.* **2004,** *79,* 352–361.

Tsuzuki, T.; M. Igarashi; T. Iwata; Y. Yamauchi-Sato; T. Yamamoto; K. Ogita; T. Suzuki; and T. Miyazawa. Oxidation Rate of Conjugated Linoleic Acid and Conjugated Linolenic Acid Is Slowed by Triacylglycerol Esterification and α-Tocopherol. *Lipids* **2004,** *39,* 475–480.

van den Berg, J.J.M.; N.E. Cook; and D.L. Tribble. Reinvestigation of the Antioxidant Properties of Conjugated Linoleic Acid. *Lipids* **1995,** *30,* 599–605.

van Gorkum, R. Manganese Complexes as Drying Catalysts for Alkyd Paints, Doctoral thesis, Leiden University. Leiden, 2005, pp. 17-24.

Wahle, K.W.J.; S.D. Heys; and D. Rotondo. Conjugated Linoleic Acids: Are They Beneficial or Detrimental to Health? *Prog. Lipid Res.* **2004,** *43,* 553–587.

Yang, L.; L.K. Leung; Y. Huang; and Z.-Y. Chen. Oxidative Stability of Conjugated Linoleic Acid Isomers. *J. Agric. Food Chem.* **2000,** *48,* 3072–3076.

Yu, L. Free Radical Scavenging Properties of Conjugated Linoleic Acids. *J. Agric. Food Chem.* **2001,** *49,* 3452–3456.

Yurawecz, M.P.; J.K. Hood; M.M. Mossoba; J.A.G. Roach; and Y. Ku. Furan Fatty Acids Determined as Oxidation Products of Conjugated Octadecadienoic Acid. *Lipids* **1995,** *30,* 595–598.

Yurawecz, M.P.; N. Sehat; M.M. Mossoba; J.A.G. Roach; and Y. Ku. Oxidation Products of Conjugated Linoleic Acid and Furan Fatty Acids. In *New Techniques and Applications in Lipid Analysis,* McDonald, R.E., and Mossoba, M.M., Eds.; AOCS Press: Champaign, 1997, pp. 183–215.

Zhang, A.; and Z.Y. Chen. Oxidative Stability of Conjugated Linoleic Acids Relative to Other Polyunsaturated Fatty Acids. *J. Am. Oil Chem Soc.* **1997,** *74,* 1611–1613.

# 5

# Oxidation of Cholesterol and Phytosterols

Afaf Kamal-Eldin and Anna-Maija Lampi
Department of Food Science, Swedish University of Agricultural Sciences, 750 07 Uppsala, Sweden; and Department of Applied Chemistry and Microbiology, 00014 University of Helsinki, Finland

## Introduction

Sterols are the major components in the unsaponifiable fractions of lipids, cholesterol in animal lipids, and a wide range of phytosterols (generally dominated by β-sitosterol) in vegetable oils and fats. The basic skeleton is similar for cholesterol and the major sterols (Fig. 5.1); they only differ in substitutions in the rings and/or the side chains. The stanols are saturated forms of the sterols and are present in a few natural sources including wheat and rye grains.

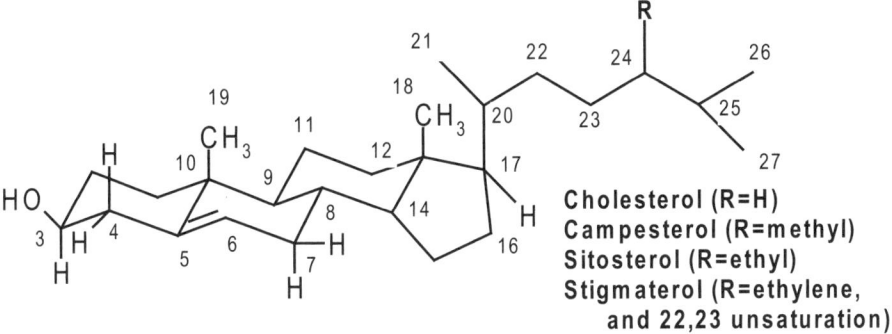

Fig. 5.1. The structures of major sterols and stanols in food products.

Some animal food products, for example egg, butter, and meat with approximately 4, 2, and 0.8 mg cholesterol/g lipid, respectively, are often subjected to high temperature treatments. Heating cholesterol was shown to induce the formation of a number of oxidation products (Osada et al., 1993a). Therefore, food products rich in cholesterol have been specifically investigated for their content of cholesterol oxidation products (COP) (Tsai and Hudson, 1984; Sander et al., 1989; Bösinger et al., 1993; Osada et al., 1993b; Paniangvait et al., 1995; Kerry et al., 2002; Stanton and Devery, 2002). Spray dried eggs, for example, are generally subjected to high temperatures during processing and leads to COP levels of up to 0.3 mg/g. Not only are the contents of total cholesterol oxide different, but the relative amounts of the various oxides are also different which indicates significant variations in other components of the egg and/or different processing and storage conditions (Galobart and Guardiola, 2002).

Phytosterol oxidation products (POP) are known to be present in vegetable oils exposed to high temperatures, for example refined oils (Bortolomeazzi et al., 2003) and in frying oils and fried food products (Dutta, 1997). The formation of POP in food depends on the sterols/stanols present (Table 5.1) and the type of food (Table 5.2).

**Table 5.1.** List of Some Characterized Oxidation Products of Major Sterols

| Trivial Name | Systematic Name |
|---|---|
| **Cholesterol** (Cholest-5-en-3β-ol) | |
| 7α-Hydroperoxycholesterol | Cholest-5-en-3β-ol-7α-peroxide |
| 7β-Hydroperoxycholesterol | Cholest-5-en-3β-ol-7β-peroxide |
| 7α-Hydroxycholesterol | Cholest-5-en-3β,5α-diol |
| 7β-Hydroxycholesterol | Cholest-5-en-3β,5β-diol |
| 7-Ketocholesterol | Cholest-5-en-3β-ol-7-one |
| 5α,6α-Epoxycholesterol | 5α,6α-Epoxycholestan-3β-ol |
| 5β,6β-Epoxycholesterol | 5β,6β-Epoxycholestan-3β-ol |
| Cholestanetriol | Cholestan-3β,5α,6β-triol |
| 20-Hydroxycholesterol | Cholest-5-en-3β,20-diol |
| 25-Hydroxycholesterol | Cholest-5-en-3β,25-diol |
| 5-Cholesten-3-one | Cholest-5-en-3-one |
| 4-Cholesten-3-one | Cholest-4-en-3-one |
| 4,6-Cholestadien-3-one | Cholest-4,6-dien-3-one |
| 3,5-Cholestadien-7-one | Cholest-3,5-dien-7-one |

*Cont. on p. 113.*

## Oxidation of Cholesterol and Phytosterols

**Table 5.1., cont.** List of Some Characterized Oxidation Products of Major Sterols

| Campesterol ((24R)-Methylcholest-5-en-3β-ol) | |
|---|---|
| 7α-Hydroperoxycampesterol | (24R)-Methylcholest-5-en-3β-ol-7α-peroxide |
| 7β-Hydroperoxycampesterol | (24R)-Methylcholest-5-en-3β-ol-7β-peroxide |
| 7α-Hydroxycampesterol | (24R)-Methylcholest-5-en-3β,5α-diol |
| 7β-Hydroxycampesterol | (24R)-Methylcholest-5-en-3β,5β-diol |
| 7-Ketocampesterol | (24R)-Methylcholest-5-en-3β-ol-7-one |
| 5α,6α-Epoxycampesterol | (24R)-5α,6α-Epoxy-24-methylcholestan-3β-ol |
| 5β,6β-Epoxycampesterol | (24R)-5β,6β-Epoxy-24-methylcholestan-3β-ol |
| Campestanetriol | (24R)-Methylcholestan-3β,5α,6β-triol |
| 25-Hydroxycampesterol | (24R)-Methylcholest-5-en-3β,25-diol |

| Sitosterol ((24R)-Ethylcholest-5-en-3β-ol) | |
|---|---|
| 7α-Hydroperoxysitosterol | (24R)-Ethylcholest-5-en-3β-ol-7α-peroxide |
| 7β-Hydroperoxysitosterol | (24R)-Ethylcholest-5-en-3β-ol-7β-peroxide |
| 7α-Hydroxysitosterol | (24R)-Ethylcholest-5-en-3β,5α-diol |
| 7β-Hydroxysitosterol | (24R)-Ethylcholest-5-en-3β,5β-diol |
| 7-Ketositosterol | (24R)-Ethylcholest-5-en-3β-ol-7-one |
| 5α,6α-Epoxysitosterol | (24R)-5α,6α-Epoxy-24-ethylcholestan-3β-ol |
| 5β,6β-Epoxysitosterol | (24R)-5β,6β-Epoxy-24-ethylcholestan-3β-ol |
| Sitostanetriol | (24R)-Ethylcholestan-3β,5α,6β-triol |
| 25-Hydroxysitosterol | (24R)-Ethylcholest-5-en-3β,25-diol |

| Stigmasterol ((24R)-Ethylcholest-5,22-dien-3β-ol) | |
|---|---|
| 7α-Hydroperoxystigmasterol | (24R)-Ethylcholest-5,22-dien-3β-ol-7α-peroxide |
| 7β-Hydroperoxystigmasterol | (24R)-Ethylcholest-5,22-dien-3β-ol-7β-peroxide |
| 7α-Hydroxystigmasterol | (24R)-Ethylcholest-5,22-dien-3β,5α-diol |
| 7β-Hydroxystigmasterol | (24R)-Ethylcholest-5,22-dien-3β,5β-diol |
| 6β-Hydroxystigmasterol | (24R)-Ethylcholest-5,22-dien-3β,5β-diol |
| 7-Ketostigmasterol | (24R)-Ethylcholest-5,22-dien-3β-ol-7-one |
| 5α,6α-Epoxystigmasterol | (24R)-5α,6α-Epoxy-24-ethylcholest-22-en-3β-ol |
| 5β,6β-Epoxystigmasterol | (24R)-5β,6β-Epoxy-24-ethylcholest-22-en-3β-ol |
| Stigmastanetriol | (24R)-Ethylcholest-22-en-3β,5α,6β-triol |
| 25-Hydroxystigmasterol | (24R)-Ethylcholest-5,22-dien-3β,25-diol |
| 6β-Hydroxy-3-keto-stigmasterol | (24R)-6β-Hydroxy-3-keto-Ethylcholest-5,22-dien-3β,5β-diol |
| 6α-Hydroxy-3-keto-stigmasterol | (24R)-6α-Hydroxy-3-keto-Ethylcholest-5,22-dien-3β,5β-diol |

**Table 5.2.** Sterol Oxides in Food

| Food | Sterol Oxides |
|---|---|
| Spray-dried eggs<br>Infant formula<br>Whole milk powder<br>Butter oils and Indian ghee<br>Grated cheeses<br>Freeze-dried meats<br>Dried fish | 7α-, 7β-, and 25-Hydroxy cholesterols,<br>7-ketocholesterol,<br>α- and β-epoxycholesterols,<br>cholesterol triol |
| Potato chips<br>French fries | 7α-, 7β-, and 25-Hydroxy sterols,<br>7-ketosterols,<br>α- and β-epoxysterols,<br>sterol triols<br>(the type of sterols depend on the frying oil) |

There is increased interest in foods supplemented with phytosterols and phytosterol esters as cholesterol-lowering functional foods (Law, 2000). Phytosterols did not oxidize significantly in milk powder, heat-treated milk, and in microcrystalline phytosterol suspensions in different fats and oils during processing and long-term storage (Soupas et al., 2006). However, some POP were found in phytosterol ester-enriched spreads (Grandgirard et al., 2004), and bread (Soupas et al., 2003) supporting the need to evaluate the safety of phytosterol-enriched foods (Lea et al., 2004). It is possible that most of the POP in these products are derived from the phytosterol ingredients. Baking was found not to induce any significant formation of POP, but several POP species were found to form during frying (Dutta, 1997; Soupas et al., 2007).

## Autoxidation of Sterols: Products and Mechanisms

Like other hydrocarbons, sterols are prone to oxidation when exposed to air under catalysis of heat, light, ionizing radiation, or certain chemicals. Accumulating experimental evidence suggests that autoxidation of sterols, akin to that of other unsaturated hydrocarbons, follows a free radical mechanism (Smith, 1981, 1987; Muto et al., 1982; Sevilla et al., 1986; Nawar et al., 1991; Kim and Nawar, 1991, 1993; Osada et al., 1993a; Rankin and Pike, 1993; Huber et al., 1995; Oehrl et al., 2001). γ-Irradiation of oxygen-free cholesterol revealed two major radicals by ESR, at C-7 (60%) and C-25 (40%) (Fig. 5.2), which were converted to peroxyl radicals when the irradiation was performed in an oxygen atmosphere (Sevilla et al., 1986). When irradiation was performed in tripalmitin, the cholesteryl radical was specifically formed at the allylic position. No radical was detected at C-20, possibly due to conformational restrictions.

The major primary autoxidation products of cholesterol were identified as 5,6-epoxysterols (5,6α-EP and 5,6β-EP) and 7-hydroperoxysterols (7α-OOH and 7β-OOH), which decompose to 7-hydroxysterols (7α-OH and 7β-OH), 7-ketosteol (7-keto) (Fig. 5.3). In most studies, the levels of the hydroperoxides 7α-OOH and

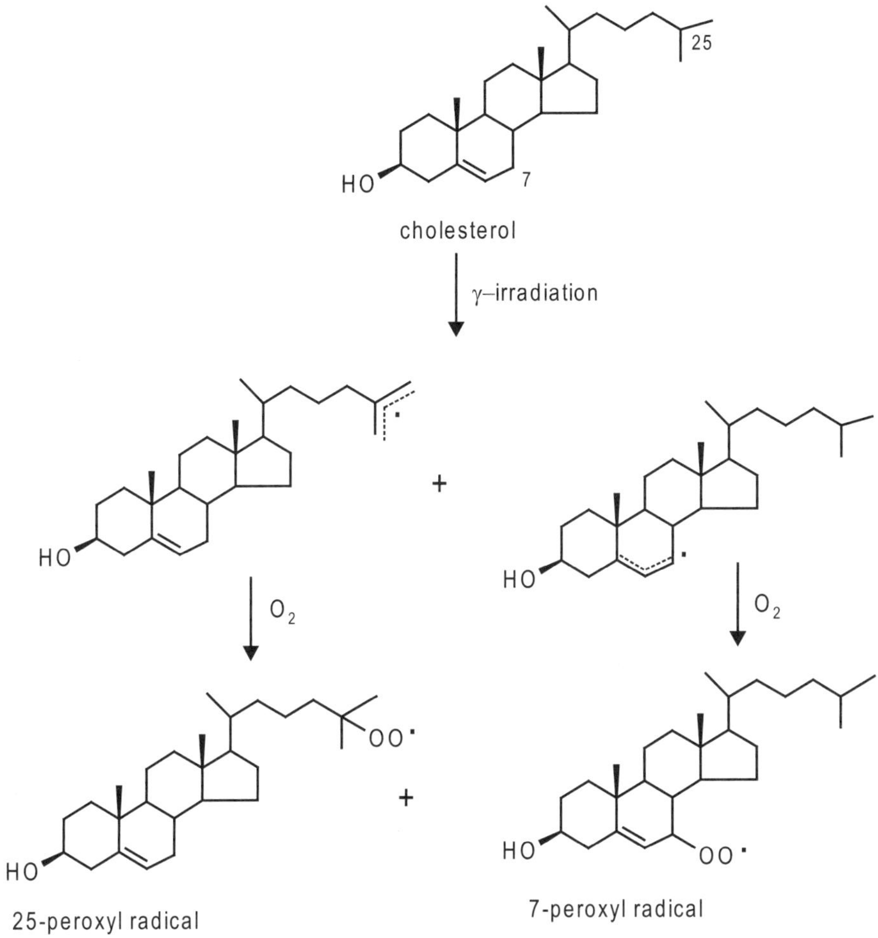

**Fig. 5.2.** Formation of radicals at C-7 and C-25 of cholesterol by γ-irradiation.

7β-OOH were not determined due to their instability and requirement of analytical methods different from the gas chromatography method often used to analyze sterol oxidation products. The best methods to analyze unstable hydroperoxides involve fluorescent detection after a specific post-column reaction of hydroperoxyl groups with non-fluorescent diphenyl-1-prenylphosphine to the fluorescent diphenyl-1-prenylphosphine oxide as shown in Fig. 5.4 (Akasaka and Ohrui, 2000; Säynäjoki et al., 2002) or electrochemical detection (Korytowski et al., 1995).

The autoxidation of sterols follows the basic rules of oxidation of hydrocarbons where peroxyl radicals act as chain carriers and react by the following pathways:

  i. Addition to the double bond at either end (C-5 or C-6) to form two epimeric epoxides, 5,6α-EP and 5,6β-EP (Aringer and Eneroth, 1974). The amount

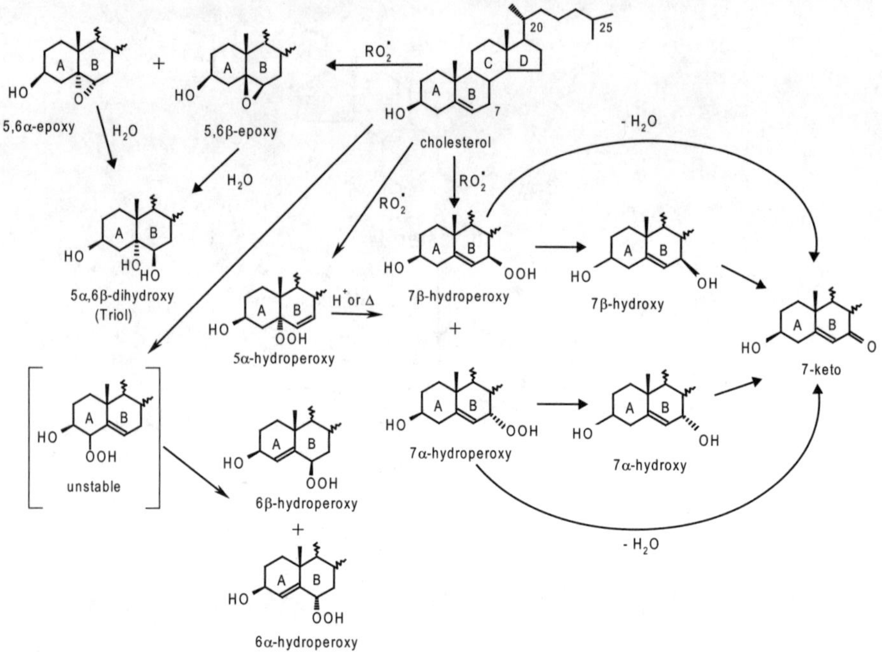


Fig. 5.3. Reaction pathways in the oxidation of cholesterol and analogous phytosterols.

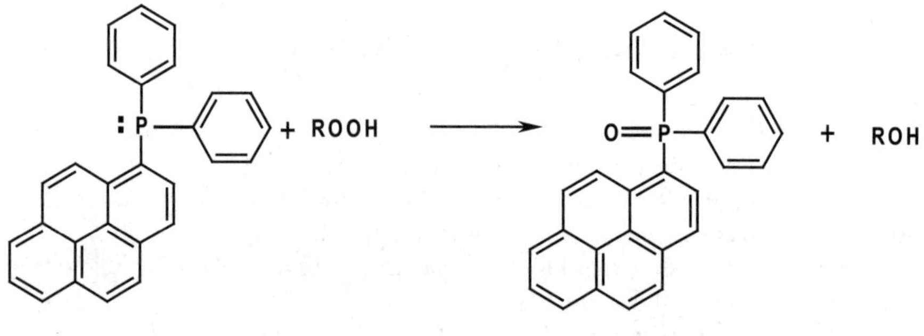


**Diphenyl-1-prenylphosphine (Non-fluorescent)**

**Diphenyl-1-prenylphosphine oxide (Fluorescent ($\lambda_{ex.}$ 352 nm, $\lambda_{em.}$ 380 nm)**

Fig. 5.4. Fluoremetric detection of hydroperoxides after reaction with diphenyl-1-prenylphosphine.

of 5,6β-EP is always higher than that of 5,6α-EP in experiments, suggesting that the formation of 5,6β-EP results from addition of the peroxyl radical to the less sterically hindered position 6 while the formation of 5,6α-EP results from peroxyl radical addition at the sterically hindered position 5. Further support for this hypothesis can be obtained from the fact that the 5-OOH formed from sterols by photoxidation (*vide supra*) has α configuration. The peroxyl radical-sterol adducts formed above decompose quickly and form epoxides by loss of a hydroxyl radical. Under hydrous conditions, both epoxides may undergo acid-catalyzed hydration to form 3β,5α,6β-sterol triol (Triol) specifically. The hydration is initiated by protonation of the epoxide oxygen followed by approach of the nucleophile, water, at the side opposite the protonated oxygen. In the β-epoxide, the dialkylated C-5 is the best to host the positive charge and thereafter the nucleophile leading to the 3β,5α,6β-sterol triol. However, steric hindrance by C-19 makes C-6 in the α-epoxide to be the position to host the positive charge leading to the 3β,5α,6β-sterol triol. Indeed hydration of the α-epoxide occurs to a much more limited extent compared to the β-epoxide (Maerker, 1987).

ii. Hydrogen abstraction primarily from allylic position C-7 forms 7α-OOH and the thermodynamically more stable 7β-OOH. These unstable hydroperoxides degrade to form more stable hydroxy and keto derivatives (Yanishlieva, 1983; Smith, 1987). Free radical abstraction rarely occurs at C-4, the other allylic position, possibly due to shielding effects of the hydroxyl group at C-3 and the tri-substituted C-5 (Maerker, 1987). Small amounts of 6α- and 6β-hydroperoxides with shifting of the $\Delta^5$ double bond to the $\Delta^4$ position (6α-OOH and 6β-OOH) have also been found under photoxidation conditions. The plausible mechanism to form these hydroperoxides and possible secondary oxidation products seems to involve primary hydrogen abstraction at the other allylic position of the $\Delta^5$ double bond, that is C-4, and fast rearrangement of the resulting unstable hydroperoxyl radical with the shift in the double bond. In fact, 4β-cholesterolhydroperoxide or 4β-hydroxycholesterol have been found in limited cases, for example heated butter and egg yolk powder (Csiky, 1982, 1985). 6β-Hydroxy derivatives of β-sitosterol, campesterol, and brassicasterol are found together with 7-ketobrassicasterol in refined, deodorized rapeseed oil (Lambelet et al., 2003). Disproportionation of sterol peroxyl radicals would lead to alkoxyl radicals resulting in singlet oxygen and hydroxy and keto oxidation products (Russel, 1957; Vardanyan et al., 1985).

iii. Very small amounts of 5α-hydroperoxide may be formed by shifting the $\Delta^5$ double bond to the $\Delta^6$ position (5α-OOH) and are followed by rearrangement to the 7α-OOH (Adachi et al., 1998). Due to the weakness of the ROO-H (BDE ~90 Kcal/mol), the rearrangement of hydroperoxide

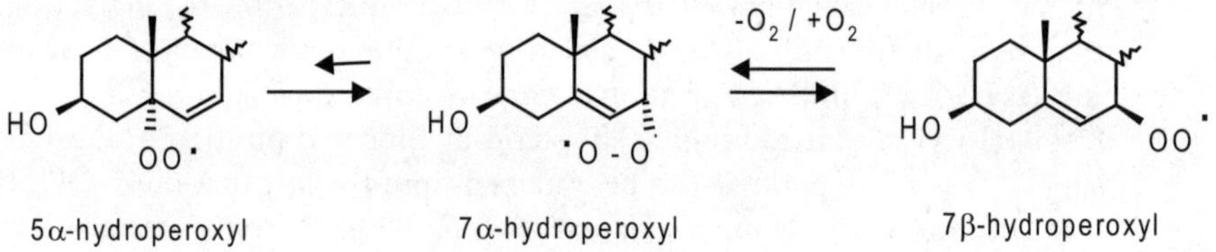

5α-hydroperoxyl      7α-hydroperoxyl      7β-hydroperoxyl

**Fig. 5.5.** Conversion of 5α-hydroperoxy sterol to 7-hydroperoxy sterols.

position with a shift in double bond (Fig. 5.5) is known to occur with monounsaturated substrates and is well known for oleic acid (Porter et al. 1994).

iv. Hydrogen abstraction from tertiary carbon atoms, which exist at C-20 and C-25 in the case of cholesterol, campesterol, campestanol, β-sitosterol, sitostanol, and brassicasterol, and C-24 in the case of campesterol, campestanol, β-sitosterol, and sitostanol. Positions C-20 and C-24, being both allylic to the $\Delta^{22\text{-}23}$ double bond and tertiary, are especially reactive in the case of stigmasterol (Blekas and Boskou, 1989). Side chain sterol hydroperoxides decompose faster than ring hydroperoxides to form the more stable hydroxy and keto derivatives (Yanishlieva, 1983). Oxidation of the sterol side chains is significant during oxidation of sterols in the solid state and is favored by their crystal structures where the aliphatic chains are allied to the outside and are more accessible to attaching free radicals than the rings (Korahani et al., 1982).

v. Further oxidation of first oxidation products has also been observed. For example, 5,6-epoxy derivatives of 7α- and 7β-OH during the oxidation of cholesteryl acetate (Lercker et al., 1999) is similar to oleate (Lercker et al., 1984).

vi. Finally, other reactions may operate at high temperatures and under anhydrous conditions, that is coupled dehydration-oxidation reactions leading to the formation of several dehydrated products of sterols (Fig. 5.6) and oligomers/polymers of unknown structures (Soupas et al., 2005). Thus, a wide range of minor oxidation products can be formed besides the major ones discussed above. Some of these products have been characterized, but many remain to be investigated since traces can be observed in different chromatograms (Lampi et al., 2002).

## Photoxidation of Sterols: Products and Mechanism

Singlet oxygen, which is formed by the reaction of triplet oxygen with photosensitizers as discussed in Chapter 1, can react rapidly with the double bonds of sterols by the ene mechanism and lead to the formation of 5-hydroperoxysterols (5-OOH) and

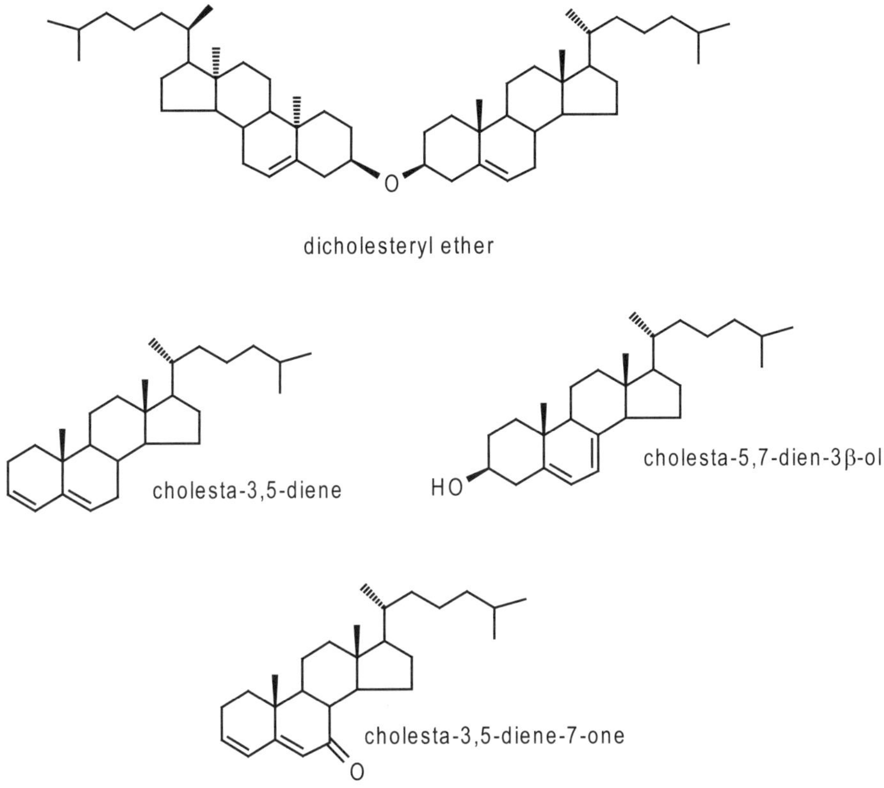

**dicholesteryl ether**

**cholesta-3,5-diene**

**cholesta-5,7-dien-3β-ol**

**cholesta-3,5-diene-7-one**

**Fig. 5.6.** Other sterol dehydration and oxidation products.

lower amounts of 6-hydroperoxysterols (6-OOH) (Kulig and Smith, 1973; Yanishlieva and Marinova, 1980; Smith, 1981, 1987, 1996). There is only one possibility for the hydroperoxide at C-5 (i.e., 5α-OOH), but two possibilities exist for hydroperoxides at C-6 (i.e., 6α-OOH and i.e. 6β-OOH) (Säynäjoki et al., 2003). Porter et al. (1994) showed that they can easily form peroxyl radicals that undergo rearrangement when they are in proximity to a double bond due to the weak ROO-H bond in hydroperoxide. They have further shown that the 5α-OOH formed in photoxidation can undergo rearrangement to for 7α-OOH, which will re-arrange further to form small amounts of 7β-OOH in agreement with other investigators (Beckwith, 1989; Bortolomeazzi et al., 1999). Beyond identification of these products and their formation pathway, studies on the photoxidation of sterols have been quite limited.

## Kinetics of Sterol Autoxidation

Oxidation of lipids starts as a pseudo-first-order reaction at peroxide values ≤20mM followed by a second-order reaction (Bateman et al., 1953). The detection and description of reaction kinetics is, however, dependent *inter alias* on the substrate, reaction temperature, reaction medium, and the monitored reaction product(s). Therefore, contradictions are often found in literature. For example, Park and Addis (1986) found that the formation of 7-ketocholesterol during heating follows a zero-order reaction, Yan and White (1990) found the formation of 7-hydroxy (OH), 7-keto, and 5,6-epoxides (EP) from cholesterol in lard fit a first-order reaction, while Chien et al. (1998) found that the formation of 7-OOH and 5,6-EP followed a second-order reaction during the oxidation of a thin film of cholesterol at 150°C, that is

$$d[CholOxid]/dt = k[CholOxid][Chol]$$

where k is the rate constant, [CholOxid] is the concentration of 7-OOH or 5,6-EP, [Chol] is the concentration of unoxidized cholesterol, and $t$ is the time.

On the other hand the formation of the secondary oxidation products, 7-OH and 7-keto compounds, followed a first-order reaction being solely dependent on the concentration of 7-OOH (Chien et al., 1998). In this study, about 50% of cholesterol underwent very fast transformation within 5 min. while the next 15% disappeared at a much slower rate between 5 and 30 min. The rate constants were approximately 1400 $h^{-1}$ for the formation of 5,6α-EP and 5,6β-EP and 1600 $h^{-1}$ for the formation of 7α-OOH and 7β-OOH from cholesterol, about 800 $h^{-1}$ for the reduction of 7-OOH to 7-OH, and 805 $h^{-1}$ for its dehydration to keto, and only 3 $h^{-1}$ for the dehydrogenation of 7-OH to keto. Like fatty acid oxidation, sterol oxidation is inhibited by antioxidants, such as α-tocopherol, (Terao et al., 1985; Li et al., 1996).

Pioneering work by Yanishlieva and Marinova (1980) showed that sitosteryl stearate oxidized much faster at the beginning of the process (PV 15-20 mM) compared to β-sitosterol in the pure state in the dark at 90-125°C. This difference was related to differences in the activation energies, calculated as 37.7 Kcal/mole for β-sitosterol and 16.3 Kcal/mole for sitosteryl stearate, but the reason for this large difference was not explained. The differences in oxidizability between sitosteryl stearate and β-sitosterol were largely diminished during the second phase of oxidation due to greater pro-oxidant effect of the oxidation products of β-sitosterol than sitosteryl stearate. It was suggested that the free hydroxyl group in β-sitosterol inhibits the initiation of oxidation in the initial stage but promotes the decomposition of hydroperoxides. The reaction products also differed in the two cases with oxidation products of β-sitosterol-3-one, β-sitosterol-3,6-dione, and β-sitosterol-3,5-diene formed from β-sitosterol while oxidation products at C-1, C-2, and C-4 formed in the case of β-sitosteryl stearate (Yanishlieva et al., 1983). Comparable results were obtained when the oxidation was performed in different triacylglycerol matrices.

Significant oxidative interactions may occur between sterols and co-existing lipids depending on their natures and other compositional and environmental factors (Nawar et al., 1991; Kim and Nawar, 1991, 1993). The effect of co-existing unsaturated fatty acids in the oxidized matrix on sterol oxidation is controversial. For example, Ohshima et al. (1993) found that cholesterol was stable in triolein but it co-oxidized rapidly in a mixture of fish oil triacylglycerols. Free radicals generated during the oxidation of highly oxidizable polyunsaturated fatty acids catalyze sterol oxidation leading to a high degree of co-oxidation (Li et al., 1994). The oxidation was found to be faster in bulk stigmasterol than when the oxidation was performed in purified rapeseed oil triacylglycerols (Lampi et al., 2002). Sitostanol, β-sitosterol, stigmasterol, and ergosterol were oxidized at 0.1-1.0% levels in purified rapeseed oil and tripalmitin at 80, 120, and 180°C. As expected, sitostanol, with a saturated ring structure, was more stable than unsaturated β-sitosterol and stigmasterol while ergosterol, with a conjugated $\Delta^{5,7}$ double bond system, was by far least stable. Stigmasterol was stable in oil matrices at 80°C for 7 days but approximately 50% was degraded after 1 day at 120°C. All sterols were more stable in rapeseed oil than in tripalmitin at 180°C (Lampi et al., 2002), which can be explained by the concept of competitive oxidation under conditions of low prevalence of oxygen. Early experiments showed that when β-sitosterol was added at a 5% level to mixtures oxidized at 120°C, it was less stable in purified sunflower oil triacylglycerols than lard triacylglycerols in tristearin (Yanishlieva and Marinova, 1986). Under these conditions, β-sitosterol oxidation was accelerated by the increased degree of unsaturation in the lipid medium which is critical for chain initiation and propagation. In addition, the rate of oxidation is also dependant on the physical state, such as melting behavior, of sterols and co-existing lipids (Kim and Nawar, 1991, 1993).

Table 5.3 presents results of stigmasterol oxidation in tripalmitin and purified rapeseed oil at 100, 140, and 180°C (Soupas et al., 2004). The results showed that at 100°C, stigmasterol oxidation was faster in purified rapeseed oil than in tripalmitin but at >140°C, stigmasterol oxidation was much slower in purified rapeseed oil than in tripalmitin. It is known that oxidizing lipids compete for the low concentration of oxygen and that other reactions, for example dehydration and polymerization, are also important at high temperatures. Unlike the unsaturated sterols (Li et al., 1994; Soupas et al., 2004), the oxidation of saturated sitostanol was enhanced by the unsaturated lipid matrix at all temperatures (Soupas et al., 2004). Similarly, phytosteryl esters oxidized faster than free phytosterols in tripalmitin at 100°C but not at 180°C due to intermolecular interactions (Soupas et al., 2005). In an early study, the percentage of oxidized cholesterol at 100°C in the presence of benzoyl peroxide in the solid phase followed the order: cholesterol linolenate < cholesterol linoleate << cholesterol stearate < cholesterol oleate < cholesterol acetate (Korahani et al., 1982). The presence of unsaturated linoleate and linolenate clearly protected cholesterol against oxidation. This is not true for highly unsaturated fatty acids like fish oil triacylglycerols (Li et al., 1994).

Depending on the composition of the matrix and the temperature of the reaction, sterols may induce a weak lowering or elevation of the oxidation rate of co-existing fatty acyls. The presence of 10% cholesterol in purified sunflower triacylglycerols was found to increase the oxidation rate at 80, 90, and 100°C (Marinova et al., 2005) in agreement with earlier results by Wu et al. (1978) in monolayers of linoleic acid. This prooxidant effect of sterols was explained by their participation in the formation of mixed micelles with initial hydroperoxides leading to a small increase in their rate of decomposition (Brimberg and Kamal-Eldin, 2003). On the other hand, at 0.1% level $\Delta^5$-avenasterol and fucosterol were found to exert a slight protective effect on the degradation of the linoleate residues of olive oil at 180°C (Gordon and Magos, 1983). This effect was not possessed by cholesterol, β-sitosterol or stigmasterol suggesting that the $\Delta^{24,28}$ double bond in the side chain was responsible for this effect. However, this effect was not found in a previous study (Lampi et al., 1999).

**Table 5.3.** Transformation of Stigmasterol to Stigmasterol Oxides (%) During Heat Treatment in Tripalmitin or Purified Rapeseed Oil Containing 1% Stigmasterol

| Treatment | Tripalmitin | | | | Purified Rapeseed Oil | | | |
|---|---|---|---|---|---|---|---|---|
| | 7α- and | 5,6-EP | 7-Keto | Total | 7α- and 7β-hydroxy | 5,6-EP | 7-Keto | Total |
| 100°C, 6h | 0.02 | 0.03 | 0.02 | 0.07 | 1.4 | 0.2 | 0.7 | 2.3 |
| 24h | 0.03 | 0.04 | 0.03 | 0.10 | 8.8 | 1.9 | 1.7 | 12.4 |
| 48h | 0.07 | 0.12 | 0.11 | 0.30 | 15.9 | 5.4 | 3.6 | 24.9 |
| 140°C, 1h | 0.05 | 0.10 | 0.10 | 0.25 | 0.3 | 0.1 | 0.2 | 0.6 |
| 3h | 0.6 | 0.9 | 1.1 | 2.6 | 1.6 | 0.8 | 0.5 | 2.9 |
| 6h | 2.0 | 3.0 | 2.5 | 7.5 | 4.4 | 2.4 | 0.8 | 7.6 |
| 180°C, 1h | 2.2 | 2.5 | 1.0 | 5.7 | 1.1 | 0.6 | 0.2 | 1.9 |
| 2h | 6.9 | 6.5 | 1.7 | 13.1 | 2.7 | 1.4 | 0.3 | 4.4 |
| 3h | 9.8 | 10.2 | 2.7 | 22.7 | 4.0 | 2.3 | 0.5 | 6.8 |

Source: Soupas et al., (2004)

## Conclusion

Sterols oxidize according to the same rules that govern other hydrocarbons including polyunsaturated fatty acids. Since they are present in complex lipid mixtures, their oxidation is affected by the other species present, including antioxidants, prooxidants, and co-oxidizable lipids. The physical status of the matrix, its non-lipid components, temperature, and other factors that influence the exposure of sterols to oxygen and their interactions affect the oxidation rate of sterols. Changes in sterol fluidity, for example by esterification, also seem to play a considerable role.

## References

Adachi, J.; M. Asano; T. Naito; Y. Ueno; and Y. Tatsuno. Chemiluminescent Determination of Cholesterol Hydroperoxides in Human Erythrocyte Membrane. *Lipids* **1998**, *33*, 1235-1240.

Akasaka, K.; and H. Ohrui. Development of Phosphine Reagents for the High Performance Liquid Chromatographic-Fluorometric Determination of Lipid Hydroperoxides. *J. Chro-*

*matogr. A.* **2000**, *881,* 159-170.

Aringer, L.; and P. Eneroth. Formation and Metabolism in vitro of 5,6-Epoxides of Cholesterol and β-Sitosterol. *J. Lipid Res.* **1974**, *15,* 389-398.

Bateman, L; H. Hughes; and A.L. Morris. Hydroperoxide Decomposition in Relation to the Initiation of Radical Chain Reactions. *Disc. Faraday Soc.* **1953** *14,* 190-199.

Beckwith, A.L.J. The Mechanism of the Rearrangements of Allylic Hydroperoxides: 5α-hydroperoxy-3β-hydroxycholest-6-ene and 7α-hydroperoxy-3β-hydroxycholest-5-ene. *J. Chem. Soc. Perkin Trans.* **1989**, *II,* 815-824.

Blekas, G. and D. Boskou. Oxidation of Stigmasterol in Heated Triacylglycerols. *Food Chem.* **1989** *33,* 301-310.

Bortolomeazzi, R.; M. De Zan; L.S. Pizzale; and L. Conte. Mass Spectrometry Characterization of the 5α-, 7α-, and 7β-hydroxy Derivatives of β-Sitosterol, Campesterol, Stigmasterol, and Brassicasterol. *J. Agric. Food Chem.* **1999**, *47,* 3069-3074.

Bortolomeazzi, R.; F. Cordaro; L. Pizzale; L.S. Conte. Presence of Phytosterol Oxides in Crude Vegetable Oils and Their Fate During Refining. *J. Agric Food Chem.* **2003**, *51,* 2394-2401.

Bösinger, S.; W. Luf; and E. Brandl. Oxysterols: Their Occurrence and Biological Effects. *Int. Dairy J.* **1993**, *3,* 1-33.

Brimberg U.; and A. Kamal-Eldin. On the Kinetics of the Autoxidation of Fats: Influence of Prooxidants, Antioxidants and Synergists. *Eur. J. Lipid Sci. Technol.* **2003**, *105,* 83-91.

Chien, J.T.; H.C. Wang; and B.H. Chen. Kinetic Model of the Cholesterol Oxidation During Heating. *J. Agric. Food Chem.* **1998**, *46,* 2572-2577.

Csiky, I. Trace Enrichment and Separation of Cholesterol Oxidation Products by Absorption High-Performance Liquid Chromatography. *J. Chromatography* **1982**, *241,* 381-389.

Csiky, I. *Extension of the Selectivity in Column Liquid Chromatography: Applications of Pre- and Post-Column Techniques for the Separation of Complex Mixtures,* Ph.D. Thesis, University of Lund, Lund, 1985.

Dutta, P.C. Studies on Phytosterol Oxides. II. Content in Some Vegetable Oils and in French Fries Prepared in These Oils. *J. Am. Oil Chem. Soc.* **1997**, *74,* 659-666.

Galobart, J.; and F. Guardiola. Formation and Content of Cholesterol Oxidation Products in Egg and Egg Products. In *Cholesterol and Phytosterol Oxidation Products: Analysis, Occurrence, and Biological Effects,* Guardiola, F.; Dutta, P.C.; Codony, R.; and Savage, G.P. Eds.; AOCS Press: Champaign, Illinois, 2002, pp. 124-146.

Gordon, M.H.; and P. Magos. The Effect of Sterols on the Oxidation of Edible Oils. *Food Chem.* **1983**, *10,* 141-147.

Grandgirard, A.; L. Martine; P. Joffre; P. Juaneda; and O. Berdeaux. Gas Chromatographic Separation and Mass Spectrometric Identification of Mixtures of Oxyphytosterol and Oxycholesterol Derivatives: Application to a Phytosterol Enriched Food. *J. Chromatogr A* **2004**, *1040,* 239-250..

Huber, K.C.; O.A. Pike; and C.S. Huber. Antioxidant Inhibition of Cholesterol Oxidation in Spray-Dried Food System During Accelerated Storage. *J. Food Sci.* **1995**, *60,* 909-916.

Kerry, J.P.; D.A. Gilroy; and N.M. O'Brien. Formation and Content of Cholesterol Oxidation Products in Meat and Meat Products. In *Cholesterol and Phytosterol Oxidation Products: Analysis, Occurrence, and Biological Effects,* Guardiola, F.; Dutta, P.C.; Codony, R.; and Savage, G.P. Eds; AOCS Press: Champaign, Illinois, 2002, pp. 162-185.

Kim, S.K.; and W.W. Nawar. Oxidative Interactions of Cholesterol with Triacylglycerols. *J. Am. Oil Chem Soc.* **1991**, *68,* 931-934.

Kim, S.K.; and W.W. Nawar. Parameters Influencing Cholesterol Oxidation. *Lipids* **1993**, *28*, 917-922.

Korahani, V.; J. Bascoul; and A. Crastes de Paulet. Autoxidation of Cholesterol Fatty Acid Esters in Solid State and Aqueous Dispersion. *Lipids* **1982**, *17*, 703-708.

Korytowski W.; P.G. Geiger; A.W. Girotti. High-Performance Liquid-Chromatography with Mercury Cathode Electrochemical Detection: Application to Lipid Hydroperoxide Analysis. *J. Chromatogr.* **1995**, *670*, 189-197.

Kulig, M.J. and L.L. Smith. Sterol Metabolism-XXV. Cholesterol Oxidation by Singlet Molecular Oxygen. *J. Org. Chem.* **1973** *38*, 3639-3642.

Lambelet, P.; A. Grandgirard; S. Gregoire; P. Juaneda; J.L. Sebedio; and C. Bertoli. Formation of Modified Fatty Acids and Oxyphytosterols During Refining of Low Erucic Acid Rapeseed Oil. *J. Agric. Food Chem.* **2003**, *51*, 4284-4290.

Lampi A-M.; L. Dimberg; and A. Kamal-Eldin. A Study on the Influence of Fucosterol on Thermal Polymerisation of Purified High Oleic Sunflower Triacylglycerols. *J. Sci. Food Agric.* **1998**, *79*: 573-579.

Lampi, A.-M.; L. Juntunen; J. Roivo; and V. Piironen. Determination of Thermo-Oxidation Products of Plant Sterols. *J. Chromatogr. B.* **2002**, *777*, 83-92.

Law, M. Plant Sterol and Stanol Margarines and Health. *Br. Med. J.* **2000**, *320*, 861-864.

Lea, L.J.; P.A. Hepburn; A.M. Wolfreys; P. Baldrick. Safety Evaluation of Phytosterol Esters. Part 8: Lack of Genotoxicity and Subcronic Toxicity with Phytosterol Oxides. *Food Chem Toxicol.* **2004**, *42*, 771-783.

Lercker, G.; P. Capella; and L.S. Conte. Thermo-Oxidative Degradation Products of Methyl Oleate. *Riv. Ital. Sost. Grasse* **1984**, *61*, 337-344.

Lercker, G.; R. Bortolmeazzi; and L. Pizzale. Formation of 5,6-Epoxy Derivatives of 7-Hydroxy-Cholesteryl-3-Acetates During Peroxidation of Cholesteryl Acetate. *Grasas y Aceites* **1999**, *50*, 193-198.

Li, N.; T. Ohshima; K. Shozen; H. Ushio; and C. Koizumi. Effects of the Degree of Unsaturation of Coexisting Triacylglycerols on Cholesterol Oxidation. *J. Am. Oil Chem. Soc.* **1994**, *71*, 623-627.

Li, S.X.; G. Cherian; D.U. Ahn; R.T. Hardin; and J.S. Sim. Storage, Heating, and Tocopherols Affect Cholesterol Oxide Formation in Food Oils. *J. Agric. Food Chem.* **1996**, *44*, 3830-3634.

Maerker, A. Cholesterol Autoxidation: Current Status. *J. Am. Oil Chem. Soc.* **1987**, *64*, 388-392.

Marinova, E.; N. Yanishlieva; and A. Toneva. Influence of Cholesterol on the Kinetics of Lipid Autoxidation and on the Antioxidative Properties of α–-Tocopherol and Quercetin. *Eur. J. Lipid Sci. Technol.* **2005**, *107*, 418-425.

Muto, T.; J. Tanaka; T. Miura; and M. Kimura. Iron-Catalyzed Autoxidation of Cholesterol in the Presence of Unsaturated Long Chain Fatty Acid. *Chem. Pharm. Bull.* **1982**, *30*, 3172-3177.

Nawar, W.W.; S.K. Kim; Y.J. Li; and M. Vadji. Measurement of Oxidative Interactions of Cholesterol. *J. Am. Oil Chem. Soc.* **1991**, *68*, 496-498.

Oehrl, L.L.; A.P. Hansed; C.A. Rohrer; G.P. Fenner; and L.C. Boyd. Oxidation of Phytosterols in a Test Food System. *J. Am. Oil Chem. Soc.* **2001**, *78*, 1073-1078.

Ohshima, T.; N. Li; and C. Koizumi. Oxidative Decomposition of Cholesterol in Fish Products. *J. Am. Oil Chem. Soc.* **1993**, *70*, 595-600.

Osada, K.; T. Kodama; K. Yamada; and M. Sugano. Oxidation of Cholesterol by Heating. *J.*

*Agric. Food Chem.* **1993a**, *41,* 1198-1202.

Osada, K.; T. Kodama; L. Cui; K. Yamada; and M. Sugano. Levels and Formation of Oxidized Cholesterol in Processed Marine Food. *J. Agric. Food Chem.* **1993b**, *41,* 1893-1898.

Paniangvait, P.; A.J. King; A.D. Jones; and B.G. German. Cholesterol Oxides in Foods of Animal Origin. *J. Food Sci.* **1995**, *60,* 1159-1174.

Park, S.W. and P.B. Addis. Identification and Quantitative Estimation of Oxidized Cholesterol Derivatives in Heated Tallow. *J. Agric. Food Chem.* **1986** *34,* 653-659.

Porter, N.A.; K.A. Mills; S.E. Cadwell; and G.R. Dubay. The Mechanism of the [3,2] Allylperoxyl Rearrangement: A Radical-Dioxygen Pair Reaction that Proceeds with Stereochemical Memory. *J. Am. Chem. Soc.* **1994,** *116,* 6697-6705.

Rankin, S.A.; and O.A. Pike. Cholesterol Autoxidation Varies Among Several Natural Antioxidants in an Aqueous Model System. *J. Food Sci.* **1993**, *58,* 653-655.

Russell, G.A. Deuterium-Isotope Effects in the Autoxidation of Alkyl Hydrocarbons: Mechanism of the Interaction of Peroxyl Radicals. *J. Am. Chem. Soc.* **1957**, *79,* 3871-3877.

Sander, B.D.; P.B. Addis; S.W. Park; and D.E. Smith. Quantification of Cholesterol Oxidation Products in a Variety of Foods. *J. Food Prot.* **1989**, *52,* 109-114.

Säynäjoki, S.; S. Sundberg; L. Soupas; A.-M. Lampi; and V. Piironen. Determination of Stigmasterol Primary Oxidation Products by High-Performance Liquid Chromatography. *Food Chem.* **2002,** *80,* 415-421.

Säynäjoki, S.; S. Sundberg; L. Soupas; A-M. Lampi; and V. Piironen. Determination of Stigmasterol Primary Oxidation Products by High Performance Liquid Chromatography. *Food Chem.* **2003**, *80,* 415-421.

Sevilla, C.L.; D. Becker; and M.D. Sevilla. An Electron Spin Resonance Investigation of Radical Intermediates in Cholesterol and Related Compounds: Relation to Solid-State Autoxidation. *J. Phys. Chem.* **1986**, *90,* 2963-2968.

Smith, L.L. *Cholesterol Autoxidation*, Plenum Press: New York, 1981.

Smith, L.L. Cholesterol Autoxidation 1981-1986. *Chem. Phys. Lipids* **1987,** *44,* 87-125.

Smith, L.L. Review of Progress in Sterol Oxidations 1987-1995. *Lipids* **1996**, *31,* 453-487.

Soupas, L.; L. Juntunen; A.-M. Lampi; K.-H. Liukkonen; K. Katina; K.-M. Oksman-Caldentey; and V. Piironen. Oxidative Stability of Phytosterols in Bread Baking. In *Strategies for Safe Foods: Analytical, Industrial and Legal Aspects: Challenges in Organization and Communication,* Proceedings of the Euro Food Chem XII Conference, Koninklijke Vlaamse Chemische Vereniging, Heverlee Belgium, 2003, pp. 317-320.

Soupas, L.; L. Juntunen; A.-M. Lampi; and V. Piironen. Effects of Sterol Structure, Temperature and Lipid Medium on Phytosterol Oxidation. *J. Agric. Food Chem.* **2004,** *52,* 6485-6491.

Soupas, L.; L. Huikko; A.-M. Lampi; and V. Piironen. Esterification Affects Phytosterol Oxidation. *Eur. J. Lipid Sci. Technol.* **2005,** *107,* 107-118.

Soupas, L.; L. Huikko; A.-M. Lampi; and V. Piironen. Oxidative Stability of Phytosterols in Some Food Applications. *Eur. Food Res. Technol.* **2006,** *222,* 266-273.

Soupas, L.; L. Huikko; A.-M. Lampi; and V. Piironen. Pan Frying May Induce Phytosterol Oxidation. *Food Chem.* **2007** *101,* 286-297.

Stanton, C.; and R. Devery. Formation and Content of Cholesterol Oxidation Products in Milk and Dairy Products. In *Cholesterol and Phytosterol Oxidation Products: Analysis, Occurrence, and Biological Effects,* Guardiola, F.; Dutta, P.C.; Codony, R.; and Savage, G.P. Eds.; AOCS Press: Champaign, Illinois, 2002, pp. 147-161.

Terao, J.; K. Sugano; and M. Matsushita. $Fe^{2+}$ and Ascorbic Acid Induced Oxidation of Cholesterol in Phosphatidylcholine Liposomes and Its Inhibition by α-Tocopherol, *J. Nutr. Sci. Vitaminol.* **1985,** *31,* 499-508.

Tsai, L.S.; and C.A. Hudson. Cholesterol Oxides in Commercial Dry Egg Products: Isolation and Identification. *J. Food Sci.* **1984,** *49,* 1245-1248.

Vardanyan, R.L.; R.L. Safiullin; and V.D. Komissarov. Absolute Values of Rate Constants of Disproportionation of Peroxyl Radicals Formed from Cholesterol. *Kinetic Katalysis* **1985,** *26,* 1140-1144.

Wu, G.-S.; R.A. Stein; and J.F. Mead. Autoxidation of Fatty Acid Monolayers Adsorbed on Silica Gel. III. Effects of Saturated Fatty Acids and Cholesterol. *Lipids* **1978,** *13,* 517-523.

Yan, P.S. and P.J. White. Cholesterol Oxidation in Heated Lard Enriched with Two Levels of Cholesterol. *J. Am. Oil Chem. Soc.* **1990** *67,* 927-931.

Yanishlieva, N.; and E. Marinova. Autoxidation of Sitosterol. I. Kinetic Studies on Free and Esterified Sitosterol. *Rivista Ital. Del Sostanzr Grasse* **1980,** *LVII,* 477-480.

Yanishlieva, N.; and E. Marinova. Effect of the Unsaturation of Lipid Media on the Autoxidation of Sitosterol. *Grasa y Aceites* **1986,** *37,* 343-347.

Yanishlieva, N.; E. Marinova; H. Schiller; and A. Seher. Comparison of Sitosterol Autoxidation in Free Form, as Fatty Acid Ester, and in Triacylglycerol Solution. Kinetics of the Process and Structure of the Products Formed. Proc. 16th ISF Congress, Fat Science, Budapest 1983, pp. 619-626.

# 6

# Tocopherol Concentrations and Antioxidant Efficacy

Afaf Kamal-Eldin[a], Hyun Jung Kim[b], Levon Tavadyan[c], and David B. Min[b]
[a]Department of Food Science, Swedish University of Agricultural Sciences, P.O. Box 7051, 75007, Uppsala, Sweden; [b]Department of Food Science and Technology, The Ohio State University, 2015 Fyffe Road, Columbus, OH 43210; [c]A. B. Nalbandyan Institute of Chemical Physics of the National Academy of Sciences of Armenia, 5/2 Sevak str. 0014 Yerevan, Armenia

## Introduction

Lipid oxidation is one of the important reactions contributing to aging in foods and biological systems, including the human body. The reaction of lipids with oxygen is thermodynamically driven but its speed is stimulated by a number of agents including temperature and other pro-oxidative catalysts (mainly trace metal ions), inhibited by antioxidants and antioxidant synergists, and variably modulated by the other chemical species in the reaction environment. The reaction between free radicals and lipids contributes to the deterioration of flavor and the presence of potential toxic products in food systems, and to membrane fragility and dysfunction in biological tissues.

Tocopherols are the major lipid-soluble antioxidants in nature (Kamal-Eldin and Appelqvist 1996; Choe and Min 2006). The inhibitory activity of tocopherols, particularly α-tocopherol, against lipid oxidation is usually associated with an optimal concentration of antioxidants after which the stabilization is lowered (Cillard et al., 1980; Blekas et al., 1945; Burlakova et al., 1998). This phenomenon was observed in triacylglycerols purified from vegetable oils, for example corn oil (Huang et al., 1995), butter oil (Lampi and Piironen. 1998), sunflower oil (Fuster et al., 1998), rapeseed oil (Lampi et al., 1999), olive oil (Deiana et al., 2002), and soybean oil (Yanishlieva et al., 2002; Kim et al., 2007). Under these conditions, the transition from the induction period to the exponential oxidation phase is not always a consequence of complete tocopherol consumption or the ratio of tocopherols to the amount of hydroperoxides present as suggested by Witting (1969). This effect was considered a "pro-oxidant effect," but this qualification is not correct considering the great stability of tocopherol-supplemented samples compared to controls void of the antioxidant. Loss of antioxidant efficacy at post-optimal concentrations was suggested to describe this phenomenon (Fuster et al., 1998).

The same phenomenon was observed in the oxidation of low-density lipoprotein particles (Bowry et al., 1992; Bowry and Stocker, 1993; Bowry and Ingold, 1999) and it was proposed that the α-tocopheroxyl radical, which is formed by the one-electron oxidation of α-tocopherol, is responsible for the observed "pro-oxidant" effect. However, the fact that other tocopherols are less pro-oxidative than α-tocopherol suggests that this is not the only mechanism involved. Nevertheless, it is generally agreed that tocopherol concentration is important whether it can be an antioxidant or a "pro-oxidant." It could be assumed that more intermediate radicals are formed during the oxidation and storage of lipids containing tocopherols at concentrations higher than certain optima. These intermediate radicals, including the tocopheroxyl radicals, can initiate the processes of lipid oxidation. In this chapter, we review the different reactions in which the tocopherol molecules and radicals are active and the possible anti- and pro-oxidant contributions of these reactions to the overall lipid oxidation reaction.

## The Antioxidant Mechanisms of α-Tocopherol

Lipid oxidation is a typical free-radical chain reaction of unsaturated fatty acids. The high reactivity of *bis*-allylic methylene groups in unsaturated fatty acids makes these molecules the primary targets for free radical reactions. The chain reaction includes initiation, propagation, degenerate branching, and termination steps.

| | | |
|---|---|---|
| LH + X• → L• + XH | (initiation) | [1] |
| L• + $O_2$ → LOO• | (propagation) | [2] |
| LOO• + LH → L• + LOOH | | [3] |
| LOOH + LH → LOO• + L• + $H_2O$ | (degenerate branching) | [4] |
| LOO• + LOO• → non-radical products | (termination) | [5] |
| LOO• + L• → non-radical products | | [6] |
| L• + L• → non-radical products | | [7] |

Where LH represents a polyunsaturated fatty acid moiety, such as linoleate or linolenate; X• is an initiating free radical; LOO• is a lipid peroxyl radical; L• is a C-centered lipid radical, LOOH is a lipid hydroperoxide molecule. Other free radicals and reactions are also involved in this extremely complex reaction (Tables 6.1 and 6.2).

The propagation step leads to the formation of hydroperoxides and is responsible for the autoinitiation nature of the reaction. Tocopherols protect polyunsaturated fatty acids from oxidative degradation by donating their hydrogen to chain carrier lipid peroxyl radicals and preventing further chain radical reaction. Tocopherols (TH) can transfer a hydrogen atom from their 6-phenoxyl group to lipid peroxyl radicals. Tocopherols, with oxidation potentials of 275-400 mV, easily donate their phenolic hydrogen to peroxyl radical (LOO•) with a reduction potential of 1000 mV and produce lipid hydroperoxide (LOOH) and tocopheroxyl radical (T•). At the same time the reaction exothermicity promotes its flow (Denisov and Denisova, 2000).

$\Delta H° = BDE(T–H) – BDE(LOO–H) = 331 – 372 = –41 kJ/mol$

LOO• + TH → T• + LOOH [8]

Table 6.1. Standard Reduction Potentials for Free Radicals of Importance in the Antioxidant Effect of Tocopherols[a]

| Compounds | Half-Cell | Standard Reduction Potential (mV) |
|---|---|---|
| •OH | $H^+/H_2O$ | 2310 |
| LO•[b] | $H^+/ROH$ | 1600 |
| LOO•[b] | $H^+/ROOH$ | 1000 |
| L• | $H^+/RH$ | 600 |
| α-Tocopheroxyl• | $H^+/α$-Tocopherol | 273 |
| β-Tocopheroxyl• | $H^+/β$-Tocopherol | 343 |
| γ-Tocopheroxyl• | $H^+/γ$-Tocopherol | 348 |
| δ-Tocopheroxyl• | $H^+/δ$-Tocopherol | 405 |

[a] Adapted from Choe and Min (2005) and Kamal-Eldin and Appelqvist (1996).
[b] LO• and LOO• forms for tocopherol are tocopheroxyl oxy radical and tocopheroxyl peroxy radical, respectively, and their structures are shown in Figures 6.1-6.4.

Tocopheroxyl radicals (T•) are resonance-stabilized structures that are more stable than lipid peroxyl radicals (LOO•) and can further scavenge another peroxyl radical by addition

$$T• + LOO• \rightarrow \text{non-radical products} \quad [9]$$

Or alternatively form non-radical products mainly by disproportionation or by combination

$$T• + T• \rightarrow \text{non-radical products} \quad [10]$$

The rate constants of the reaction of α-tocopherol with lipid peroxyl radical (Eq. 8) is $10^6$ to $10^7$ $M^{-1}$ $sec^{-1}$ (Niki et al., 1984; Denisov and Denisova, 2000; Choe and Min, 2005) and is $10^4$ to $10^5$ times higher than that of unsaturated lipid with lipid peroxyl radical (Niki et al. 1984; Denisov and Denisova, 2000; Naumov and Vasil'ev, 2003). The antioxidant activity of tocopherols was found to depend on temperature, pH, the degree and number of unsaturated fatty acids, the availability of oxygen and transition metal ions (Roginsky, 1990; Verleyen, et al., 2002; Kamal-Eldin et al., 2002; Reische et al., 2002).

## The Pro-Oxidant Mechanisms of α-Tocopherol

To understand the nature of the pro-oxidative side reactions affecting the antioxidant potency of α-tocopherol, one needs to examine the effect of the initial α-tocopherol concentration on the length of the induction period as well as on the rate of peroxidation during the induction period. Yanishlieva and co-workers used two terms to

**Table 6.2.** Rates Constants of Important Autoxidation Reactions Involving Unsaturated Fatty Acids (LH), $\alpha$-Tocopherol (TH), and Their Oxidation Products

| N | Reaction Steps | Reaction Rate Constant[a] |
|---|---|---|
| 1 | $LH + O_2 \rightarrow L\bullet + HOO\bullet$ | $2.24 \times 10^{-10}$ |
| 2 | $L\bullet + O_2 \rightarrow LOO\bullet$ | $8.75 \times 10^8$ |
| 3 | $LOO\bullet + LH \rightarrow L\bullet + LOOH$ | 76.3 |
| 4 | $LO\bullet + LH \rightarrow L\bullet + LOH$ | $1.26 \times 10^7$ |
| 5 | $HOO\bullet + LH \rightarrow L\bullet + H_2O_2$ | 228 |
| 6 | $LOO\bullet \rightarrow NRP + HOO\bullet$ | $5.49 \times 10^{-2}$ |
| 7 | $LOO\bullet + LOO\bullet \rightarrow L(-H)=O + LOH + O_2$ | $2.01 \times 10^7$ |
| 8 | $LOO\bullet + HOO\bullet \rightarrow L(-H)=O + H_2O + O_2$ | $10^8$ |
| 9 | $LOOH \rightarrow LO\bullet + \bullet OH$ | $1.23 \times 10^{-8}$ |
| 10 | $LOO\bullet + TH \rightarrow T\bullet + LOOH$ | $1.85 \times 10^6$ |
| 11 | $HOO\bullet + TH \rightarrow T\bullet + H_2O_2$ | $5.55 \times 10^6$ |
| 12 | $T\bullet + LH \rightarrow TH + L\bullet$ | $3.97 \times 10^{-2}$ |
| 13 | $T\bullet + LOOH \rightarrow TH + LOO\bullet$ | 10.6 |
| 14 | $T\bullet + H_2O_2 \rightarrow TH + HOO\bullet$ | 10.6 |
| 15 | $T\bullet + O_2 \rightarrow TMQ + HOO\bullet$ | $1.16 \times 10^{-2}$ |
| 16 | $T\bullet + T\bullet \rightarrow TMQ + TH$ | $1.28 \times 10^3$ |
| 17 | $T\bullet + HOO\bullet \rightarrow TH + O_2$ | $0.75 \times 10^8$ |
| 18 | $LOO\bullet + TMQ \rightarrow NRP$ | $8.35 \times 10^2$ |
| 19 | $LOOH + TH \rightarrow LO\bullet + T\bullet + H_2O$ | $6.6 \times 10^{-6}$ |
| 20 | $HOOH + TH \rightarrow HO\bullet + T\bullet + H_2O$ | $6.6 \times 10^{-6}$ |
| 21 | $\bullet OH + LH \rightarrow L\bullet + H_2O$ | $10^{10}$ |
| 22 | $T\bullet + LOO\bullet \rightarrow o\text{-}T\text{-}OOL$ | $1.23 \times 10^8$ |
| 23 | $T\bullet + LOO\bullet \rightarrow p\text{-}T\text{-}OOL$ | $0.27 \times 10^8$ |
| 24 | $o\text{-}T\text{-}OOL \rightarrow LO\bullet + o\text{-}TE\bullet$ | $5.54 \times 10^{-8}$ |
| 25 | $o\text{-}TE\bullet + LOO\bullet \rightarrow o\text{-}TE\text{-}OOL$ | $1.5 \times 10^8$ |
| 26 | $o\text{-}TE\text{-}OOL \rightarrow o\text{-}TE\text{-}O\bullet + LO\bullet$ | $6.34 \times 10^{-10}$ |
| 27 | $o\text{-}TE\text{-}O\bullet + LH \rightarrow o\text{-}TEQ + L\bullet$ | $1.26 \times 10^7$ |
| 28 | $p\text{-}T\text{-}OOL \rightarrow LO\bullet + TQ\text{-}O\bullet$ | $6.34 \times 10^{-10}$ |
| 29 | $TQ\text{-}O\bullet + LH \rightarrow TQ\text{-}OH + L\bullet$ | $1.26 \times 10^7$ |
| 30 | $T\bullet + HOO\bullet \rightarrow o\text{-}T\text{-}OOH$ | $2.7 \times 10^8$ |
| 31 | $T\bullet + HOO\bullet \rightarrow p\text{-}T\text{-}OOH$ | $0.6 \times 10^8$ |

[a] The rate constants of the reactions are given in M and sec at 40°C. The designation of some reaction species corresponds to those shown in Fig. 6.1. Source: Tavadyan et al., 2007

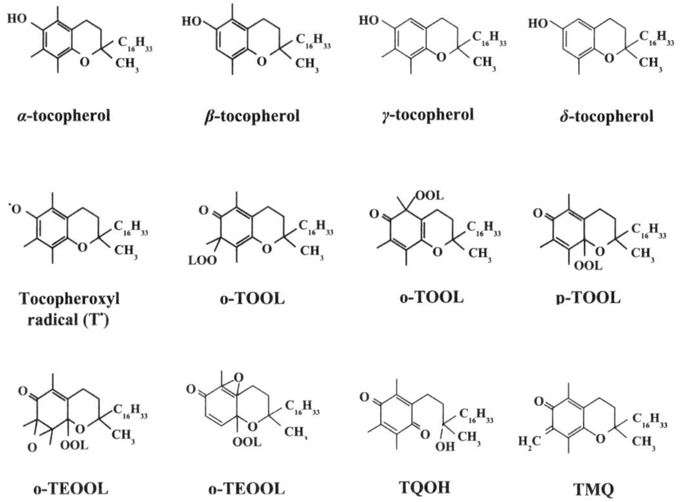

**Fig. 6.1.** Chemical structures of α-, β-,γ-, δ- tocopherols and the main molecular products of α- tocopherol oxidation according the scheme presented in Table 6.2.

describe the effects of antioxidant inhibitors on lipid peroxidation

1. Effectiveness, representing the ability of the antioxidant to block the radical chain reactions by interaction with peroxyl radicals, which is responsible for the duration of the induction period, and

2. Strength, expressing the ability of the inhibitor moieties to participate in side reactions, which may lead to a change in oxidation rate during the induction period and the magnitude of the reaction induction period (Yanishlieva and Marinova, 1992; Kasaikina et al., 1999; Yanishlieva et al., 2002).

The Yanishlieva-Marinova model (1992) uses the stabilization factor (F), which is the ratio between the induction period in the presence of inhibitor and its absence to measure effectiveness and uses the oxidation rate ratio (ORR), the ratio of the rate of lipid oxidation reactions in the presence of inhibitor and its absence, as a measure of the strength. A combining factor, antioxidant activity (A), is calculated as the ratio F/ORR. If F is not a linear function of the concentration of the antioxidant, it indicates that the antioxidant molecule participates in reactions other than the main reaction of chain interception. When the rate of inhibited reaction is equal to the reciprocal of the initial concentration of antioxidant, then the antioxidant radical does not participate in side reactions. Using these principles, it was found that the molecule of α-tocopherol readily participates in one side reaction with hydroperoxides leading to branching to produce two radicals, while the tocopheroxyl radical(s) seemed to participate in a number of pro-oxidant side reactions (Yanishlieva and Marinova, 1992; Yanishlieva et al., 2002).

Results of the kinetic numerical analysis (Tavadyan et al., 2007) of the reaction mechanisms of the methyl linoleate oxidation inhibited by α-tocopherol (Table 6.2) showed that the length of the induction period initially rises with increased initial tocopherol concentration followed by an arena of practical independence and ends with a decrease. In accordance with this, the rate of peroxidation during the induction period starts with a fast drop, remains stable up to about 20 mM α-tocopherol in methyl linoleate and then starts to increase proportionally to the initial tocopherol concentration. Such complex dependencies of both the length of the induction period and the rate of the inhibited oxidation on the initial tocopherol concentration are connected to the net product of both antioxidant and pro-oxidant activities of α-tocopherol (Kamal-Eldin and Appelqvist 1996; Mukai et al. 2005).

The kinetic analysis by Hamiltonian systematization method (Tavadyan and Martoyan, 2005) of the pro-oxidant mechanism of α-tocopherol in methyl linoleate (Tavadyan et al., 2007) revealed that three types of side reaction might significantly contribute to the manifestation of pro-oxidant effects of α-tocopherol,

1. Chain transfer reactions by the tocopheroxyl radical leading to the abstraction of a hydrogen atom from bis-allylic methylene groups of fatty acid molecules and/or from hydroperoxide,

2. An autoinitiation reaction between α-tocopherol with hydroperoxide leading to alkoxyl radical formation, and

3. Reactions of the oxidation products of tocopherol

### The Chain Transfer Reactions by α-Tocopheroxyl Radical

The "pro-oxidant effect" of tocopherols in bulk lipid solutions was first attributed to its tocopheroxyl radical (T•). When this species is present at high concentration, a number of undesirable side reactions that can initiate or enhance the rate of lipid oxidation are possible. It was first proposed by Mukai et al. (1993) that T• might abstract a hydrogen atom from the bis-allylic position of an unsaturated fatty acid moiety or the peroxide hydrogen from lipid hydroperoxides,

T• + LH → TH + L•   [11]
T• + LOOH → TH + LOO•   [12]

These chain transfer steps were shown by kinetic modeling to have the most essential pro-oxidant activity during the induction period when the formation of peroxyl radicals is minimized by the presence of α-tocopherol. Thus, the ratio [T•]/[LOO•] increases with increasing initial tocopherol concentration leading to the increase in the

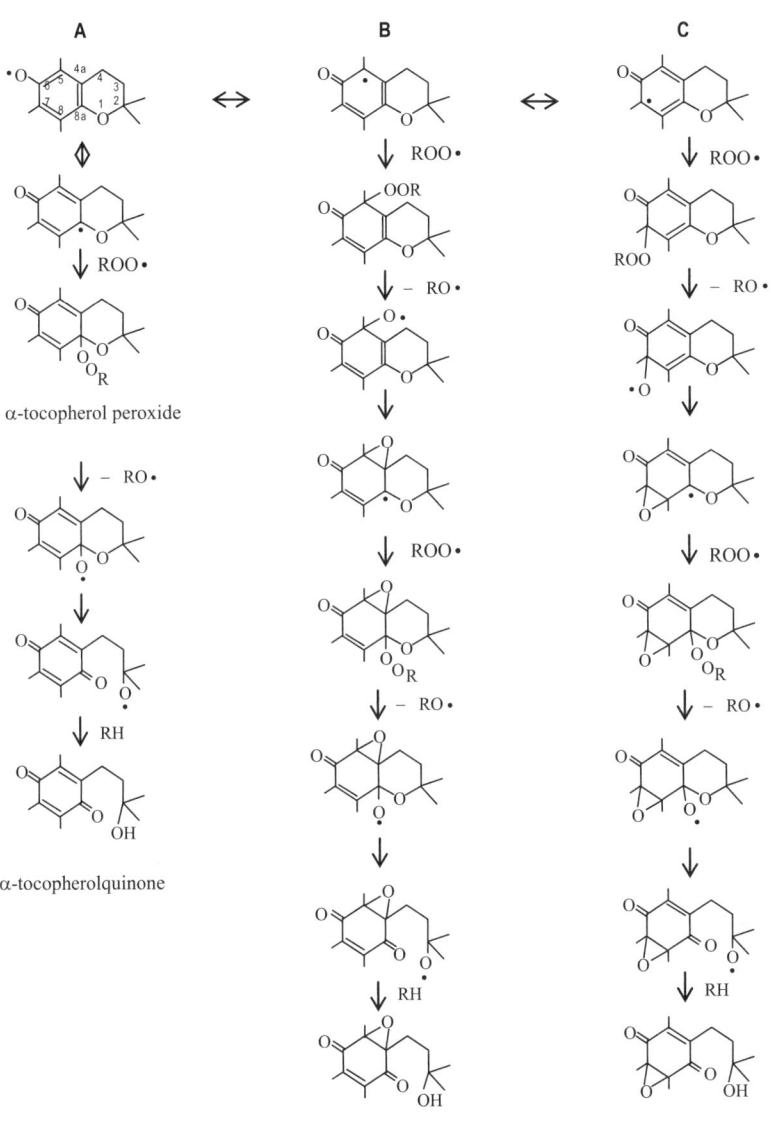

**Fig. 6.2.** Possible formation of α-tocopherolquinone (a), 4α,5-epoxy-α-tocopherolquinone (b), and 7,8-epoxy-α-tocopherolquinone (c) from α-tocopheroxyl radical with lipid peroxyl radical.

reaction rate of chain transfer compared with the rate of the main reaction of chain termination (eq 9). This leads to the non-linear dependence of the length of induction period on tocopherol concentration. Under these conditions, oxidation proceeds with T• acting as the main chain-carrier (Tavadyan et al., 2007). This phenomenon was named tocopherol-mediated peroxidation (TMP) and was used to describe the loss of antioxidant activity of α-tocopherol in the oxidation of low-density lipoprotein (Bowry et al., 1992; Bowry and Stocker, 1993; Bowry and Ingold, 1999).

## Autoinitiation Due to the Decomposition of Hydroperoxides by α-Tocopherol

One of the main reasons for the complex non-linear dependence of the length of the induction period and the rate of oxidation during the induction period on the initial concentration of α-tocopherol is the autoinitiation reaction involving the molecules of α-tocopherol and hydroperoxides (Denisov and Denisova, 2000; Denisov and Azatyan, 2000)

$$TH + LOOH \rightarrow T\bullet + LO\bullet + H_2O$$

[13]

This reaction is as important for the manifestation of loss of α-tocopherol antioxidant efficacy as the TMP reaction discussed previously. This is the reaction that also plays a substantial role in the nonlinear behavior and the extreme dependence of both the induction period and the rate of inhibited reaction on the initial α-tocopherol concentration. The rates of participation of different tocopherols (α-, β-, γ-, and δ-) in this reaction are inversely related to the bond dissociation enthalpies of their phenolic hydroxyl groups.

## Pro-Oxidant Effects Due to the Oxidation Products of Tocopherol

Tocopherols are degraded in the presence of oxygen and free radicals and produce oxidized products resulting in the loss of antioxidant activity (Jung and Min, 1990; Evans et al., 2002). The oxidized tocopherol products, such as peroxide compounds, act as pro-oxidants in lipids. The addition of oxidized α-, γ-, and δ-tocopherols to soybean oil lowered the oxidative stability of soybean oil (Jung and Min, 1992). The more oxidized tocopherols in lipids, the more oxidation of lipids occurs (Choe and Min, 2006). The oxidation of tocopherols could be induced by strong oxidizing agents, such as chromic acid, nitric acid, and ferric chloride, to produce lactones, quinines, and many degradation products (Kamal-Eldin and Appelqvist, 1996). α-Tocopherolquinone, α-tocopherolhydroquinone, 4a,5-epoxy-α-tocopherolquinone, and 7,8-epoxy-α-tocopherolquinone have been reported as the oxidation products of α-tocopherol in beef and bovine muscle microsomes (Faustman et al., 1999; Liebler et al., 1996), in triolein (Verleyen et al., 2002), triolein and tripalmitin mixtures (Verleyen et al., 2002), and in fish muscle (Pazos et al., 2005). During the formation of α-tocopherolquinone, α-tocopherolhydroquinone, 4a,5-epoxy-α-tocopherolquinone, and 7,8-epoxy-α-tocopherolquinone from α-tocopherol oxidation, many intermediate radical species could be produced. Rietjens et al. (2002) suggested that the increased levels of oxidized α-tocopherol could result in increased levels of α-tocopheroxyl radicals, which can initiate lipid oxidation by themselves.

The formation of α-tocopherolquinone, α-tocopherolhydroquinone, epoxy-α-tocopherolquinone, and α-tocopherol hydroperoxide during the oxidation of α-tocopherol is known (Faustman et al.. 1999; Verleyen et al.. 2002a,b; Pazos et al., 2005). The mechanisms for the formation of these oxidation products are shown in Fig. 6.2. The formations of α-tocopherol peroxide, α-tocopherolquinone, 4a,5-epoxy-α-tocopherolquinone, and 7,8-epoxy-α-tocopherolquinone as a result of the combination of α-tocopheroxy radical and lipid peroxy radical (ROO•) are also shown in Fig. 6.2. Figure 6.2a shows the formation of α-tocopherolquinone from α-tocopheroxyl radical. The α-tocopheroxyl radical rearranges by resonance to form the 8a-carbon centered α-tocopheroxyl radical. This radical reacts with lipid peroxy radical (ROO•) to form α-tocopherol lipid peroxide at the rate constant of $1.5 \times 10^8$ (Table 6.2). The α-tocopherol lipid peroxide can produce α-tocopherol oxy radical at the 8a carbon and an alkoxyl radical (RO•) by cleavage of the peroxide bond. The α-tocopherol oxy radical, at the 8a position, forms α-tocopherolquinone oxy radical by cleavage of the bond between 1 and 8a position leading to the formation of α-tocopherolquinone (Fig. 6.2a).

Figure 6.2b shows the formation of 4a,5-epoxy-α-tocopherolquinone from the carbon-centered α-tocopheryl radical at the 5 position. The 5-carbon-centered α-tocopheryl radical reacts with lipid peroxyl radical (ROO•) to form α-tocopherol peroxide. The breakdown of α-tocopherol peroxide forms α-tocopherol oxy radical at the carbon 5 and alkoxyl radical (RO•). The α-tocopherol oxy radical is cyclized to form 4a,5-epoxy-α-tocopheryl radical at the 8a position. The 4a,5-epoxy-α-tocopheryl radical can react with lipid peroxyl radical (ROO•) to form 4a,5-epoxy-α-tocopherol peroxide. The 4a,5-epoxy-α-tocopherol peroxide produces 4a,5-epoxy-α-tocopherol oxy radical at the carbon 8a and alkoxyl radical (RO•). The 4a,5-epoxy-α-tocopherol oxy radical at the carbon 8a forms 4a, 5-epoxy-α-tocopherolquinone oxy radical at the carbon 2 and then produces 4a,5-epoxy-α-tocopherolquinone by abstracting a hydrogen from unsaturated lipids (Fig. 6.2b). Similar to this mechanism, Fig. 6.2c shows the formation of 7,8-epoxy-α-tocopherolquinone from the carbon-centered α-tocopheryl radical at the carbon 7. The tocopheroxyl radical might re-arrange to give a carbon-centered radical (Fig. 6.3) but this transformation, which leads to the formation of tocopherol hydroperoxide, is insignificant and was shown to lack effect on the activity of α-tocopherol (Doba et al., 1984).

Figure 6.4 shows that singlet oxygen might also be formed from the carbon-centered α-tocopheroxyl radicals if they add oxygen and form α-tocopherol peroxy radicals. Two molecules of α-tocopherol peroxy radical form dimerized α-tocopherol peroxide and singlet oxygen at the rate constant of $10^5$ $M^{-1}sec^{-1}$ (Barclay et al., 1989; Min and Boff, 2002). Singlet oxygen is a very reactive oxygen species and a strong pro-oxidant (Bradley and Min, 1992; Min and Boff, 2002). Its reactions with lipids

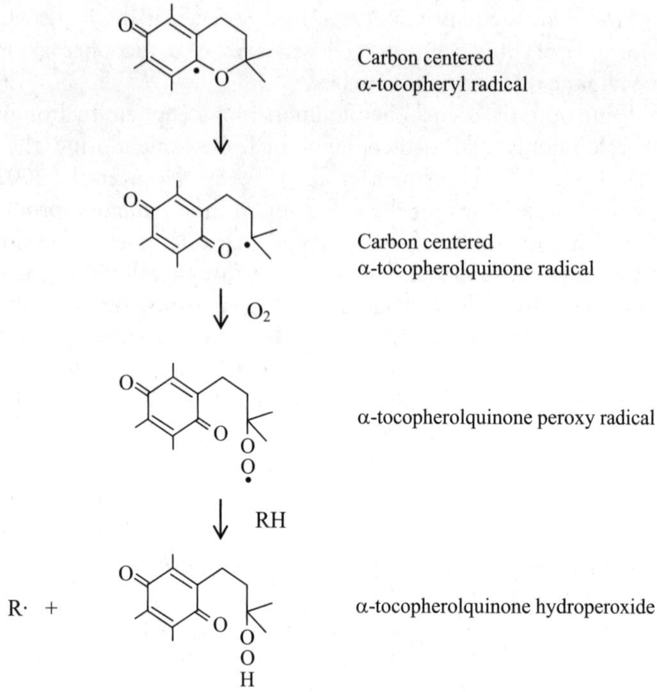

**Fig. 6.3.** Possible formation of α-tocopherolquinone hydroperoxide from the combination of α-tocopheryl radical with oxygen.

are very fast and occur even at very low temperature (Min and Boff, 2002). It was recently shown that when α-tocopherol is excited, it can sensitize the formation of singlet oxygen (Dad et al., 2006).

The tocopheroloxyl radical, α-tocopherolquinone oxy radical, α-tocopherolquinone peroxyl radical, alkoxyl radical (RO•), hydroxyl radical (•OH), and singlet oxygen ($^1O_2$ ($^1\Delta g$)) formed from the oxidation of tocopherol can all act as pro-oxidants (Table 6.1). Oxidized α-tocopherol compounds have polar hydroxyl and nonpolar hydrocarbon groups. Yoon et al. (1988) and Mistry and Min (1988a, 1988b) reported that thermally oxidized lipid compounds with polar hydroxyl and nonpolar hydrocarbons in the same molecule were pro-oxidants in soybean oil during storage. They reported that the oxidized lipids with hydroxyl and/or carbonyl groups were less soluble in the soybean oil and moved to the surface of the oil. The oxidized oils having polar and nonpolar groups decreased the surface tension between air and oil and increased the transportation of oxygen from air to oil to accelerate the oxidation of oil (Yoon et al., 1988; Mistry and Min, 1988a,b). The oxidized α-tocopherol compounds with the polar groups and nonpolar hydrocarbons in the same molecule reduce the surface

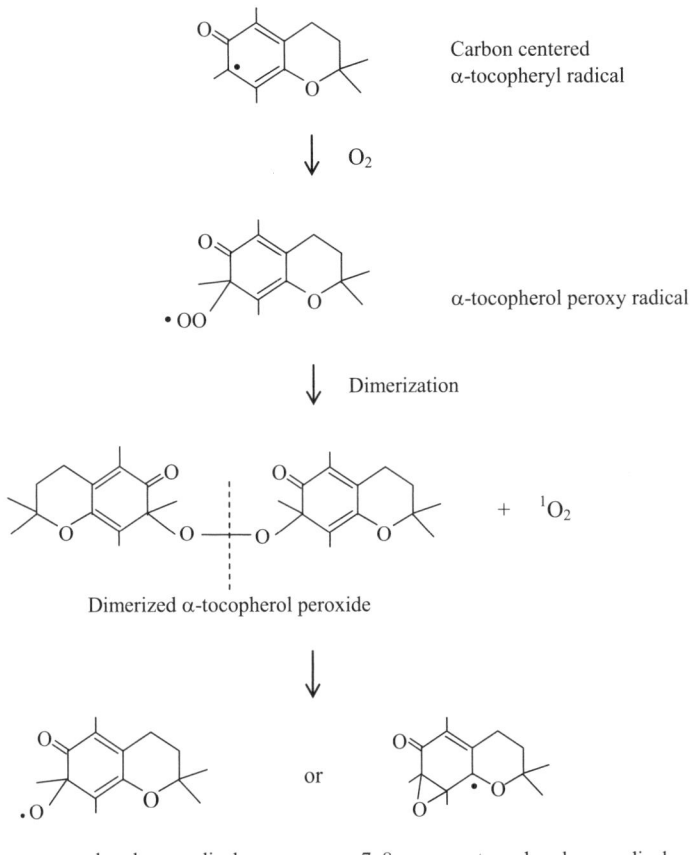

**Fig. 6.4.** Possible formation of singlet oxygen and a dimeric α-tocopherol peroxide, which dissociates to form α-tocopheryl alkoxy radical. The formation of singlet oxygen is determined by the rule of spin conservation.

tension between headspace air and oil and accelerate the oxidation of oil.

The pro-oxidant mechanisms of oxidized α-tocopherol may be mainly due to α-tocopherol peroxyl radical, α-tocopherol oxy radical, α-tocopherolquinone oxy radical, and hydroxyl radical which have high reduction potential and singlet oxygen formed from the oxidation of tocopherols during storage in foods. The oxidized α-tocopherol compounds with polar hydroxyl and nonpolar hydrocarbons in the same molecule may contribute to the oxidation of oil by reducing the surface tension of the oil and increasing the diffusion of oxygen from air to the oil (Denisov and Khudyakov, 1987).

Moreover, it is possible that two tocopherol peroxyl radicals would combine to form a tetraoxy intermediate that would dissociate by the Russel mechanism to yield a molecule of singlet oxygen (Fig. 6.4). In fact, singlet oxygen was detected by 1270 nm luminescence after pulsed laser excitation (308 nm) of vitamin E and an its analog, 2,2,5,7,8-pentamethyl-6-hydroxy-chroman (PMHC) (Dad et al., 2006).

## Effect of Temperature on Tocopherol Antioxidant Activity

The effect of temperature on the antioxidant activity of α- and δ-tocopherol was compared in pork lard using the Oxipres apparatus in the temperature range from 80 to 150°C (Réblová, 2006). The antioxidant activity of both tocopherols decreased with increasing temperature although at different rates. At 100 ppm, δ-tocopherol had about double the antioxidant activity of α-tocopherol at 80°C but its activity was the same at 130°C and none of the tocopherols was effective at 150°C. It was also found that γ-tocopherol is a more efficient antioxidant than α-tocopherol at 180°C in purified triacylglycerols from high oleic sunflower oil (Lampi and Kamal-Eldin, 1998). Heating at high temperatures causes rapid tocopherol degradation. Interestingly, it was previously shown that the rate of tocopherol degradation in triacylglycerols purified from nine vegetable oils at higher temperatures (180 and 240°C) decreased as their degree of unsaturation increased (coconut, tallow, palm, olive, high oleic sunflower, maize, sunflower, soybean, and flaxseed oil) (Verleyen et al., 2002c). Even if the overall efficacy of tocopherols decrease at high temperatures, the negative effect of high tocopherol concentrations becomes insignificant when the rate of initiation (Ri) is high. Only at low Ri, does the contribution of side reactions leading to loss of antioxidant efficacy become most important. This is especially true for the reaction between tocopherols and hydroperoxides (eq 13). Other reactions, for example, polymerization, become important with high-temperature heating. Mass balance (Verleyen et al., 2002b). Future research needs to identify the different oxidation/polymerization products of tocopherols at different temperatures.

## References

Barclay, L.R.C.; K.A. Baskin; S.J. Locke; and M.R. Vinquist. Absolute Rate Constants for Lipid Peroxidation and Inhibition in Model Biomembranes. *Can. J. Chem.* **1989**, *68*, 2258-2269.

Blekas G.; M. Tsimidou; and D. Boskou. Contribution of α-Tocopherol to Olive Oil Stability. *Food Chem.* **1995**, *52*, 289-294.

Bowry, V.W.; and K.U. Ingold. The Unexpected Role of Vitamin E (α-Tocopherol) in the Peroxidation of Human Low–Density Lipoprotein. *Acc. Chem. Res.* **1999**, *32*, 27–34.

Bowry, V.W.; and R. Stocker. Tocopherol-Mediated Peroxidation: The Prooxidant Effect of Vitamin E on the Radical-Initiated Oxidation of Human Low Density Lipoprotein. *J. Am. Chem. Soc.* **1993**, *115*, 6029-6044.

Bowry, V.W.; K.U. Ingold; and R. Stocker. Vitamin E in Human Low–Density Lipoprotein. When and How this Antioxidant Becomes a Pro–Oxidant. *Biochem J.* **1992**, *288*, 341–344.

Bradley, D.G.; and D.B. Min. Singlet Oxygen Oxidation of Foods. *Crit. Rev. Food Sci. Nutri.*

1992, *31*, 211-236.

Burlakova, E.B.; S.A. Karshakov; and N.G. Khrapova. The Role of Tocopherols in Biomembranes' Lipids Peroxidation. *Biol. Membranes* **1998**, *15*, 137–167 (*in Russian*).

Choe, E.; and D.B. Min. Chemistry and Reactions of Reactive Oxygen Species in Foods. *J. Food Sci.* **2005**, *70*, R142-159.

Choe, E.; and D.B. Min. Mechanisms and Factors for Edible Oil Oxidation. *Comp. Rev. Food Sci. Food Safety* **2006**, *5*, 169-186.

Cillard, J.; P. Cillard; and M. Cornier. Effect of Experimental Factors on the Prooxidant Behavior of α-Tocopherol. *J. Am. Oil Chem. Soc.* **1980**, *57*, 255–261.

Dad, S.; R.H. Bisby; I.P. Clark; and A.W. Parker. Formation of Singlet Oxygen from Solutions of Vitamin E. *Free Rad. Res.* **2006**, *40*, 333-338.

Deiana, M.; A. Rosa; C.F.Q. Cao; F.M. Pirisi; G. Bandino; and M.A. Dessi. Novel Approach to Study Oxidative Stability of Extra Virgin Olive Oils: Importance of Alpha-Tocopherol Concentration. *J. Agric. Food Chem.* **2002**, *50*, 4342-4346.

Denisov, E.T.; and V.V. Azatyan. *Chain Reactions Inhibition.* Taylor and Francis: London, 2000.

Denisov, E.T.; and T.G. Denisova. *Handbook of Antioxidants: Bond Dissociation Energies, Rate Constants, Activation Energies and Enthalpies of Reactions.* 2$^{nd}$ edn.; CRC Press: Boca Ration, 2000.

Denisov, E.; and I. Khudyakov. Mechanism of Action and Reactivities of the Free Radicals of Inhibitors. *Chem. Rev.* **1987**, *87*, 1313-1357.

Doba, T.; G.W. Burton; K.U. Ingold; and M. Matsuo. α-Tocopheroxyl Delay: Luck of Effect of Oxygen. *J. Chem. Soc. Chem. Commun.* **1984**, 461–462.

Evans, J.C.; D.R. Kodali; and P.B. Addis. Optimal Tocopherol Concentrations to Inhibit Soybean Oil Oxidation. *J. Am. Oil Chem. Soc.* **2002**, *79*, 47-51.

Faustman, C., D.C. Liebler; and J.A. Burr. α-Tocopherol Oxidation in Beef and in Bovine Muscle Microsomes. *J. Agric. Food Chem.* **1999**, *47*, 1396-1399.

Fuster, M.D.; A.-M. Lampi; A. Hopia; and A. Kamal-Eldin. Effects of α- and γ-Tocopherols on the Autoxidation of Purified Sunflower Triacylglycerols. *Lipids* **1998**, *33*, 715-722.

Huang, S.-W.; E.N. Frankel; and J.B. German. Effects of Individual Tocopherols and Tocopherol Mixtures on the Oxidative Stability of Corn Oil Triglycerides. *J. Agric. Food Chem.* **1995**, *43*, 2345-2350.

Jung, M.Y.; and D.B. Min. Effects of α-, γ-and δ–Tocopherols on the Oxidative Stability of Purified Soybean Oil. *J. Food Sci.* **1990**, *55*, 1464-1465.

Jung, M.Y.; and D.B. Min. Effects of Oxidized α-, γ- and δ-Tocopherols on the Oxidative Stability of Purified Soybean Oil. *Food Chem.* **1992**, *45*, 183-187.

Kamal-Eldin, A.; and L. Appelqvist. The Chemistry and Antioxidant Properties of Tocopherols and Tocotrienols. *Lipids* **1996**, *31*, 671–701.

Kamal-Eldin, A.; M. Makinen; A. Lampi; and A. Hopia. A Multivariate Study of α-Tocopherol and Hydroperoxide Interaction during the Oxidation of Methyl Linoleate. *Eur. Food Res. Technol.* **2002**, *214*, 52–57.

Kasaikina, O.T.; V.D. Kortenska; and N.V. Yanishlieva. Effect of Chain Transfer and Recombination/Disproportionation of Inhibitor Radicals on Inhibited Oxidation of Lipids. *Russ. Chem. Bull.* **1999**, *48*, 1891-1896.

Kim, H.J.; H.O. Lee; and D.B. Min. Effects and Prooxidant Mechanisms of Oxidized Alpha-Tocopherol on the Oxidative Stability of Soybean Oil. *J. Food Sci.* **2007**, *72*, C223-C230.

Lampi, A.M.; and A. Kamal-Eldin. Effect of Alpha- and Gamma-Tocopherols on Thermal Polymerization of Purified High-Oleic Sunflower Triacylglycerols. *J. Am. Oil Chem. Soc.* **1998,** *75,* 1699-1703.

Lampi, A.-M.; and V. Piironen. α- and γ-Tocopherols as Efficient Antioxidants in Butter Oil Triacylglycerols. *Fett/Lipid* **1998,** *100,* 292-295.

Lampi, A.-M.; L. Katja; A. Kamal-Eldin; and P. Vieno. Antioxidant Activities of α- and γ-Tocopherols in the Oxidation of Rapeseed Oil Triacylglycerols. *J. Am. Oil Chem. Soc.* **1999,** *76,* 749-755.

Liebler, D.C.; J.A. Burr; L. Philips; and A.J.L. Ham. Gas Chromatography-Mass Spectrometry Analysis of Vitamin E and Its Oxidation Products. *Anal. Biochem.* **1996,** *236,* 27-34.

Min, D.B.; and J.M. Boff. Chemistry and Reaction of Singlet Oxygen in Foods. *Comprehensive Rev. Food Sci. Food Safety* **2002,** *1,* 58-72.

Mistry, B.S.; and D.B. Min. Prooxidant Effects of Monoglycerides and Diglycerides in Soybean Oil. *J. Food Sci.* **1988a,** *53,* 1896-1897.

Mistry, B.S.; and D.B. Min. Isolation and Identification of Minor Components and Their Effects on Flavor Stability of Soybean Oil. In *Frontiers of Flavor,* Elsevier Science: Holland, 1988b, pp. 499-519.

Mukai, K.; H. Morimoto; Y. Okauchi; and S. Nagaoka. Kinetic Study of Reactions between Tocopheroxyl Radicals and Fatty Acids. *Lipids* **1993,** *28,* 753-756.

Mukai, K.; S. Noborio; and S.I. Nagaoka. Why Is the Order Reversed? Peroxyl–Scavenging Activity and Fats and Oils Protecting Activity of Vitamin E. *Int. J. Chem. Kinetics* **2005,** *37,* 605–610.

Naumov, V.V.; and R.F. Vasil'ev. Antioxidant and Prooxidant Effects of Tocopherol. *Kinetics Catalysis* **2003,** *44,* 101-105.

Niki, E.; T. Satio; A. Kawakami; and Y. Kamiya. Inhibition of Oxidation of Methyl Linoleate in Solution by Vitamin E and Vitamin C. *J. Biol. Chem.* **1984,** *259,* 4177-4182.

Pazos, M.; L. Sanchez; and I. Medina. α-Tocopherol Oxidation in Fish Muscle during Chilling and Frozen Storage. *J. Agric. Food Chem.* **2005,** *53,* 4000-4005.

Réblová, Z. The Effect of Temperature on the Antioxidant Activity of Tocopherols. *Eur. J. Lipid Sci. Technol.* **2006,** *108,* 858-863.

Reische, D.W.; D.A. Lillard; R.R. Eitenmiller. Antioxidants. In: C. Akoh and D.B. Min (Eds.), *Food Lipids* (2nd ed.), pp. 489–516, New York: Marcel Dekker, 2002.

Rietjens, I.M.C.M.; M.G. Boersm; L. de Haan; B. Spenkelink; H.M. Awad; N.H.P. Cnubben; J.J. van Zanden; H. van der Woude; G.M. Alink; and J.H. Koeman. The Prooxidant Chemistry of the Natural Antioxidants Vitamin C, Vitamin E, Carotenoids and Flavonoids. *Environ. Toxicol. Pharmacol.* **2002,** *11,* 321-333.

Roginsky, V.A. Kinetics of Polyunsaturated Fat Acid Ethers Oxidation, Inhibited by Substituted Phenols. *Kinetics Catalysis* **1990,** *31,* 475–481.

Tavadyan, L.A.; and G.A. Nartoyan. Analysis of Kinetic Models of Chemical Reaction Systems. Value Approach, Gitutiun, Yerevan Armenia, (In *Russian*) **2005**.

Tavadyan, L.A.; A.A. Khachoyan; G.A. Martoyan; and A. Kamal-Eldin. Numerical Revelation of the Kinetic Significance of Individual Steps in the Reaction Mechanism of Methyl Linoleate Peroxidation Inhibited by α-Tocopherol. *Chem. Phys. Lipids* **2007,** *147,* 30-45.

Verleyen, T.; R. Verhe; A. Huyghebaert; K. Dewettinck; and W. De Grey. Identification of α-Tocopherol Oxidation Products in Triolein at Elevated Temperature. *J. Agric. Food Chem.* **2002a,** *49,* 1508-1511.

Verleyen, T.; A. Kamal-Eldin; C. Dobarganes; R. Verhe; R. Dewettinck; and A. Huyghebaert.

Modeling of α-Tocopherol Loss and Oxidation Products Formed during Thermoxidation in Triolein and Tripalmitin Mixture. *Lipids* **2002b,** *36,* 719-726.

Verleyen, T.; A. Kamal-Eldin; R. Mozuraityte; R. Verhe; K. Dewettinck; A. Huyghebaert; and W. De Greyt. Oxidation at Elevated Temperatures: Competition between Alpha-Tocopherol and Unsaturated Triacylglycerols. *Eur. J. Lipid Sci. Technol.* **2002c,** *104,* 228-233.

Witting, I.A. The Oxidation of α-Tocopherol during the Autoxidation of Ethyl Oleate, Linoleate, Linolenate and Arachidonate. *Arch. Biochem. Biophys.* **1969,** *129,* 142-151.

Yanishlieva, N.V.; and E.M. Marinova. Inhibited Oxidation of Lipids I. Complex Estimation and Comparison of the Antioxidative Properties of Some Natural and Synthetic Antioxidants. *Fat Sci. Technol.* **1992,** *94,* 374-379.

Yanishlieva, N.V.; A. Kamal-Eldin; E.M. Marinova; and A.G. Toneva. Kinetics of Antioxidant Action of Alpha- and Gamma-Tocopherols in Sunflower and Soybean Triacylglycerols. *Eur. J. Lipid Sci. Technol.* **2002,** *104,* 262-270.

Yoon, S.H.; M.Y. Jung; and D.B. Min. Effects of Thermally Oxidized Triglycerides on the Oxidative Stability of Soybean Oil. *J. Am. Oil Chem. Soc.* **1988,** *65,* 1652-1656.

# 7

# Carotenoids and Lipid Oxidation Reactions

Afaf Kamal-Eldin
Department of Food Science, Swedish University of Agricultural Sciences, P. O. Box 7051, 750 07 Uppsala, Sweden

## Introduction

Carotenoids are yellow-orange pigments that are widely distributed in the plant and animal kingdoms. They are common in leaves, flowers, and fruits of plants even if co-occurring chlorophyll and flavonoids often mask their colors. The main function of carotenoids in plants seems to be the protection of the photosynthetic apparatus from the harmful effects of chlorophyll-photosensitized oxidation. Carotenoids are also important colorants in nature, for example they are responsible for the colors of animals especially birds and fish (Goodwin, 1986). Marine algae contribute large amounts to the commercial production of carotenoids estimated as 108 tons per year with the main commercial carotenoids being fucoxanthin, axtaxanthin, lutein, violaxanthin, neoxanthin, lycopene, and α- and β- carotenes (Harborne and Baxter, 1993). Commercial carotenoids are used as food colorants and as precursors for vitamin A (Bauernfeind, 1981; Simpson, 1983).

During the past couple of decades, carotenoids have received considerable attention because of their anticipated potential to contribute to protect against degenerative diseases, such as atherosclerosis, some types of cancer, compromised immunity, and aging (Cutler, 1984; Prabhala et al., 1990, 1993; Mathews-Roth, 1991; Gaziano and Hennekens, 1993; Olson and Krinsky, 1995; Biyani and Sheorey, 1994; Gaziano et al., 1995; Nishino, 1995). The main mechanisms believed to be involved in the protective role of carotenoids include production of nutritional vitamin A retinoids, antioxidant effects, and signaling mechanisms. A great disappointment, however, was witnessed after results of the Finnish ATBC study showing that β-carotene supplementation led to 18% increased incidence of lung cancers and 8% increased overall mortality (ATBC, 1994). This study was one of the first to indicate the complexity of the health contribution of phytochemicals and its dependence of the test situation, level of different compounds, such as other antioxidants that may synergize the carotenoids or even protect them against oxidative degradation.

β-Carotene and other carotenoids are believed to play very important roles as antioxidative substances both in vivo and in vitro. They have shown antioxidant activities in homogeneous solutions, liposomes, microsomal membranes, and lipoproteins (Krinsky and Denke, 1982; Burton and Ingold, 1984; Vile and Winterbourn, 1988a,b; Terao, 1989; Jialal et al., 1991; Palozza and Krinsky, 1991, 1992a,b; Kennedy and Liebler, 1992; Lim et al., 1992; Palozza et al., 1992). Two mechanisms are believed to be involved in carotenoid antioxidant reactions, first quenching reactive singlet oxygen and excited photosensitizers, and second scavenging peroxyl and other reactive free radicals (*vide infra*). While the mechanisms of singlet oxygen quenching by carotenoids remain fairly well understood, those of their reaction with free radicals are highly controversial and subject to experimental set up. Burton and Ingold (1984) present an early thesis on this complexity. This chapter aims to review literature on the implications and roles of carotenoids in lipid oxidation reactions and to highlight areas in need of further research.

## Structure and Nomenclature of Carotenoids

As β-carotene is a highly visible pigment in carrot root, the name carotenoids was derived from the name of this plant (*Daucus carota*). The carotenoid structure is basically composed of a long chain of eight isoprenoid units (40 carbons) joined head-to-tail to give a conjugated system of alternate double bonds (Fig. 7.1). The conjugated double bond structure is the main chromophore responsible for their colors and oxidative properties.

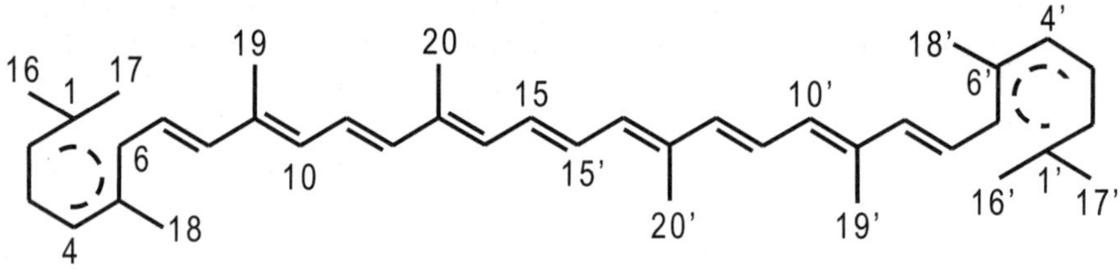

**Fig. 7.1.** Basic skeleton for carotenoid structures.

Different carotenoids are derived from the acyclic long-chain conjugated structure shown in Fig. 7.1 by hydrogenation, dehydrogenation, cyclization, oxidation, and combination of these reactions (IUPAC, 1975). Carotenoids can be classified into two major groups: (i) carotenes, which are simple non-polar unsaturated hydrocarbons based on the lycopene structure, and (ii) xanthophylls, which are more polar carotenoids with oxygen functions at one or both ends of the molecule. The structures, sources, and trivial and systematic names of 563 carotenoids have been published by Pfander (1987), and the rules for systematic nomenclature are published by the IUPAC (1975). Thus, the structures and semi-systematic names given in this chapter (Fig. 7.2) are only for a few carotenoids with significance to the discussion.

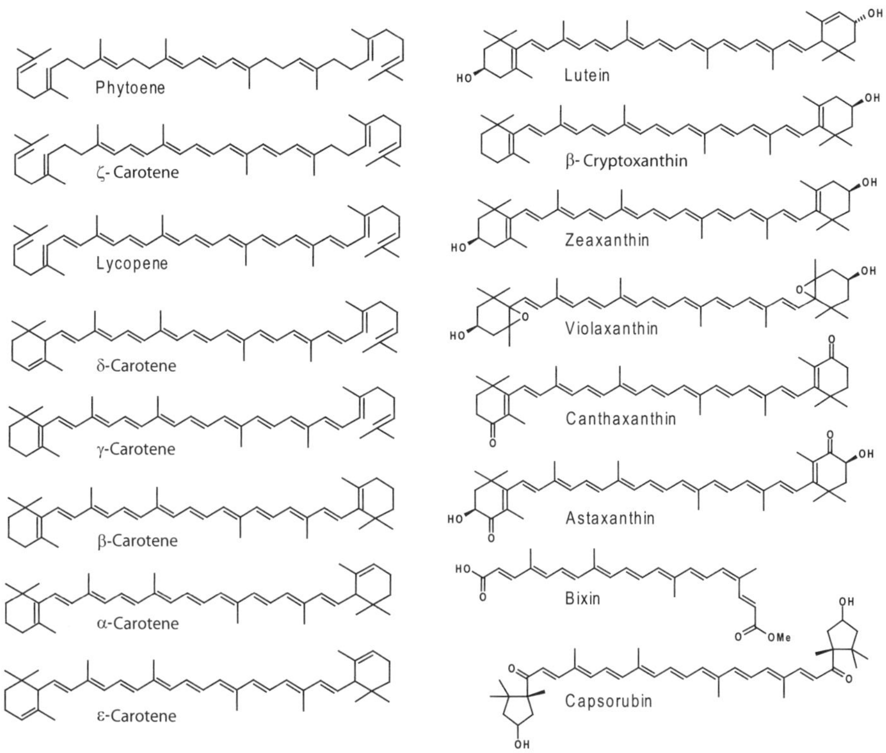

Fig. 7.2. Structures of selected carotenes (right hand) and xanthophylls (left hand).

## Structural Features Relevant to The Reactivity of Carotenoids

Four different structural features that influence the electron density along the conjugated double bond system of carotenoids dictate their properties and reactivity. These factors are explained in subsequent sections.

### The Degree of Conjugation

The high degree of double bond conjugation is the main chemical feature responsible for the distinct physical and chemical properties of carotenoids compared to other polyenes with isolated double bonds or to phenolic compounds with aromatic rings. Each carbon atom in the acyclic conjugated polyene chain has a trigonal arrangement, due to sp$^2$-hybridization, with one electron in the p$_z$ orbital. The p$_z$ orbitals of all conjugated carbons overlap forming a π-molecular orbital covering all conjugated carbons. As a result electron pairs are no longer confined to regions between nuclei, and bonds are called delocalized bonds. Thus, the degree of conjugation in the carotenoid molecules (9-13 conjugated double bonds) is perhaps the most important feature that determines their oxidative and antioxidative properties (Mathews-Roth et al., 1974; Foote, 1976; Terao, 1989; Woodall et al., 1995).

An important detail in this respect is the variation in electron density within the delocalized electronic system. This is illustrated by considering butadiene, $CH_2=CH-CH=CH_2$, the simplest conjugated hydrocarbon. According to molecular orbital theory, four $p_z$ atomic orbitals combine to form four π-molecular orbitals; two fully occupied bonding π-orbitals and two non-occupied antibonding π*-orbitals (Fig. 7.3). Due to the localization of the π-electrons over the whole conjugated structure, each central bond has a bond order close to 1.5. Since each of the two outer carbons in the conjugated polyene structure shares its π-electrons with only one other carbon, the bond order of these two outer bonds are close to 2. Thus, the electron density is not evenly distributed within the conjugated system of carotenoids with the terminal double bonds having the highest densities (Zechmeister et al., 1943). This difference is important when considering products of reactions with electrophiles such as oxygen (*vide infra*).

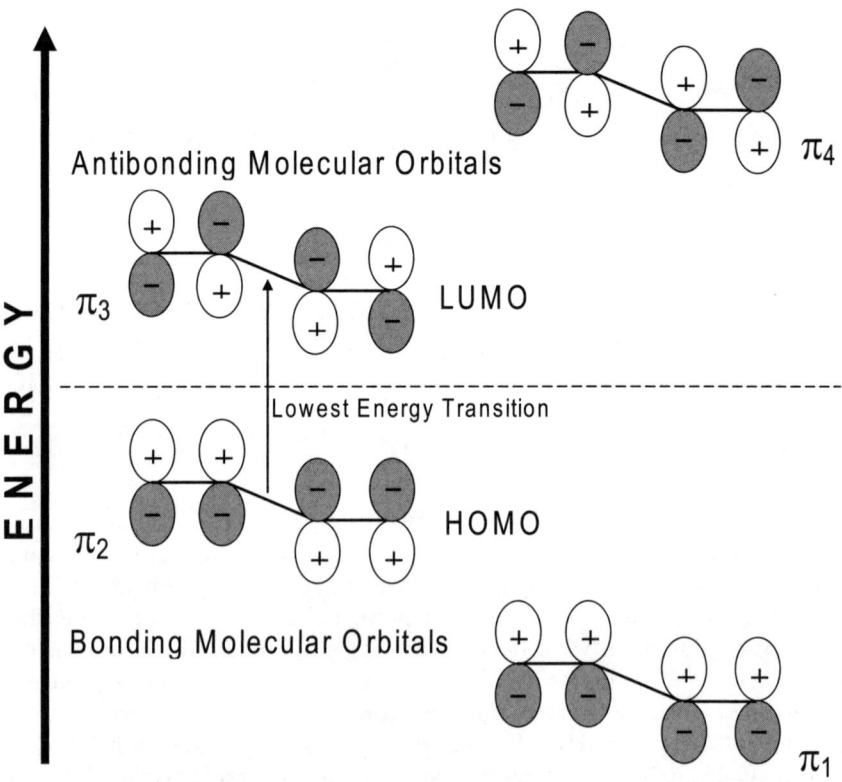

**Fig. 7.3.** Molecular orbitals of a conjugated double bond system.

## Geometrical Isomerism

The second important structural feature is the presence of a number of *cis* and *trans* geometrical isomers of the chain double bonds (Zechmeister et al., 1943; Saleh and Tan, 1991). There are 272 theoretical possibilities for *cis/trans* isomers of β-carotene, but only 20 unhindered isomers are possible to obtain (Zechmeister et al., 1943; Zechmeister, 1960). According to the Pauling rules for steric hindrance in isoprenoid systems, such as carotenoids, only the double bonds at *cis*-9, *cis*-13, and *cis*-15 are unhindered. Although Pauling rules approve di-*cis* and tri-*cis* isomers that may occur in an equilibrium mixture as a result of spontaneous or catalytic isomerization, energy calculations predict a very low probability for their formation. Thus, the all-*trans* and the three mono-*cis* isomers (*cis*-9, *cis*-13, and *cis*-15) are the predominant isomers of β-carotene in an equilibrium mixture (Fig. 7.4). The *all-trans* isomer, which is energetically the most stable, can undergo *cis*-isomerization upon exposure to heat or light. Indeed, *cis*-isomerization affects the planarity of the molecule and the evenness of the relative electron density in the delocalized electron cloud.

## Configuration of Cyclic Ends of Carotenoids

A third structural feature that determines the planarity of the carotenoid molecule is the possible forms resulting from rotation about the 6,7-single bond (Orlandi et al., 1991). There are two planar conformations where all the $p_z$ orbitals forming the

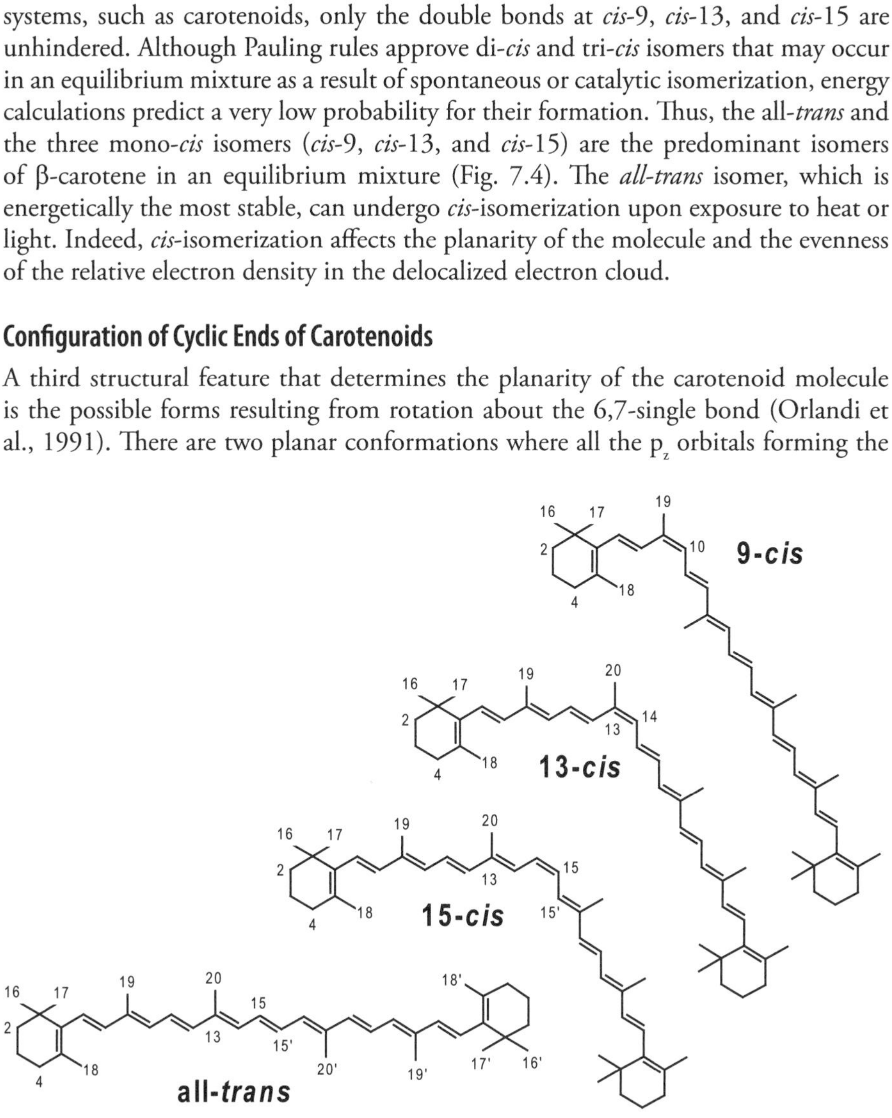

**Fig. 7.4.** Natural all-*trans* β-carotene and its *cis* isomers.

conjugated system are parallel and give maximum overlap. These two conformations are called s-*cis* and s-*trans* referring to the fact that the geometrical isomerism is in respect to a single bond (Fig. 7.5). Besides these two extremes, other conformations are found in which the rings and the chain are not coplanar. UV and NMR spectroscopy showed that the β-ionone rings in all-*trans*-β-carotene are not coplanar with the chain while X-ray analysis of some all-*trans* carotenoids in crystalline states demonstrated that their conformation is close to the s-*cis* form. The non-planarity in carotenoids is ascribed to steric interactions between the ring methyl groups and hydrogens at positions 7 and 7' resulting in an incomplete conjugation of the ring double bonds with the extended polyene system. This may explain why β-carotene with two β-ionone rings has a shorter absorption wavelength ($\lambda_{max}$ = 466 nm in $CHCl_3$) than γ-carotene with one β-ionone ring ($\lambda_{max}$ = 475 nm in $CHCl_3$) than lycopene with no β-ionone rings ($\lambda_{max}$ = 480 nm in $CHCl_3$) (Finar, 1981). This difference in planarity, and consequently the degree of conjugation, was also thought to explain why lycopene has twice the activity of β-carotene as a singlet oxygen quencher (DiMascio et al., 1989, 1991, 1992; Conn et al., 1991; Devasagayam et al., 1992).

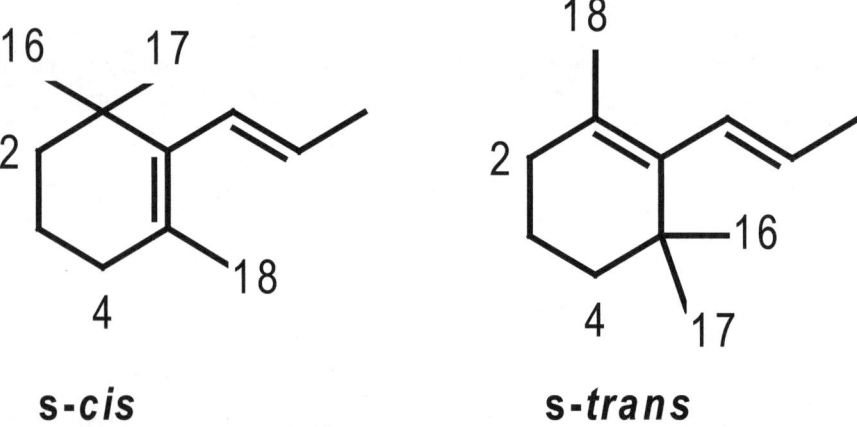

**Fig. 7.5.** s-*cis* and s-*trans* forms of the cyclic ends of carotenoids.

## Oxygenated Groups at Cyclic Ends of Carotenoid

The presence of an oxy or hydroxyl substituent in the cyclic part of the carotenoids modifies their polarity characteristics and thereby their solubility, stability, and distribution in different media. Moreover, oxygenation of the cyclic rings may positively contribute to the antioxidant properties of carotenoids (Terao, 1989; DiMascio et al., 1989, 1991, 1992; Miki, 1991; Jorgensen and Skibsted, 1993; Oshima et al., 1993). The presence of oxygenated functions in xanthophylls makes them markedly different from simple carotenes, especially in solubility, light absorption, reactivity with other chemicals, and biological potency as determined by the kinetics of transfer and retention within membranes. The relative solubility, stability, and absorptivity of lutein and β-carotene were compared in 18 different solvents, among which tetrahydrofuran solutions had the highest absorptivity (Craft and Soares, 1992). Lutein was least soluble in hexane, while β-carotene was least soluble in methanol and acetonitrile. Lutein was more stable than β-carotene in all solvents; both carotenoids were least stable in cyclohexanone and β-carotene was quite unstable in dichloromethane.

## Light Absorption and Quenching of Photosensitizers and Singlet Oxygen

### Light Absorption

Carotenoids display a broad absorption band in the visible region, mainly 400-500 nm, due to an allowed $\pi$-$\pi^*$ transition in the conjugated polyene structure (Jaffé and Orchin, 1962). Typically, all-*trans*-β-carotene has an absorption maximum around 460 nm with a very high molar absorption coefficient, $\varepsilon_{max}$ is approximately $1.5 \times 10^5$ $M^{-1}cm^{-1}$. Carotenoids with a *cis* double bond show somewhat reduced $\varepsilon_{max}$ values but show an extra band, called the *cis* band, in the UV region. Thus, 15-*cis*-β-carotene has $\varepsilon_{max}$ of $0.9 \times 10^5$ $M^{-1}cm^{-1}$ at 460 nm and a *cis*-band at approximately 340 nm with $0.5 \times 10^5$ $M^{-1}cm^{-1}$. As the number of conjugated double bonds increases, the carotenoids show a bathochromic shift with increasing $\lambda_{max}$ and $\varepsilon_{max}$ values (Truscott, 1990). These light absorption properties of carotenoids are of relevance to their colors, photoquenching as well as antioxidant and proxidant properties.

Upon absorption of light energy, ground state carotenoid molecules ($S_o$, $^1Car$) are excited to higher singlet energy states ($^1Car^*$). For symmetrical all-*trans*-carotenoids, quantum chemical calculations identified the state arising from the excitation of a single electron from the highest occupied $\pi$-molecular orbital (HOMO) to the lowest unoccupied $\pi^*$-molecular orbital (LUMO) of all-*trans*-β-carotene as the $S_2$ ($^1Bu^+$) singlet state (Suzuki, 1967; Hudson et al., 1982). The transition to this excited singlet state ($S_2$, $^1Bu^+$) from the ground state ($^1Ag^-$) is fully allowed, but the excited $^1Bu^+$ state is very unstable and it decays rapidly to a more stable lower energy singlet excited state ($S_1$, $^1Ag^*$) by loss of heat. The latter state then relaxes back to the ground state either directly by fluorescence (Thrash et al., 1979) or indirectly through vibrational, non-radiating, intersystem crossing (ISC, [1]) to an equivalent-energy triplet state ($T_1$, $^3Bu^+$), which can relax back to the ground state by phosphorescence (Fig. 7.6).

$^1Car$ ($S_1$, $^1Ag^*$) → $^3Car(T_1, ^3Bu^+)$    intersystem-crossing    [1]

The relaxation of carotenoids from higher singlet and triplet states to the ground state is not 100% efficient and the carotenoids are subject to different amounts of loss during this process. Evidence for the incomplete relaxation of excited carotenoids comes from the finding that triplet state β-carotene ($^3Car^*$) shows a strong absorption band in the 500-550 nm range due to electronic transition from the lower ($T_1$, $^3Bu^+$) to a higher excited triplet state (Wolff and Witt, 1969; Land et al., 1971). Moreover,

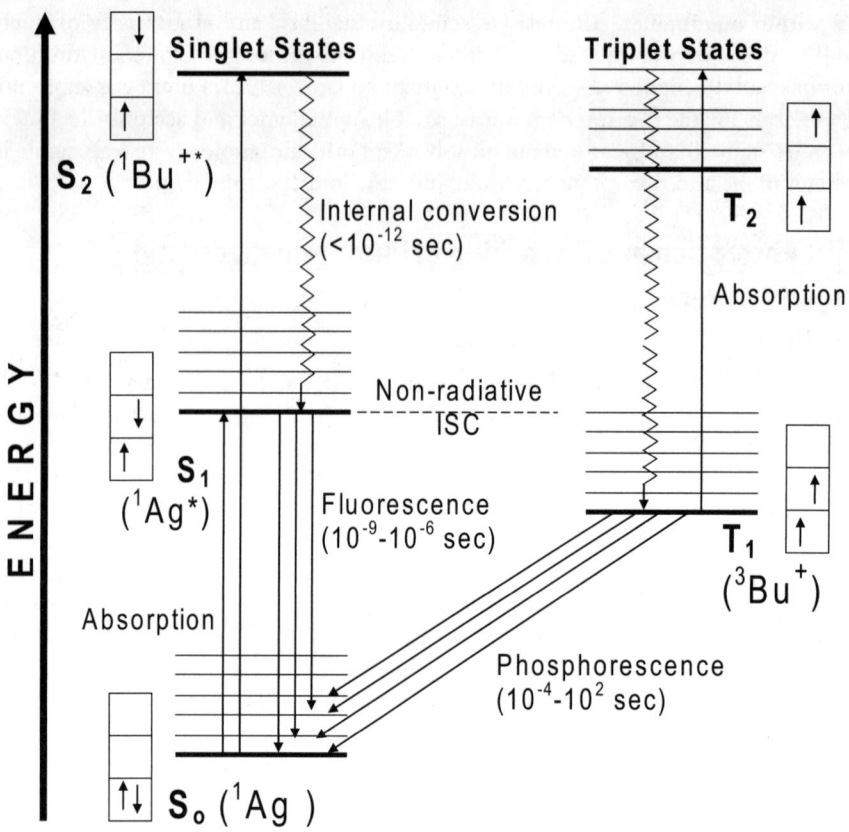

**Fig. 7.6.** Light absorption, excitation, and relaxation of carotenoid molecules.

photoradiation in the absence of photosensitizers causes carotenoid losses in the following order lycopene>>β-carotene=α-carotene>>astaxanthin (Oshima et al., 1993). The carotenoids may also be excited by heat and/or mechanical agitation. As a result of excitement, carotenoids undergo isomerization and oxidative destruction and may act as pro-oxidants for co-existing lipids.

## Physical Quenching of Photosensitizers and Singlet Oxygen by Carotenoids

Ground state photosensitizers ($S_o$), typically chlorophyll, are excited by absorption of a quantum of radiation and lead to the formation of short-life, $10^{-11}$ s, singlet state, excited photosensitizers ($^1S^*$). These excited photosensitizers dissipate their energy either by photon emission via fluorescence or by intersystem crossing to form excited triplet state photosensitizers ($^3S^*$) with relatively long lifetimes, $10^{-4}$ s, during which they may return to the ground state by phosphorescence or deactivation by carotenoids. Depending on the light intensity and the nature of other reactive molecules in the assay system, particularly oxygen concentration, triplet photosensitizers may either react with compounds or initiate photochemical reactions. Gollnick and Schenk (1967) classified these reactions into type I reactions, redox reactions involving abstraction of a hydrogen atom or electron from the substrate forming free radicals that lead *inter alias* to the degradation of the photosensitizer, and type II reactions involving exchange of energy with triplet molecular oxygen ($^3O_2$) generating the most reactive singlet oxygen ($^1O_2$) (Fig. 7.7).

Singlet oxygen, produced by type II photosensitized reactions, reacts with linoleic acid approximately 1500 times faster than triplet oxygen and therefore was thought to be involved in the initiation of lipid oxidation (Rawls and Van Santen, 1970). In addition, the relative oxidizability of oleic and linoleic acid was 1:1.7 in reactions with singlet oxygen (Terao and Matsushita, 1977) compared to 1:12 in reactions with triplet oxygen (Gunstone and Hilditch, 1945) making oleic acid an important reactant in lipid oxidation reactions catalyzed by singlet oxygen. Moreover, singlet oxygen is also known to oxidize proteins, amino acids, and DNA bases at much higher rates than hydroperoxides produced from fatty acids. Amino acids containing heterocyclic rings or sulfur atoms appear to be attacked by singlet oxygen with the following second order rate constants ($M^{-1}s^{-1}$): tryptophan ($2.5 \times 10^8$), histidine ($1.3 \times 10^8$), tyrosine ($0.27 \times 10^8$), methionine ($0.22 \times 10^8$), and alanine ($0.02 \times 10^8$) (Wilkinson and Brummer, 1981). The finding that the effects of singlet oxygen on protein oxidation were more significant than its effects on lipid oxidation made Wilkinson and Brum-

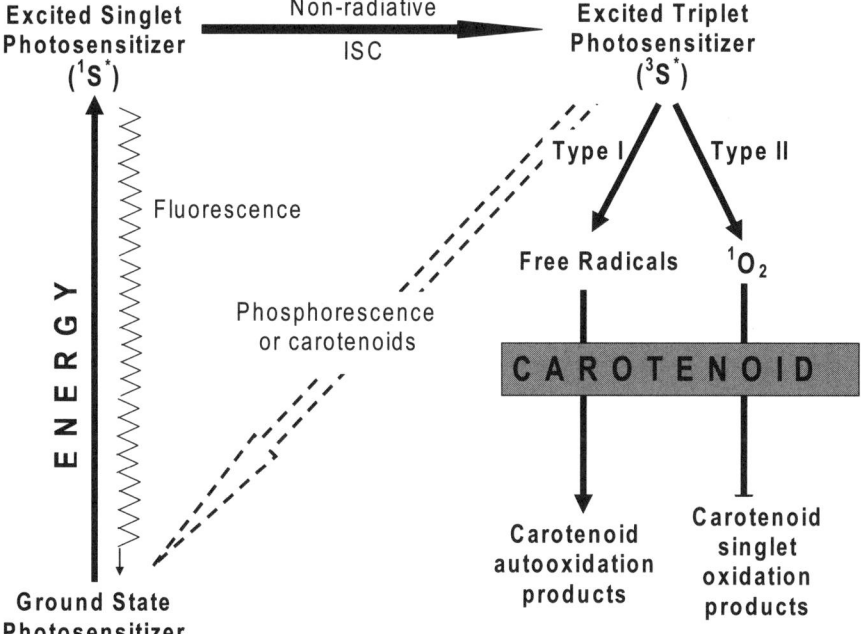

**Fig. 7.7.** Effect of photosensitizers on carotenoid photooxidation.

mer (1981) associate the anticarcinogenic properties of carotenoids to protection of the genetic material against singlet oxygen.

The carotenoids are recognized as the most efficient protective agents against the harmful effects of excited pigments and singlet oxygen generated by photochemical reactions (Foote, 1979; Foote and Denny, 1968; Foote et al., 1970a,b,c; Mathews-Roth, 1987). The protective effects of carotenoids against photosensitized reactions are due to their ability to quench both excited photosensitizers and singlet oxygen at a diffusion-controlled rate without being consumed in the process, mainly by a harmless energy transfer physical quenching mechanism (pathway C), that is

$$^3Sen^* + {}^1Car \rightarrow {}^1Sen + {}^3Car^* \qquad [2]$$
$$^1O_2 + {}^1Car \rightarrow {}^3O_2\ (^3\Sigma_g^-) + {}^3Car^* \qquad [3]$$

The excitation energy is dissipated to the solvent system as thermal energy through rotational and vibrational intersystem relaxation of the C-C and C=C bonds in the polyene chain to recover the ground state carotenoid

$$^3Car^* \rightarrow {}^1Car + \text{Thermal energy} \qquad [4]$$

The ability of conjugated polyene systems to quench photosensitizers and singlet oxygen largely depends on the number of their conjugated double bonds (Foote, 1976; Lee and Min, 1988, 1990; Jung and Min, 1991). To be able to accept extra energy from singlet oxygen, carotenoids need to have their singlet states below 22.4 Kcal/mole. Energy transfer from $^1O_2$ to carotenoids with nine (singlet state 22 Kcal/mole) or more conjugated double bonds is exothermic and these carotenoids are efficient oxygen quenchers. The singlet oxygen quenching rate constants for selected carotenoids in dichloromethane were lutein ($5.72 \times 10^9$ $M^{-1}s^{-1}$), zeaxanthin ($6.79 \times 10^9$ $M^{-1}s^{-1}$), lycopene ($6.93 \times 10^9$ $M^{-1}s^{-1}$), and isozeaxanthin ($7.39 \times 10^9$ $M^{-1}s^{-1}$) (Lee and Min, 1990). Carotenoids with seven conjugated double bonds (29 Kcal/mole) and retinol with five conjugated double bonds (34 Kcal/mole) are much less efficient or even inactive as singlet oxygen quenchers (Mathews-Roth et al., 1974; Foote, 1976; Krinsky, 1979). On the other hand, the efficiency of carotenoids with seven conjugated double bonds was found to be about 75% of that of β-carotene in quenching chlorophyll, and retinol was about half efficient as β-carotene in quenching methylene blue (Foote et al., 1970a; Mathis and Klero, 1973). The relative ability of compounds with conjugated polyene structures to quench singlet oxygen and chlorophyll is shown in Fig. 7.8. The fact that C50 and C60 carotenoids, with 15 and 19 conjugated double bonds, exhibit essentially the same quenching rates as β-carotene suggests that the reaction rate approaches diffusion-controlled rates after about 11 conjugated double bonds (Foote et al., 1970a).

The singlet oxygen quenching rate constant for β-carotene was reported to vary from $5 \times 10^9$ to $13 \times 10^9$ $M^{-1}s^{-1}$ depending on the solvent and other experimental conditions (Carlsson et al., 1972; Farmilo and Wilkinson, 1973; Fahrenholtz et al.,

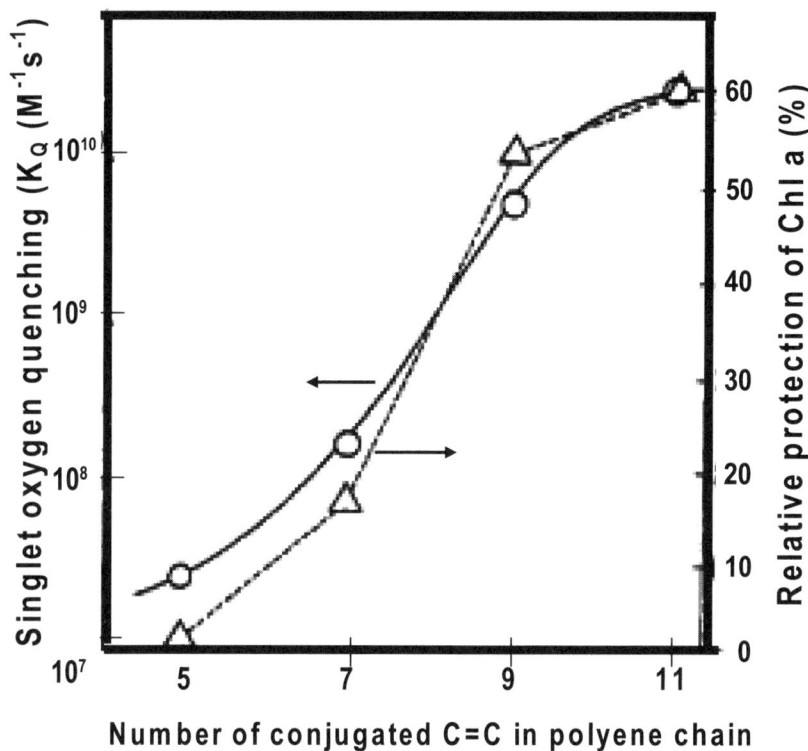

**Fig. 7.8.** Effect of the number of conjugated double bonds of carotenoids on the quenching of singlet oxygen and chlorophyll-a (Adapted from Foote et al. 1970a with permission from the American Chemical Society Press).

1974; DiMascio et al., 1989). The presence of oxygenated functions, carbonyl or hydroxyl groups, in the carotenoids enhances their singlet oxygen quenching abilities, for example astaxanthin ($9.79 \times 10^9$ $M^{-1}s^{-1}$ in dicholomethane) is more efficient than the previously mentioned carotenoids with 11 double bonds. Further, γ-carotene and lycopene are more efficient singlet oxygen quenchers than β-carotene indicating that an open chain has a higher quenching ability than a β-iononre ring (DiMascio et al., 1989, 1991, 1992; Conn et al., 1991; Devasagayam et al., 1992; Sundqvist et al., 1994; Hirayama et al., 1994), an effect that may be related to the differences in molecular planarity.

## Chemical Quenching of Photosensitizers and Singlet Oxygen by Carotenoids

Although most of the protective effects of the carotenoids against the harmful effects of excited photosensitizers and singlet oxygen are due to the physical quenching mechanisms discussed previously, some chemical transformations of the carotenoids occur. For example, the carotenoids may isomerize (Stahl and Sies, 1993) or undergo concomitant chemical reactions with free radicals generated during the process (pathway A) or with singlet oxygen (pathway B) (Foote, 1979; Foote and Denny, 1968; Foote et al., 1970a,b,c). An ESR study showed that when β-carotene was added to a mixture of the singlet oxygen generator triphenyl phosphate ozonide (TPPO) and the spin trap α-phenyl-*N-tert*-butylnitrone (PBN), it inhibited the formation of PBN-singlet oxygen adducts (Pryor and Govindan, 1981). However, a triplet of doublet ($a^N$ = 13.5 and $a^H$ = 1.5) was observed in the ESR spectrum indicating the presence of a PBN-adduct with a peroxyl radical that is believed to result from the oxidation of β-carotene. Although chemical reactions account for only 0.05% of the quenching activity of β-carotene (Krasnovskii and Paramonova, 1983), they can still lead to oxidative destruction of the conjugated chromophore and the formation of carotenoid oxidation products (Foote, 1979; Hasegawa et al., 1969).

Carotenoid + $^1O_2$ → Singlet oxygen oxidation products [5]
$^1$Carotenoid + $^1O_2$ → $^3$Carotenoid + $^3O_2$ → Triplet oxygen oxidation products [6]

Singlet oxygen reacts differently with isolated and conjugated double bonds. The latter reactions can be classified according to three mechanisms:

i. The "ene" reaction characteristic of isolated double bonds producing hydroperoxides (Gollnick and Kuhn, 1979)

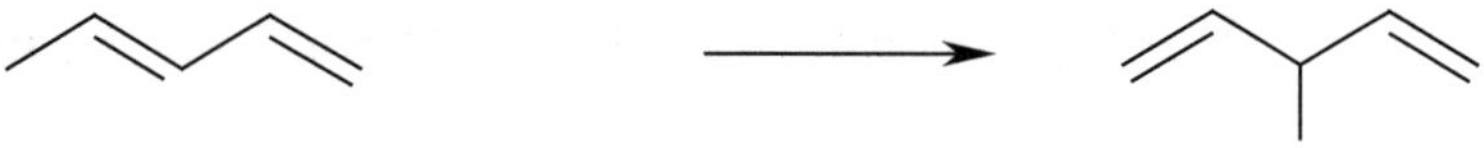

[7]

ii. 1,2-Cycloaddition for both isolated and conjugated C=C bonds producing dioxetanes,

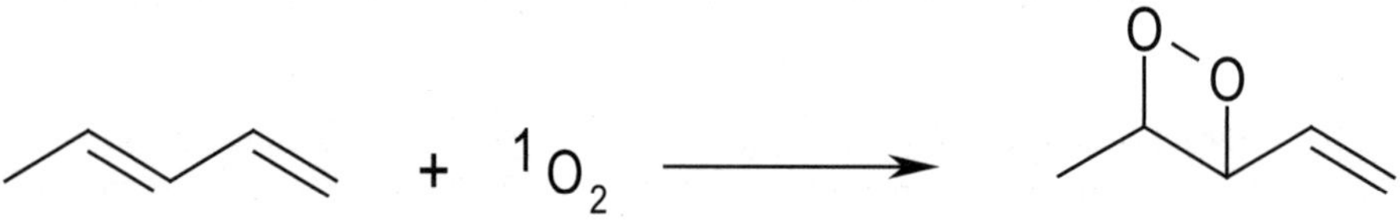

[8]

iii. 1,4-Cycloaddition restricted to conjugated *cis* dienes producing endoperoxides (Gollnick and Schenk, 1967)

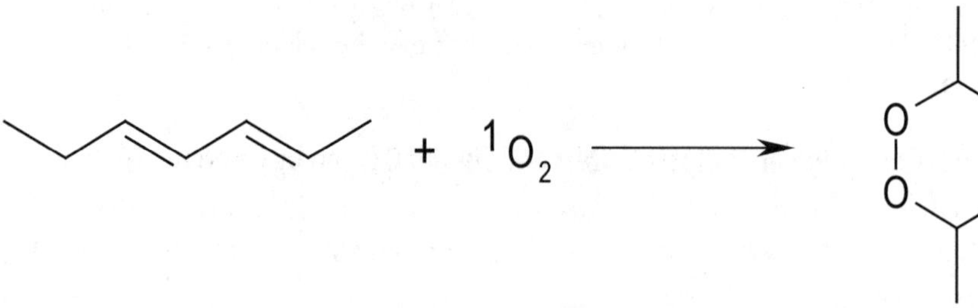

[9]

Singlet oxygen oxidation products of carotenoids include β-carotene-5,8-endoperoxides as well as β-apo-15-carotenal, β-apo-14'-carotenal, β-apo-12'-carotenal, β-apo-10'-carotenal, and β-apo-8'-carotenal resulting from dioxetane intermediates (Schenck and Schade, 1970). As expected, no allylic hydroperoxides characteristic of the "ene" reaction of singlet oxygen with isolated double bonds were found (Fig. 7.9).

## Peroxyl Radical Reactions Involving Carotenoids
### Kinetics and Possible Mechanisms

Although the oxidation of carotenoids can occur in the absence of other oxidizable substrates, it is generally enhanced by the presence of free radical generators or unsaturated fatty acids where a coupled oxidation occurs. Elahi and Cole (1964) showed that in the presence of *tert*-butyl hydroperoxides in chloroform, β-carotene oxidized in a concentration-dependent manner. Free radicals may be generated from the mono- or bimolecular decomposition of the hydroperoxide or from its reaction with chloroform, that is

*tert*-ButOOH → *tert*-ButO• + OH•  [10]
2 *tert*-ButOOH → *tert*-ButO• + *tert*-ButOO• + $H_2O$  [11]
*tert*-ButOOH + $CHCl_3$ → *tert*-ButO• + $CCl_3$• + $H_2O$  [12]

The reaction is significantly catalyzed by traces of transition metal ions that can split the hydroperoxides to generate initiating hydroxyl radicals.

The long conjugated double bond system of carotenoids makes them excellent substrates for radical attack. For example, peroxyl and alkoxyl radicals react with carotenoids at much higher rates than with unsaturated fatty acids (Weber and Grosch, 1976). Carotenoids react with electron-deficient peroxyl radicals by adding them to their conjugated polyene system; this results in carbon-centered radicals (ROO-β-Car•) that are stabilized by resonance (Scott, 1992; Mayo, 1968; Pryor et al., 1972; Weber and Grosch, 1976; Bors et al., 1981; Burton and Ingold, 1984; Krinsky and Denke, 1982; Burton, 1989; Terao, 1989; Yamauchi et al., 1993; Everets et al., 1995). Intermediate radical adducts formed from β-carotene and astaxanthin are shown in Fig. 7.10. The stability of these radicals is dependent on the addition site of the peroxyl radical, but it is generally low and leads to the collapse of the hydroperoxyl group mainly to an epoxide with carbon-centered radical or an alkoxyl radical (Mayo, 1968),

[13]

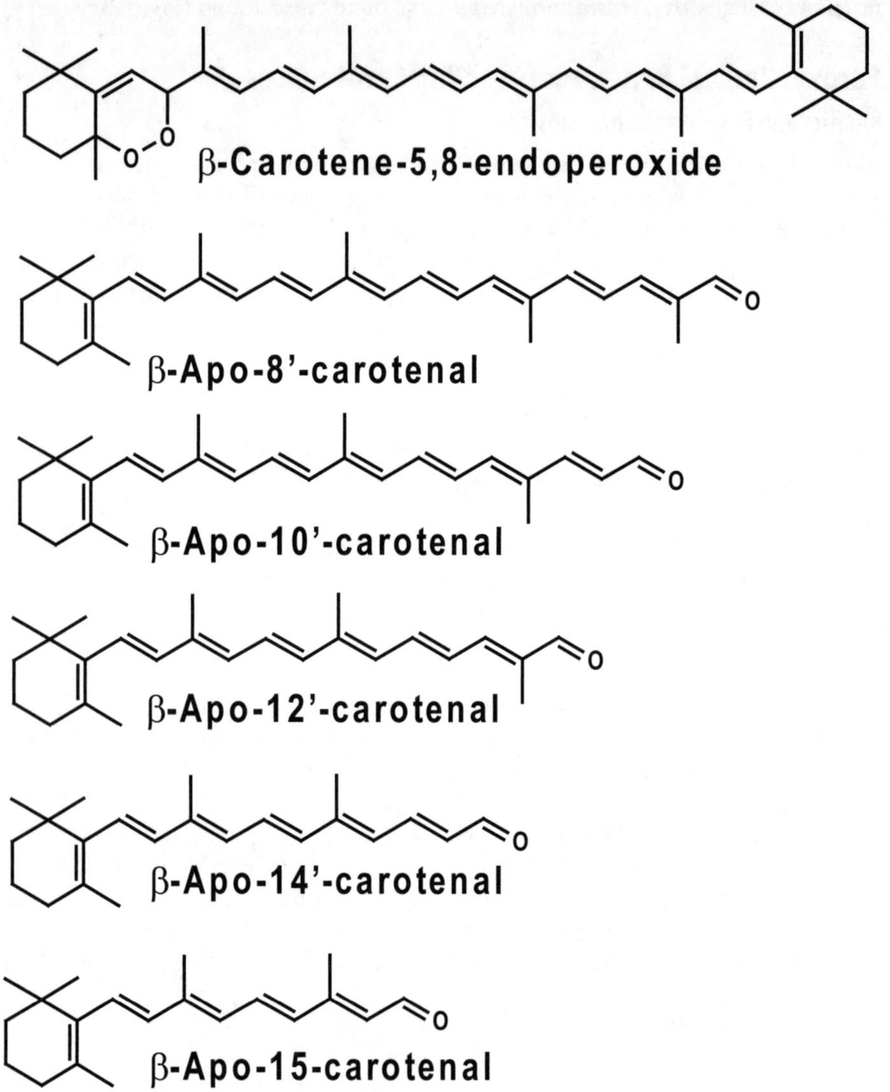

**Fig. 7.9.** Endoperoxide and carbonyl oxidation products of β-carotene.

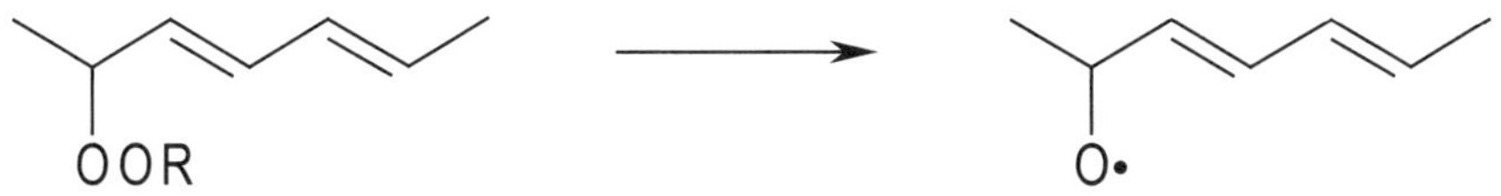

[Reaction scheme 14: OOR-substituted diene rearranging to O• radical with diene]

[14]

Krinsky (1989) speculated that at high peroxyl radical-to-carotenoid concenrations, the ROO-β-Carotene• adds on other peroxyl radicals by opening double bonds so that one carotenoid molecule can act as an antioxidant by adding several peroxyl radicals.

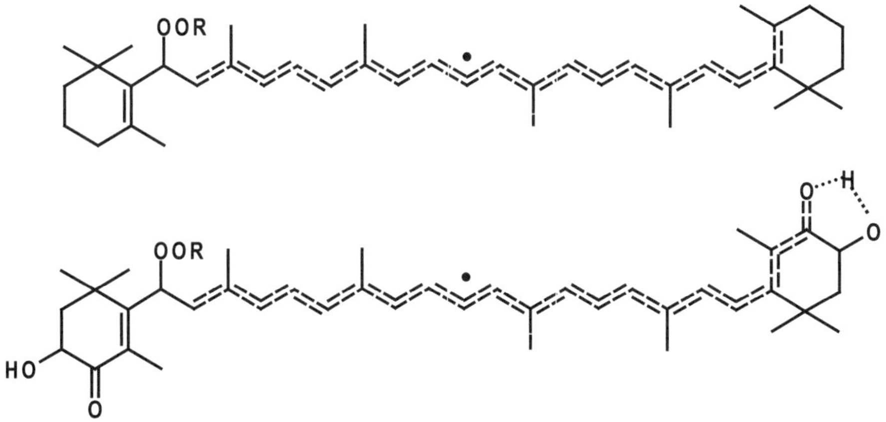

**Fig. 7.10.** Intermediate roo-carotenoid radical adducts formed with β-carotene and astaxanthin.

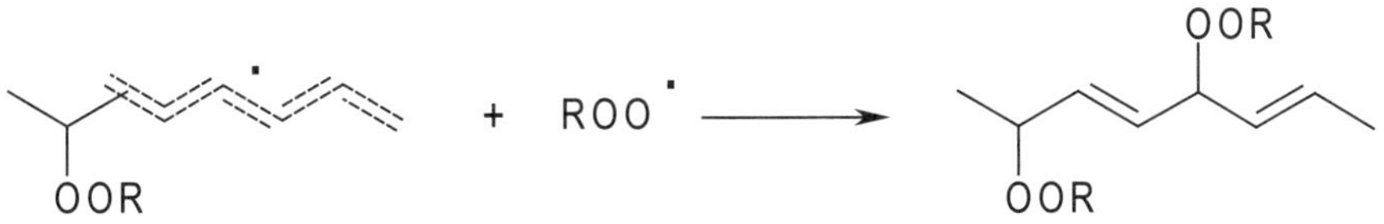

[Reaction scheme 15: carotenoid radical + ROO• → bis-OOR adduct]

[15]

Liebler and McClure (1996), however, discussed that the addition of peroxyl radicals to carotenoid molecules is an inherently inefficient antioxidant mechanism for two reasons; first, the antioxidant effectiveness depends on the extent to which released alkoxyl radicals are scavenged; and second, the resulting ROO-β-Carotene• may reversibly trap oxygen to form reactive peroxyl radicals,

$$\text{OOR-CH=CH-CH=CH-CH=CH(·)} + O_2 \longrightarrow \text{OOR-CH=CH-CH(OO·)-CH=CH-CH}$$

[16]

Studying products of the reaction between β-carotene and radicals generated from azo*bis*-2,4-dimethylvaleronitrile (AMVN), Liebler and McClure (1996) suggested an alternative mechanism by which peroxyl radicals react with β-carotene by hydrogen abstraction in an analogous way to phenolic antioxidants

β-Car + ROO• → β-Car• + ROOH     [17]
β-Car• + ROO• → β-Car-OOR     [18]

Other researchers supported this alternative mechanism. For example, Takahashi et al. (1999, 2001, 2003a,b) constructed a kinetic model based on the classical free radical oxidation mechanism of hydrocarbons that fits data describing the autoxidation of β-carotene, its co-oxidation with oleic acid, and its protection by α-tocopherol. Although an excellent fit of data was obtained in these studies, kinetic modeling cannot be considered conclusive for suggested mechanisms and needs to be supported by analytical data on reaction products.

The reaction conditions might play a significant role in the reaction mechanism. For example, under high radical fluxes generated by high temperatures or azo-initiators, the oxidation of carotenoids may follow simple first order kinetic models (Chou and Breene, 1972; Arya et al., 1979; Ramakrishnan and Francis, 1979a; Henry et al., 1998). Generally, the oxidation of β-carotene follows a sigmoidal curve and involves initiation, propagation, and termination stages (Kasaikina et al., 1975, 1981; Papadopolou and Ames, 1994; Ozhogina and Kasaikina, 1995). As discussed in Chapter 4, the activation energy for the oxidation of conjugated linoleic acid is significantly higher but the oxidation rate constants were greater than that for linoleic acid. This plus the fact that that β-carotene bleaches quickly under co-oxidation conditions (Budowski and Bondi, 1960; Ramakrishnan and Francis, 1979b; Tsuchihashi et al., 1995) indicates that carotenoids are more favorable for radical addition than for hydrogen abstraction. The solvent might affect the mechanism since hydrogen transfer may be a favored oxidative mechanism in non-polar solvents while oxidation in more polar solvents may entail electron transfer (Weber and Grosch, 1976; Grant et al., 1988; Jovanovic et al., 1992).

## Carotenoid Structure and Antioxidant Activity

Xanthophylls containing oxygenated functions (e.g., astaxanthin, canthaxanthin, zeaxanthin, and β-cyptoxanthin) seem to be more effective as peroxyl radical scavengers than β-carotene and lycopene (Kurashige et al., 1989; Terao, 1989; Miki, 1991; Boey et al., 1992; Sies et al., 1992; Terao et al., 1992; Jorgensen and Skibsted, 1993; Oshima et al., 1993; Woodall et al., 1995). Astaxanthin and canthaxanthin contain 13 conjugated double bonds compared to 11 in β-carotene and their conjugated polyene system terminates with carbonyl groups. These features seem to offer them abilities to form more stable adducts with peroxyl radicals than β-carotene. Moreover, further stabilization of the peroxyl radical adduct through an intramolecular hydrogen bond between the hydroxyl hydrogen and the conjugated electron pairs of the carbonyl oxygen may explain why astaxanthin is a slightly more powerful antioxidant than canthaxanthin (Jorgensen and Skibsted, 1993). In polar media, astaxanthin may also benefit from the 1-keto-2-hydroxy functional feature, which may chelate trace metal ions,

[19]

Zeaxanthin and β-cryptoxanthin, although having 11 conjugated double bonds like β-carotene, were more effective in protecting phosphatidyl liposomes against peroxidation induced by both water- and lipid-soluble azo-initiators (Woodall et al., 1995). Since the hydroxyl groups in these xanthophylls are not adjacent to the chromophore, they were not believed to influence the reactivity of the carotenoids with free radicals. Rather, the effect of these functions was related to biophysical interactions between the phospholipid and the carotenoid. As shown in Fig. 7.11, xanthophylls are believed to locate in membranes between the fatty acid moieties with their β-rings close to the hydrophilic/hydrophobic interface while non-polar β-carotene and lycopene are believed to occur between the two halves of the bilayer (Jezowska et al., 1994; Strzalka and Gruszecki, 1994).

## Oxidation Products of Carotenoids

Although the site of addition of peroxyl radicals to the carotenoid molecule is not yet known with certainty, it is expected to be a function of the inherent reactivity (i.e. electron density) of the carotenoid and the size and reactivity of the attaching radical. Zechmeister et al. (1943) and El-Tinay and Chichester (1970) suggested that the C5=C5 and C5'=C6' terminal double bonds of β-carotene, having the highest electron density, serve as the preferential sites of radical attachment. Alternatively, Pullman and Pullman (1963) and Marty and Berest (1990) agreed that the C7=C8 and

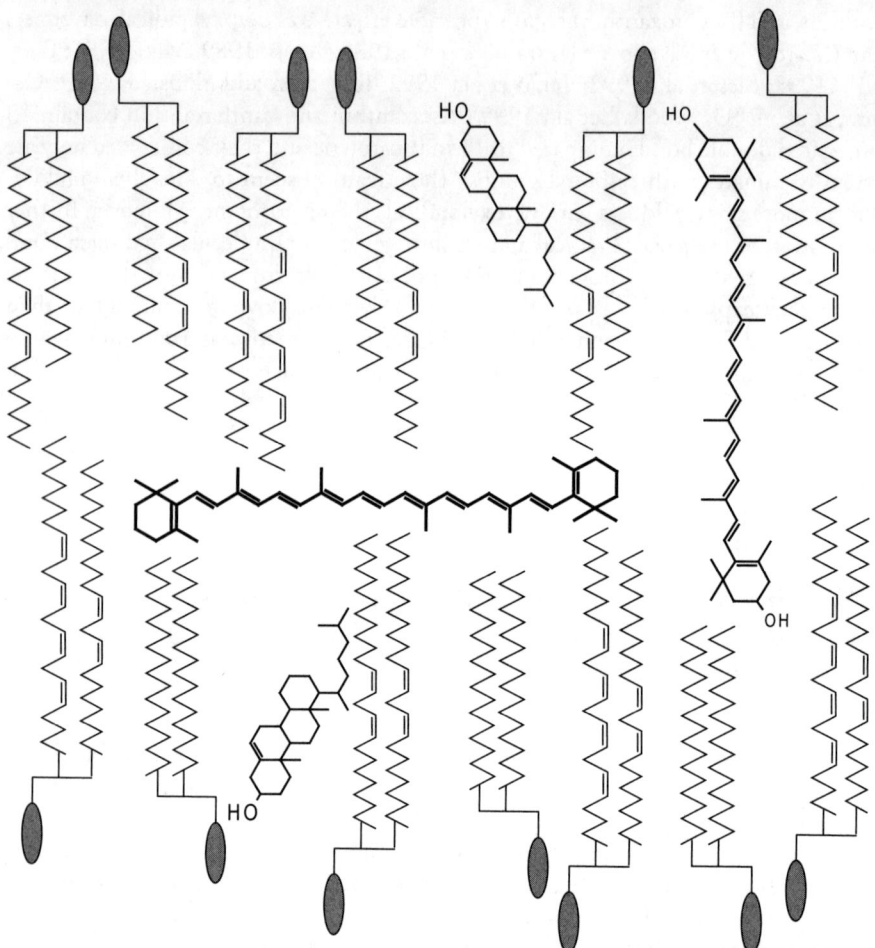

**Fig. 7.11.** Proposed differences in the localization of carotenes and xanthophylls in biological membranes.

C7'=C8' are the positions with the highest mobility index, 0.731 obtained from molecular orbital calculations, and they favored radical addition to these double bonds. On the other hand, the central double bonds have less energy and therefore are easier to open. Naturally, the addition of a certain radical to a specific double bond within the conjugated polyene depends on the steric nature of the radical, its reactivity, and the conformation and isomerization characteristics of the carotenoid molecule as determined by the amount of energy available (i.e., temperature and light).

Epoxides (5,6-, 5',6'-, 5,8-, and 5',8'-) and diepoxides (5,6,5',6'-, 5,6,5',8'-, and 5,8,5',8'-), shown in Fig. 7.12, were detected as autoxidation products of β-carotene (Hunter and Krakenberger, 1947; Friend, 1958a; El-Tinay and Chichester, 1970; Kennedy and Liebler, 1991; Mordi et al., 1991, 1993; Yamauchi et al., 1993). Also,

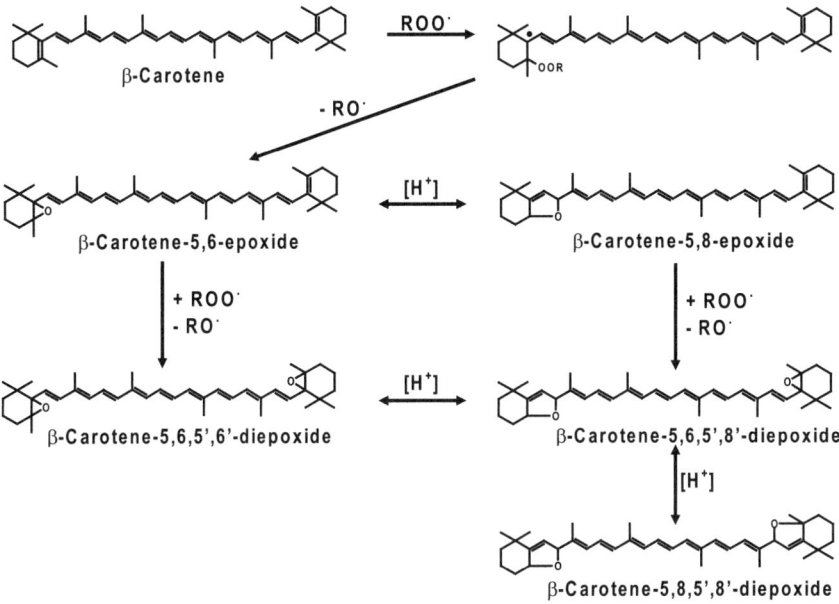

**Fig. 7.12.** Epoxides formed by the reaction of β-carotene with peroxyl radical.

three cycloepoxy oxidation products, 13,15'-epoxyvinyleno-13,15'-dihydro-dinor-β-carotene, 15',13-epoxyvinyleno-13,15'-dihydro-dinor-β-carotene (which is present in four isomers; 13R,15'R, 13S,15'S, 13R,15'S, and 13S,15'R), and 11,15'-cyclo-12,15-epoxy-11,12,15,15'-tetrahydro-β-carotene (Fig. 7.13), were identified (Yamauchi et al., 1993; McClure and Liebler, 1995). These products suggest that radicals can add to the terminal as well as the central double bonds of β-carotene.

### Enzyme-Catalyzed Co-Oxidation with Unsaturated Fatty Acids

The co-oxidation of carotenoids with polyunsaturated fatty acids was found to be highly enhanced by the presence of lipoxygenase since a large proportion of peroxyl radicals is not converted to hydroperoxides by the enzyme (Sumner, 1942; Smith and Sumner, 1948; Friend, 1956, 1958b; Blain, 1970; Tookey et al., 1958; Ben-Azziz et al., 1971; Zinsou, 1971; Grosch et al., 1977; Ikediobi and Snyder, 1977; Cohen et al., 1985; Klein et al., 1985; Katusin-Rasem and Razem, 1994). Interestingly, the co-oxidation of β-carotene by soybean lipoxygenase in the presence of lipids and lipid hydroperoxides was faster under anaerobic than aerobic conditions (Klein et al., 1984). This effect was explained by the formation of an enzyme-fatty acid radical complex,

$Enz(Fe^{II}) + LOOH \rightarrow Enz(Fe^{III}) + LO\bullet + {}^-OH$ [20]
$Enz(Fe^{III}) + LH \rightarrow Enz(Fe^{II})\text{-}L\bullet + {}^+H$ [21]
$Enz(Fe^{II})\text{-}L\bullet + Car \rightarrow Enz(Fe^{II}) + \text{Oxidized Car}$ [22]

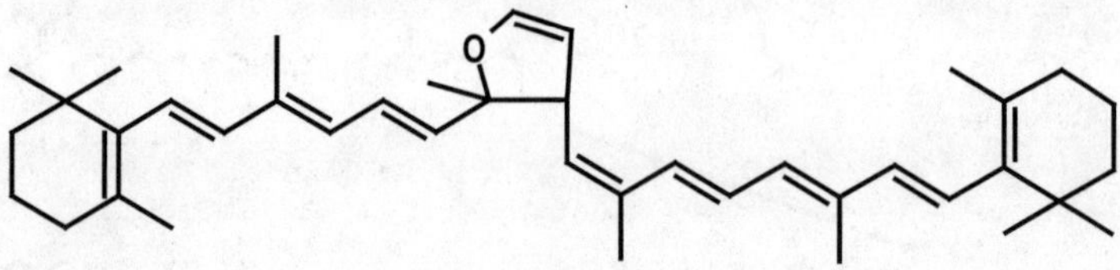

**13,15'-epoxyvinyleno-13,15'-dihydro-14,15-dinor-β,β-carotene**

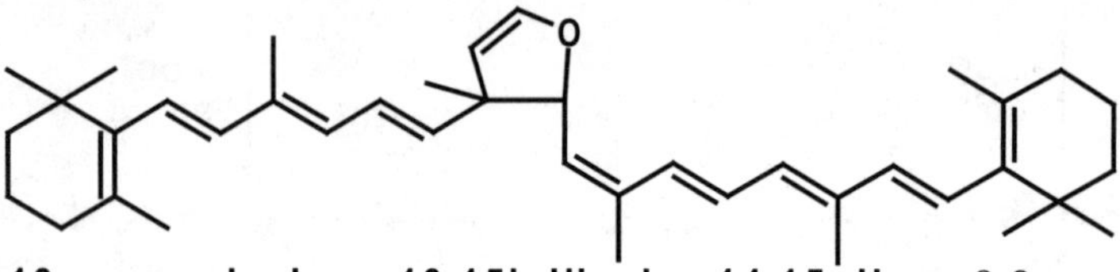

**15',13-epoxyvinyleno-13,15'-dihydro-14,15-dinor-β,β-carotene**

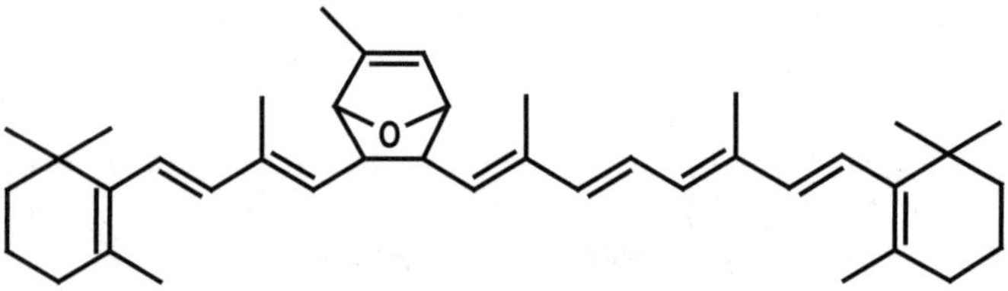

**11,15'-cyclo-12,15-epoxy-11,12,15,15'-tetrahydro-β,β-carotene**

**Fig. 7.13.** Heterocyclic oxidation products of β-carotene.

Under aerobic conditions, the last reaction would proceed as follows

$Enz(Fe^{II})$-L• + $O_2$ + LH → $Enz(Fe^{II})$-LOOH + H•     [23]
$Enz(Fe^{II})$-LOOH → $Enz(Fe^{II})$-LO• or HO•     [24]
$Enz(Fe^{II})$-LOO• or $Enz(Fe^{II})$-LO• + Car → Oxidized Car     [25]

Barimalaa and Gordon (1988) calculated the activation energies for the co-oxidation of linoleic acid and β-carotene by soybean lipoxygenase as 4.97 and 4.78 Kcal/mole, respectively. This indicates that they can compete for peroxyl radical equally well and that the competition of linoleate offers protection to the carotenoid [Eqn. 25]. This competition is also confirmed by finding that carotenoid depletion increased and linoleic acid oxidation decreased with increased carotenoid concentration. The presence of butylatedhydroxytoluene and α-tocopherol inhibited the oxidation of β-carotene to a greater extent than the oxidation of linoleic acid.

# The Pro-oxidant Effects of Carotenoids
## Reactions with Triplet Molecular Oxygen

The interaction of carotenoids with lipid oxidation is complex and subject to many factors (Vile and Winterbourn, 1988a; Kennedy and Liebler, 1991, 1992; Palozza and Krinsky, 1991) It was shown by Burton and Ingold (1984) that β-carotene acts as an effective antioxidant at low oxygen pressure (*ca* 15 torr or 2% oxygen) however its activity will decrease with increased oxygen tension (*ca* 150 torr or 20% oxygen) until it becomes a pro-oxidant at high oxygen pressure (towards 750 torr or 100% oxygen).

The fact that carotenoids can easily absorb light or heat energies and get excited to higher energy states may be involved in their pro-oxidant effect observed by many researchers (Burton and Ingold, 1984; Vile and Winterbourn, 1988b; Palozza and Krinsky, 1991; Kennedy and Liebler, 1992; Haila and Heinonen, 1994). β-Carotene was found to exert antioxidant effects in the absence of light and pro-oxidant effects in its presence (Stanescu and Eisenburger, 1969). It is known that ground-state carotenoids ($^1$Car) absorb light energy in the blue region of the spectrum (400-500 nm) and become excited to triplet states ($^3$Car*) that might initiate or propagate lipid oxidation (Rodgers and Bates, 1980),

$$^1Car + h\upsilon\ (450\ nm) \rightarrow\ ^3Car^* \qquad [26]$$
$$^3Car^* +\ ^3O_2(^3\Sigma_g^-) \rightarrow\ ^1Car +\ ^1O_2 \qquad [27]$$

The blue light represent about 20% of the energy emitted from light sources (Paul et al., 1972a). In the absence of sensitizers, photooxidation of polyunsaturated fatty acids is catalyzed more strongly by short light wavelengths, 325-460 nm (Parker et al., 1952; Radtke et al., 1970; Chahine and deMan, 1971; Paul et al., 1972b,c; Satter et al., 1976a,b). Actually, Schenck and Schade (1970) showed that β-carotene can act as a sensitizer in type II photooxidation reactions, generating singlet oxygen via excited carotene-$O_2$ complexes.

Excitation of a π-electron to higher singlet states may cause partial opening of a double bond in the conjugated structure (Fig. 7.14). Consequently, the bond order decreases and the double bonds in the carotenoid molecule elongate and give some carbon atoms a partial biradical character (Wasielewski et al., 1989; Christophersen et al., 1991). Molecular orbital calculations showed that the central carbon-carbon double bonds are elongated more than threefold in the higher energy state $^1Bu^+$ than in the lower energy state $S_1$ (Wasielewski et al., 1989). The biradical intermediate, formed by this reaction

$$^3Car^* +\ ^3O_2(^3\Sigma_g^-) \rightarrow\ ^3Car\bullet-OO\bullet \qquad [28]$$

Depending on oxygen concentration, the *Doering's diradical* may undergo *cis*-isomerization or oxidation generating a wide range of oxidation products, including per-

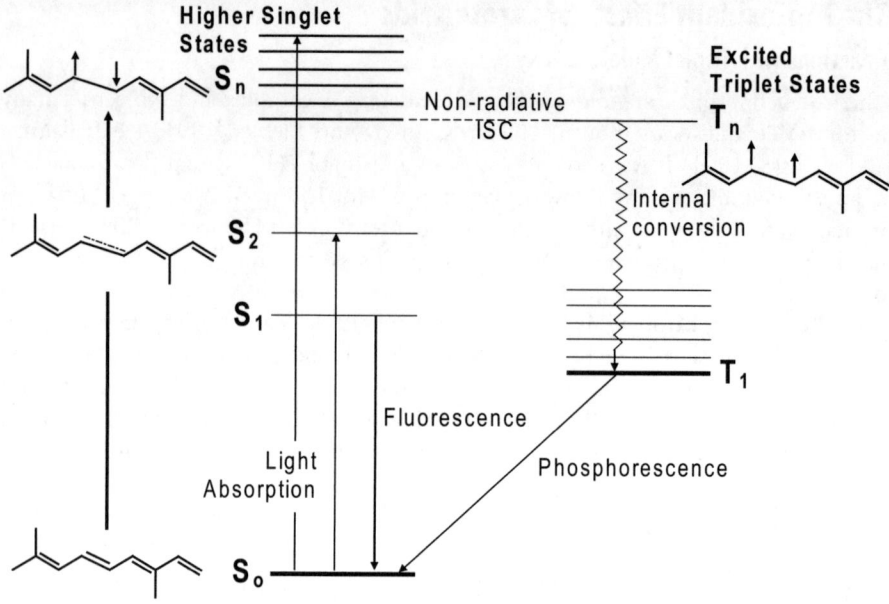

**Fig. 7.14.** Excitation of conjugated double bonds by light to form an excited triplet state (tn) that can react with molecular oxygen.

oxides, epoxides, aldehydes, and ketones (Fig. 7.15). Oxidation products are more pronounced for the central double bonds while the rings and adjacent double bonds are largely unchanged (Isoe et al., 1969). This reaction pathway leads to products different from those of oxidation by singlet oxygen (Fig. 7.9) or peroxyl radicals (Fig. 7.12). Excited triplet state carotenoids ($^3$Car*) can also react with triplet oxygen by 1,2-addition forming dioxetanes, which decompose to short-chain carbonyl compounds, mainly apocarotenals similar to those shown in Fig. 7.9. In the presence of energy-acceptors, excited carotenoids can exchange their extra energy and relax back to the ground state. Since relaxation is an exothermic process, lowering the temperature of the system can enhance it.

Canthaxanthin and other carbonyl-containing carotenoids are more stable towards photodegradation than carotenes (Terao, 1989; Nielsen et al., 1996). This difference was attributed to the nature of the excited states having some n,π* character in canthazanthin compared to only π, π* states in the case of β-carotene. Thus, the excited states of xanthophylls may have less of a biradical character and may be less labile than those of the carotenes. Xanthophylls are regarded as pure antioxidants in contrast to carotenes, which exhibit both anti- and pro-oxidant effects as shown in Fig. 7.16 (Martin et al., 1999).

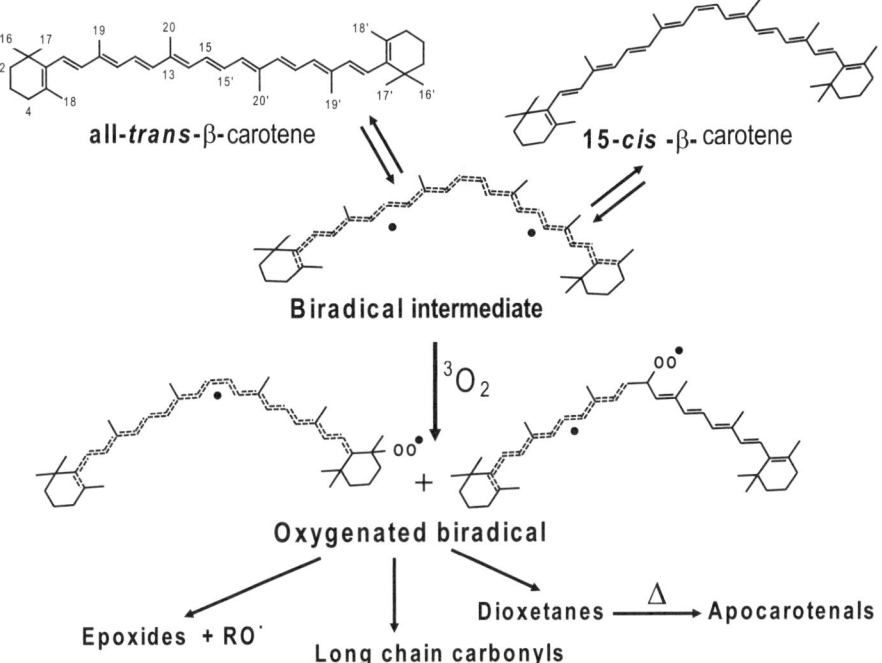

Fig. 7.15. Opening of a carotenoid double bond and formation of biradical intermediates that can react with molecular oxygen.

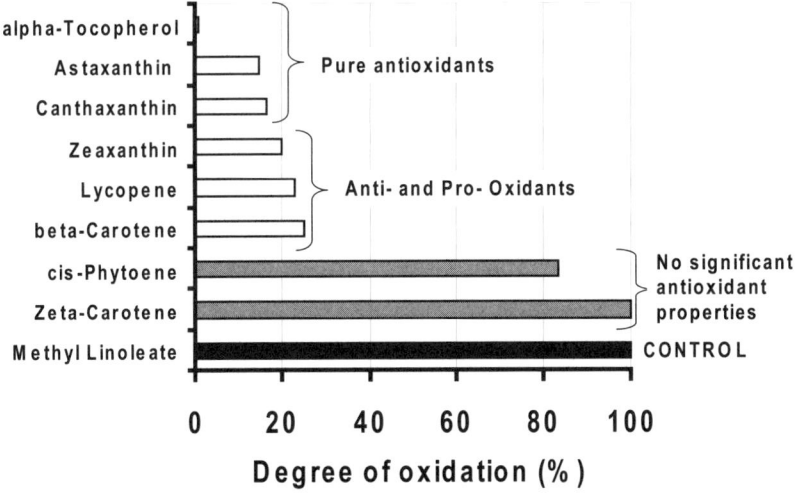

Fig. 7.16. Classification of carotenoids according to their anti- and pro-oxidant reactions (data from Martin et al. 1999).

## Effect of Heat on Carotenoid Oxidation and Cleavage

Heat causes isomerization of the carotenoids and enhances their reaction with oxygen and degradation to undesirable compounds (McKeown, 1965; Scotter, 1995). Heat-excited carotenoids ($^3$Car*) undergo *cis→trans* isomerization or react with oxygen to form the Car•-OO• diradical species discussed previously (Mordi, 1993). The Car•-OO• may rearrange to form dioxetanes, which degrade to apocarotenals, or react with radicals or other species to form peroxides that degrade to epoxides and a wide range of volatile compounds (Cole and Kapur, 1957; Day and Erdman, 1963; Mader, 1964; Isoe et al., 1969; La Roe and Shipley, 1970; Demole and Berthet, 1972; Schreier et al., 1979; Kawakani, 1982; Marty and Berest, 1986a,b; Kanasawud and Crouzet, 1990a,b; Crouzet and Kanasawud, 1992). Heat-excited carotenoids ($^3$Car*) isomerize to a number of mono- and poly- *cis*-isomers at C-9, C-13, and C-15 (Claes and Nakayama, 1959; Zechmeister, 1960; Claes, 1961; Sweeney and Marsh, 1970; Ogunlesi and Lee, 1979; Tsukida and Saiki, 1983). Carotenoids with *cis* configuration have a higher steric interaction between the two parts of the molecule and possess a higher potential energy than their *trans* isomers. In effect, *cis* carotenoids are less stable and are more susceptible to various oxidation reactions than their *trans* counterparts (Scotter, 1995). For example, 9-*cis*-β-carotene oxidizes preferentially and protects both all-*trans*-β-carotene and methyl linoleate from oxidation (Levin and Mokedy, 1994).

The thermal degradation of β-carotene under sufficient oxygen follow zero order kinetics (El-Tinay and Chichester, 1970; Kanasawud and Crouzet, 1990a). All-*trans*-β-carotene was resistant to degradation at 180°C and in the absence of mechanical mixing favored oxygen diffusion (Marty and Berest, 1986a). Non-volatile oxidation products of β-carotene include the five mono- or diepoxy compounds shown in Fig. 7.12, as well as five apocarotenals (β-apo-15-carotenal, β-apo-14'-carotenal, β-apo-12'-carotenal, β-apo-10'-carotenal, and β-apo-8'-carotenal), one polyene ketone (β-carotene-4-one), one monohydroxy (3 or 4)-5,8,5',8'-diepoxy derivative, and one dihydroxy derivative (*trans*-β-carotene-3,3'-diol) shown in Fig. 7.17 (Marty and Berest, 1986a,b; Ouyang et al., 1980, 1986; Vecchi et al., 1981; Kanasawud, 1984; Berest and Marty, 1992). In addition, a number of volatile oxidation products are formed, including β-ionone, 5,6-epoxy- β-ionone, dihydroactinidiolide, 2-hydroxy-2,6,6-trimethylcyclo-hexanone, 2.6,6-trimethylcyclohexen-1-one, 2,6,6-trimethyl-cyclohexanone, 1-carboxaldehyde-2,6,6-trimethyl-1-cyclohexene, and β-cyclocitral (Fig. 7.18) (Kanasawud and Crouzet, 1990a). At 30°C, only dihydroactinidiolide was produced and most of the other compounds appeared when temperature reached 50°C.

Fig. 7.19 shows the main thermal degradation products of all-*trans*-lycopene including 2-methyl-2-hepten-6-one (cleavage of C6-C7), geranial and neral (cleavage of C8-C9), and pseudoionone and 6-methyl-3,5-heptadien-3-one (cleavage of C10-C11) (Cole and Kapur, 1957; Schreier et al., 1977; Sieso and Crouzet, 1977; Drawert et al., 1981; Kanasawud and Crouzet, 1990b; Crouzet and Kanasawud, 1992). The main thermal degradation product of 9'-*cis*-bixin was identified as the *trans*-mono-

## Carotenoids and Lipid Oxidation Reactions 167

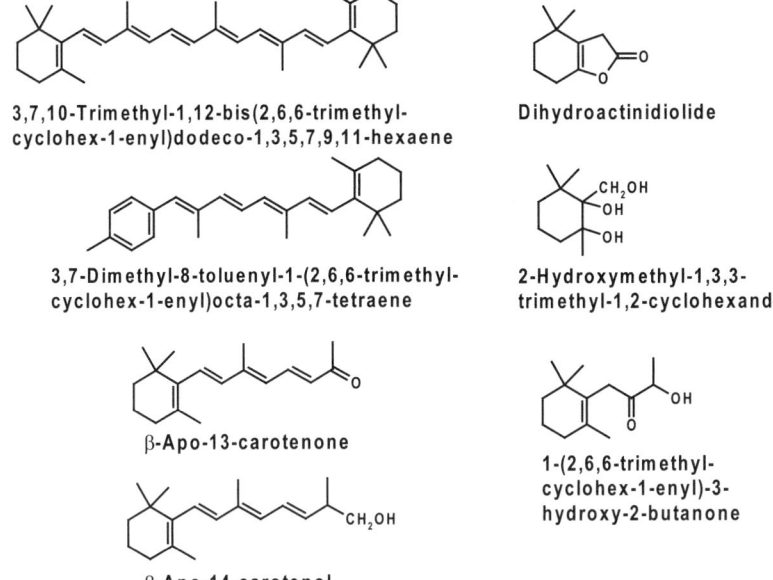

Fig. 7.17. Diverse oxidation products of β-carotene.

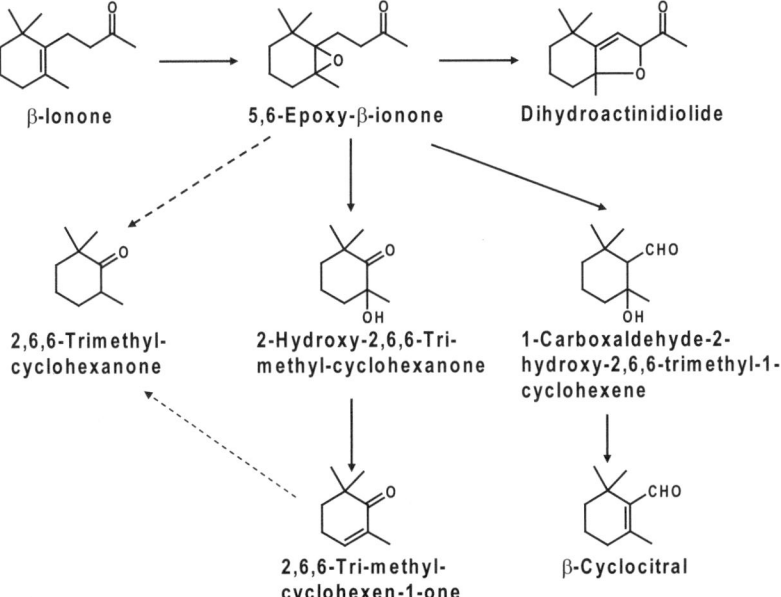

Fig. 7.18. Formation of some volatile oxidation products of β-carotene.

methyl ester of the 17-carbon polyene 4,8-dimethyl-tetradecahexaenedioic acid accompanied by the release of m-xylene (Fig. 7.20). The mechanism of this elimination consists of three steps; an eight electron conrotatory, a six electron disrotatory electrocyclic reaction, and opening of the four-membered ring (Scotter, 1995). Under thermal conditions, 9'-*cis*-bixin degrades to a much lesser extent to yield toluene and an 18-carbon polyene and to dimethyldihydronaphthalene and a 13-carbon tetraene.

## Synergism between Carotenoids and Tocopherols in Lipid Oxidation Reactions

Lipid oxidation reactions are often a combination of free radical and singlet oxygen reactions varying in proportion as dictated by the physicochemical conditions of the reaction(s). In these reactions, β-carotene acts mainly as a physical quencher of singlet oxygen and photosensitizers, having roughly 100 times the rate constant for physical quenching of $^1O_2$ and being much less reactive towards this species than α-tocopherol (Terao et al., 1980; DiMascio et al., 1989; Miki, 1991). Moreover, carotenoids are much more efficient as scavengers of alkoxyl radicals than tocopherols (Saran et al., 1980; Bors et al., 1982, 1984). For example, Trolox C, the water-soluble analog of α-tocopherol, is only about 30% as reactive towards alkoxyl radicals as the water-soluble carotenoid crocetin or canthaxanthin (Krinsky, 1989). The reaction kinetics of α-tocopherol and β-carotene in suppressing azo*bis*-isobutyronitrile (AIBN)-induced formation of malondialdehyde from unsaturated fatty acids in hexane differed significantly (Palozza et al., 1995). The addition of the chain inhibtor α-tocopherol led to a lag phase corresponding to the time of consumption of tocopherols but the propagation rate, which is dependent on the degree of fatty acid unsaturation, did not change. The addition of β-carotene, on the other hand, does not cause a lag phase but decreased the rate of propagation in a competitive way with fatty acids.

Carotenoids and tocopherols act synergistically to inhibit lipid oxidation reactions (Leibovitz et al., 1990; Palozza and Krinsky, 1992a). As mentioned *vide supra*, β-carotene reacts with peroxyl radicals and oxygen-forming propagative radicals, that is Car•-OOL and $^3$Car•-OO•, respectively, that would enhance the oxidation of the carotenoid itself and co-existing lipids leading to a pro-oxidant effect. The presence of an efficient peroxyl radical scavenger, such as α-tocopherol, would inhibit such reactions (Budowski and Bondi, 1960; El-Tinay and Chichester, 1970; Stratton et al., 1993). α-Tocopherol effectively retarded the co-oxidation of β-carotene and linoleate catalyzed by soybean lipoxygenase (Barimalaa and Gordon, 1988) and of β-carotene and oleate catalyzed by *tert*-butyl hydroperoxide and ferrous ions (Tsuchiya et al., 1992). On the other hand, by scavenging singlet oxygen generated from the decomposition of peroxyl radicals by the Russell mechanism (Russell, 1957), the carotenoids may help increase the antioxidant potency of tocopherols. In the presence of trace metal ions, able to dissociate hydroperoxides, the antioxidant activity of tocopherols is compromised and some carotenoids are more effective antioxidants (Miki, 1991; Kurashige et al., 1990; Esterbauer et al., 1991). For example, astaxanthin was 100-500

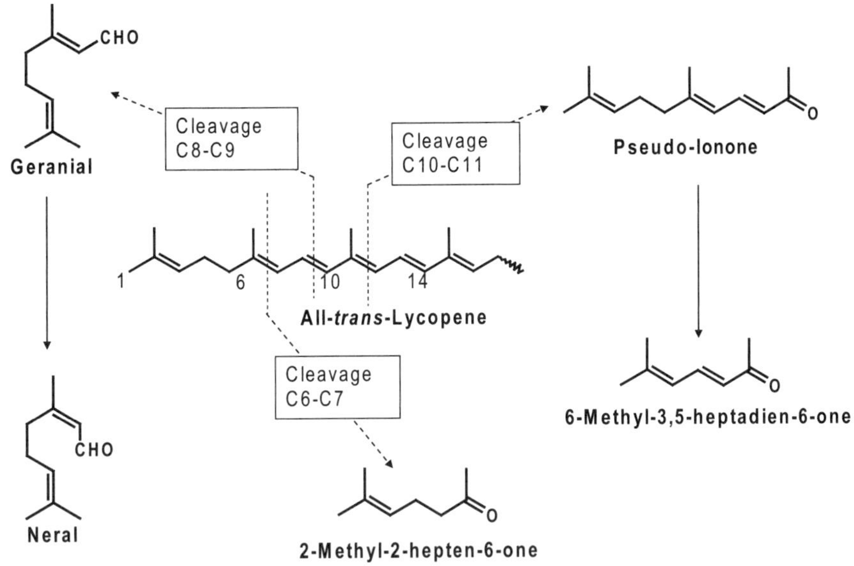

**Fig. 7.19.** Some decomposition products of lycopene.

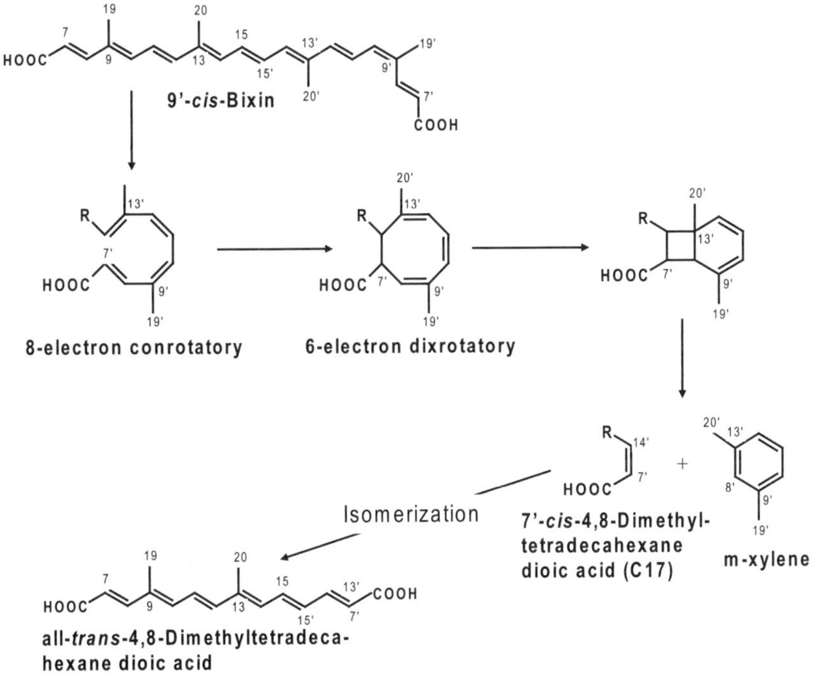

**Fig. 7.20.** Some decomposition products of 9'-cis-bixin.

times stronger than α-tocopherol in protecting rat liver mitochondria against ferrous ion-induced peroxidation (Marty and Berest, 1986a).

When tocopherols and carotenoids co-exist during lipid oxidation, the tocopherols preferentially scavenge hydroperoxyl radicals and get consumed (Tsuchihashi et al., 1995). For example, the antioxidant consumption during the oxidation of low-density lipoprotein particles was found to follow the following sequence: α-tocopherol > γ-tocopherol > lycopene > β-carotene (Esterbauer et al., 1989). However, when α-tocopherol and β-carotene were incorporated into soybean liposomal membranes, α-tocopherol was consumed faster when radicals were generated outside the membranes while β-carotene was consumed faster when radicals were generated within the membranes (Tsuchihashi et al., 1995). This might be explainable by differences in localization with α-tocopherol being localized at the surface of membrane (Buettner, 1993) and β-carotene being localized in the central hydrophobic region (Milon et al., 1986).

## Conclusions

The carotenoids, with their extended double bond system, are special lipid substrates when it comes to oxidation. Compared to a fair realization of the protective effects of carotenoids against photosensitized oxidations, the understanding of the chemistry of oxidation of carotenoid molecules and its relevance to the free radical oxidation of co-existing lipids is preliminary due to the wide range of reactions in which the extended system of conjugated double bonds can participate. This chapter provides a review of the state of the art of current knowledge in this area with the aim to serve, with other relevant literature, as a base for future research pertinent to the roles played by carotenoids in lipid oxidation reactions as well as food quality and food safety.

## References

Arya, S.S.; V. Natesan; D.B. Psrihar; and P.K. Vijayaraghavan. Stability of Beta-Carotene in Isolated Systems. *J. Food Technol.* **1979** *14*, 571-578.

ATBC, The Alpha-Tocopherol Beta Carotene Cancer Prevention Study Group The Effect of Vitamin E and Beta Carotene on The Incidence of Lung Cancer and Other Cancers in Male Smoker. *N. Engl. J. Med.* **1994** *330*, 1029-1035.

Bauernfeind, J.C. *Carotenoids as Colorants and Vitamin A Precursors: Technical and Nutritional Applications*, Academic Press: New York, 1981, pp. 48-319.

Barimalaa, I.S.; and M.H. Gordon. Cooxidation of β-Carotene by Soybean Lipoxygenase. *J. Agric. Food Chem.* **1988** *36*, 685-687.

Ben-Azziz, A.; S. Grossman; I. Ascarelli; and P. Budowski. Carotene Bleaching Activities of Lipoxygenase and Heme Proteins as Studied by a Direct Spectrophotometric Method. *Phytochemistry* **1971** *10*, 1445-1452.

Berest, C.; and C. Marty. Formation of Non-Volatile Compounds by Thermal Degradation of β–Carotene: Protection by Antioxidants. *Methods Enzymol.* **1992** *213*, 129-142.

Biyani, M.K.; and D.S. Sheorey. Carotenes, A Ray of Hope in Prevention of Cardiovascular Disorders, Cancers, and Cataract. *J. Assoc. Physicians India* **1994** *42*, 899-903.

Blain, J.A. Carotene Bleaching Activity of Plant Tissue Extract. *J. Sci. Food Agric.* **1970** *21*, 35-38.

Boey, P.L.; A. Nagao; J. Terao; K. Tanaka; T. Suzuki; and K. Takama. Antioxidant Activity of Xanthophylls on Peroxy Radical-Mediated Phospholipid Peroxidation. *Biochim. Biophys. Acta* **1992** *1126*, 178-184.

Bors, W.; C. Michel; and M. Saran. Organic Oxygen Radicals in Biology: Generation and Reactions. In *Oxygen and Oxy-Radicals in Chemistry and Biology,* Rodgers, M.A.; and Powers, E.L. Eds; Academic Press: New York, 1981, pp. 75-81.

Bors, W.; M. Saran; and C. Michel. Radical Intermediates Involved in Bleaching of the Carotenoid Crocin: Hydroxyl Radicals, Superoxide Anions, and Hydrated Electrons. *Int. J. Radiat. Biol. Relat. Stud. Phys. Chem. Med.* **1982** *41*, 493-501.

Bors, W.; C. Michel; and M. Saran. Inhibition of the Bleaching of the Carotenoid Crocin: A Rapid Test for Quantifying Antioxidant Activity. *Biochim. Biophys. Acta* **1984** *796*, 312-319.

Budowski, P.; and A. Bondi. Autoxidation of Carotenes and Vitamin A: Influence of Fat and Antioxidant. *Arch. Biochem. Biophys.* **1960** *89*, 66-73.

Buettner, G.R. The Pecking Order of Free Radicals and Antioxidants: Lipid Peroxidation, -Tocopherol, and Ascorbate. *Arch. Biochem. Biophys.* **1993** *300*, 535-543.

Burton, G.W. Antioxidant Action of Carotenoids. *J. Nutr.* **1989** *119*, 109-111.

Burton, G.W.; and K.U. Ingold. β-Carotene: An Unusual Type of Antioxidant. *Science* **1984** *224*, 569-573.

Carlsson, D.J.; T. Mendenhall; T. Suprunchuk; and D.M. Wiles. Singlet Oxygen Quenching in the Liquid Phase by Metal Chelates. *J. Am. Chem. Soc.* **1972** *94*, 8960-8962.

Chahine, M.H.; and J.M. deMan. Autoxidation of Corn Oil Under The Influence of Fluorescent Light. *Food Sci. Technol.* **1971** *4*, 24-28.

Chou, H.; and W. Breene. Oxidative Decoloration of Beta-Carotene in Low Moisture Model System. *J. Food Sci.* **1972** *37*, 66-68.

Christophersen, A.G.; H. Jun; K. Jorgensen; and L.H. Skibsted. Photobleaching of Astaxanthin and Canthaxanthin: Quantum Yields Dependence of Solvent, Temperature, and Wavelength of Irradiation in Relation to Packaging and Storage of Carotenoids Pigmented Salmonoids. *Z. Lebensm. Unters Forsch.* **1991** *192*, 433-439.

Claes, H. Energieübertragung von Angergtem Chlorophyll auf C40-Polyene mit Verschiedenen Chromophoren Gruppen. *Z. Naturforschg* **1961** *16B*, 445-454.

Claes, H.; and T.O.M. Nakayama. Das Photooxydative Ausbleichen von Chlorophyll in vitro in Gedenwart von Carotinen mit Verschiedenen Chromophoren Gruppen. *Z. Naturforschg* **1959** *14B*, 746-747.

Cohen, B.; S. Grossman; B.P. Klein; and A. Pinsky. Pigment Bleaching by Soybean Lipoxygenase Type-2 and the Effect of Specific Chemical Modifications. *Biochim. Biophys. Acta* **1985** *837*, 279-287.

Cole, E.R.; and N.S. Kapur. The Stability of Lycopene: Degradation by Oxygen and Oxidation During Heating of Tomato Pulps. *J. Sci. Food Agric.* **1957** *8*, 360-368.

Conn, P.F.; W. Schalch; and T.G. Truscott. The Singlet Oxygen and Carotenoid Interaction. *J. Photochem. Photobiol.* **1991** *B 11*, 41-47.

Craft, N.E.; and J.H. Soares. Relative Solubility, Stability, and Absorptivity of Lutein and β-Carotene in Organic Solvents. *J. Agric. Food Chem.* **1992** *40*, 431-434.

Crouzet, J.; and P. Kanasawud. Formation of Volatile Compounds by Thermal Degradation of Carotenoids. *Methods Enzymol.* **1992** *213*, 54-62.

Cutler, R.G. Carotenoids and Retinol: Their Possible Importance in Determining Longevity of Primate Species. *Proc. Natl. Acad. Sci. U.S.A.* **1984** *87*, 7627-7631.

Day, W.C.; and J.G. Erdman. Thermal Degradation Products of β-Carotene. *Science* **1963** *141*, 808-810.

Demole, E.; and D. Berthet. A Chemical Study of Burley Tobacco Flavour (*Nicotina tabacum*, L.). I. Volatile to Medium-Volatile Constituents (b.p. 84 ºC/0.001 Torr). *Helv. Chim. Acta* **1972** *55*, 1866-1882.

Devasagayam, T.P.A.; T. Werner; H. Ippendorf; H.D. Martin; and H. Sies. Synthetic Carotenoids, Novel Polyene Polyketones, and New Capsorubin Isomers as Efficient Quenchers of Singlet Molecular Oxygen. *Photochem. Photobiol.* **1992** *55*, 511-514.

DiMascio, P.; S. Kaiser; and H. Sies. Lycopene as the Most Efficient Biological Carotenoid Singlet Oxygen Quencher. *Arch. Biochem. Biophys.* **1989** *274*, 532-538.

DiMascio, P.; M.E. Murphy; and H. Sies. Antioxidant Defense Systems: The Role of Carotenoids, Tocopherols and Thiols. *Am. J. Clin. Nutr.* **1991** *53*, 194S-200S.

DiMascio, P.; A.R. Sundquist; T.P.A. Devasagayam; and H. Sies. Assay of Lycopene and Other Carotenoids As Singlet Oxygen Quenchers. *Methods Enzymol.* **1992** *213*, 429-438.

Drawert, F.; P. Schreier; S. Bhiwapurkar; and I. Heindze. Chemical-Technological Aspects for Concentration of Plant Aromas. In *Flavour'81,* Schreier, P. Ed., W. de Gruyter: Berlin, New York, 1981, pp. 649-663 .

Elahi, M.; and E.R. Cole. Oxidation of β-Carotene by Hydroperoxides. *Nature* **1964** *203*, 186-187.

Esterbauer, H.; G. Striegl; H. Puhl; and M. Rotheneder. Continuous Monitoring of in vivo Oxidation of Low-Density Lipoprotein. *Free Rad. Res. Commun.* **1989** *6*, 67-75.

Esterbauer, H.; M. Rotheneder; G. Striegl; and G. Waeg. The Role of Vitamin E in Preventing the Oxidation of Low-Density Lipoprotein. *Am. J. Clin. Nutr.* **1991** *53*, 314S-321S.

El-Tinay, A.H.; and C.O. Chichester. Oxidation of β-Carotene: Site of Initial Attack. *J. Org. Chem.* **1970** *35*, 2290-2293.

Everets, S.A.; S.C. Kundu; S. Maddix; and R.L. Willson. Mechanisms of Free Radical Scavenging by the Nutritional Antioxidant β-Carotene. *Biochem. Soc. Trans.* **1995** *23*, 230S-233S.

Fahrenholtz, S.R.; F.H. Doleiden; A.M. Trozollo; and A.A. Lamolla. On the Quenching of Singlet Oxygen by α-Tocopherol. *Photochem. Photobiol.* **1974** *20*, 505-509.

Farmillo, A.; and F. Wilkinson. On the Mechanism of Quenching of Singlet Oxygen in Solution. *Photochem. Photobiol.* **1973** *18*, 447-450.

Finar, I.L. *Organic Chemistry, vol 2,* 5th Edn,; Longman: London, U.K., 1981, pp. 463-491.

Foote, C.S. Photosensitized Oxidation and Singlet Oxygen: Consequencies in Biological Systems. In *Free Radicals in Biology, vol. 2,* Pryor, W.A. ed.; Academic Press: New York, 1976, pp. 85-133.

Foote, C.S. Quenching of Singlet Oxygen. In *Singlet Oxygen,* Wasserman, H.H.; and Murray, R.W. Eds., Academic Press: New York, 1979, pp. 139-171.

Foote, C.S.; and R.W. Denny. Chemistry of Singlet Oxygen. VIII. Quenching by β-Carotene. *J. Am. Chem. Soc.* **1968** *90*, 6233-6235.

Foote, C.S.; Y.C. Chang; and R.W. Denny. Chemistry of Singlet Oxygen. X. Carotenoid Quenching Parallels Biological Protection. *J. Am. Chem. Soc.* **1970a** *92*, 5216-5218.

Foote, C.S.; Y.C. Chang; and R.W. Denny. Chemistry of Singlet Oxygen. XI. cis/trans Isomerization of Carotenoids by Singlet Oxygen and a Probable Quenching Mechanism. *J. Am. Chem. Soc.* **1970b** *92*, 5218-5219.

Foote, C.S.; R.W. Denny; L. Weaver; Y.C. Chang; and J. Peters. Quenching of Singlet Oxygen. *Ann. N.Y. Acad. Sci.* **1970c** *171*, 130-148.

Friend, J. The Oxidation of β-Carotene by a Lipoxidase Linoleate System. *Biochem. J.* **1956** *64*, 19-20.

Friend, J. The Coupled Oxidation of β-Carotene by a Linoleate Lipoxidase System and by Autoxidizing Linoleate. *Chem. Ind.* **1958a** *20*, 597-598.

Friend, J. The Biochemical Oxidation of β-Carotene. *Qual. Plant Mat. Veg.* **1958b** *4-4*, 354-359.

Gaziano, J.M.; and C.H. Hennekens. The Role of Beta-Carotene in the Prevention of Cardiovascular Disease. *Ann. N.Y. Acad. Sci.* **1993** *691*, 796-799.

Gaziano, J.M.; J.E. Manson; L.G. Branch; G.A. Colditz; W.C. Willett; and J.E. Buring. Prospective Study of Consumption of Carotenoids in Fruits and Vegetables and Decreased Cardiovascular Mortality in the Elderly. *Ann. Epidemiol.* **1995** *5*, 255-260.

Gollnick, K.; and H. Kuhn. Ene-Reactions with Singlet Oxygen. In *Singlet Oxygen,* Wasserman, H.H.; and Murray, R.W. Eds., Academic Press: New York, 1979, pp. 287-427.

Gollnick, K.; and G.O. Schenk. Oxygen as Dienophile. In *1,4-Cycloaddition Reactions,* Hamer, J. Ed., Academic Press: New York, 1967, pp. 255-344.

Goodwin, T.W. Metabolism, Nutrition and Function of Carotenoids. *Ann. Rev. Nutr.* **1986** *6*, 273-297.

Grant, J.L.; V.J. Kramar; R. Ding; and L.D. Kispert. Carotenoid Cation Radicals: Electrochemical, Optical and ESR Study. *J. Am. Chem. Soc.* **1988** *110*, 2152-2157.

Grosch, W.; F. Weber; and K.H. Fischer. Bleaching of Carotenoid by the Enzyme Lipoxygenase. *Ann. Technol. Agric.* **1977** *26*, 133-137.

Gunstone, F.D.; and T.P. Hilditch. The Union of Gaseous Oxygen with Methyl Oleate, Linoleate and Linolenate. *J. Chem. Soc.* **1945** 836-841.

Haila, K.; and M. Heinonen. Action of β-Carotene on Purified Rapeseed Oil During Light Storage. *Lebensm.-Wiss. U., Technol.* **1994** *27*, 573-577.

Harborne J.B.; and H. Baxter. *Phytochemical Dictionary: A Handbook of Bioactive Compounds from Plants,* Taylor and Francis: London, U.K., 1993, pp. 745-746

Hasegawa, K.; J.D. Macmillan; W.A. Maxwell; and C.O. Chichester. Photosensitized Bleaching of β-Carotene with Light at 632.8 nm from a Continuous Gas Laser. *Photochem. Photobiol.* **1969** *9*, 165-169.

Henry, L.K.; G.L. Catignani; and S.J. Schwartz. Oxidative Degradation Kinetics of Lycopene, Lutein, and 9-*cis*- and All-*trans*-β-Carotene. *J. Am. Oil Chem. Soc.* **1998** *75*, 823-829.

Hirayama, O.; K. Nakamura; S. Hamda; and Y. Kobayasi. Singlet Oxygen Quenching Ability of Naturally Occuring Carotenoids. *Lipids* **1994** *29*, 149-150.

Hudson, B.S.; B.E. Kohler; and K.S. Schulten. Linear Polyene Electronic Structure and Potential Surfaces. In *Excited States, vol. 6;* Lim E.C. Ed., Academic Press, New York, 1982, pp. 1-95.

Hunter, R.F.; and R.M. Krakenberger. The Oxidation of β-Carotene in Solution by Oxygen. *J. Chem. Soc.* **1947** 1-4.

Ikediobi, C.O.; and H.E. Synder. Cooxidation of β-Carotene by an Isozyme of Soybean Lipoxygenase. *J. Agric. Food Chem.* **1977** *25*, 124-127.

Isoe, S.; S.B. Hyeon; S. Katsunura; and T. Sakan. Photo-Oxygenation of Carotenoids. I. The Formation of Dihydrpactinidiolide and β-Ionone from β-Carotene. *Tetrahedron Lett.* **1969** *10*, 279-281.

IUPAC, Commission on Nomenclature of Organic Chemistry and IUPAC-IUB Commission on Biochemical Nomenclature, Nomenclature of Carotenoids, *Pure Appl. Chem.* **1975** *41*, 407-431.

Jaffé, H.H.; and M. Orchin. *Theory and Applications of Ultraviolet Spectroscopy*, Wiley and Sons: New York, 1962, 624 pages.

Jezowska, I.; A. Wolak; W.I. Gruszecki; and K. Strzalka. Effect of Beta-Carotene on Structural and Dynamic Properties of Model Phosphatidylcholine Membranes. II. A $^{31}$P-NMR and $^{13}$C-NMR Study. *Biochim. Biophys. Acta* **1994** *1194*, 143-148.

Jialal, I.; E.P. Norkus; L. Cristol; and S.M. Grundy. β-carotene Inhibits the Oxidative Modification of Low Density Lipoprotein. *Biochim. Biophys. Acta* **1991** *1081*, 134-138.

Jorgensen, K.; and L.H. Skibsted. Carotenoid Scavenging of Radicals: Effects of Carotenoid Structure and Oxygen Partial Pressure on Antioxidant Activity. *Z. Lebensm. Unters Forsch.* **1993** *196*, 423-429.

Jovanovic, S.V.; I. Jovanovic; and L. Josimovic. Electron Transfer Reactions of Alkyl Peroxyl Radicals. *J. Am. Chem. Soc.* **1992** *114*, 9018-9021.

Jung, M.M.; and D.B. Min. Effects of Quenching Mechanisms of Carotenoids on the Photosensitized Oxidation of Soybean Oil. *J. Am. Oil Chem. Soc.* **1991** *68*, 653-658.

Kanasawud, P. *Identification des Composes Volatils Produits per Degradation des Carotenoids en Miliew Aqueux. Mecanismw de Formation*, Ph.D. thesis, Universite des Science et Techniques du Languedoc, Montpollier, France, 1984.

Kanasawud, P.; and C, Crouzet. Mechanism of Formation of Volatile Compounds by Thermal Degradation of Carotenoids in Aqueous Medium. 1. β-Carotene Degradation. *J. Agric. Food Chem.* **1990a** *38*, 237-243.

Kanasawud, P., and Crouzet, C. Mechanism of Formation of Volatile Compounds by Thermal Degradation of Carotenoids in Aqueous Medium. II. Lycopene Degradation. *J. Agric. Food Chem.* **1990b** *38*, 1238-1242.

Kasaikina, O.T.; A.B. Gagarina; and N.M. Emanuel. Reactivity of β-Carotene in the Interaction with Free Radicals. *Izv. Akad. Nauk. USSR Ser. Khim.* **1975** *10*, 2243-2246.

Kasaikina, O.T.; Z.S. Kartasheva; and A.B. Gagarina. Polyene Compounds as Free Radical Acceptors Free Radicals. *Izv. Akad. Nauk. USSR Ser. Khim.* **1981** *3*, 536-540.

Katusin-Rasem, B.; and D. Razem. Activity of Antioxidants in Solution and in Irradiated Heterogeneous System. *J. Am. Oil Chem. Soc.* **1994** *71*, 519-523.

Kawakani, M. Ionone Species Compounds from β-Carotene by Thermal Degradation in Aqueous Medium. *Nippon Nogekagaky Kaisi* **1982** *56*, 917-921.

Kennedy, T.A.; and D.C. Liebler. Peroxy Radical Oxidation of β-Carotene: Formation of β-Carotene Epoxides. *Chem. Res. Toxicol.* **1991** *4*, 290-295.

Kennedy, T.A.; and D.C. Liebler. Peroxy Radical Scavenging by β-Carotene in Lipid Bilayers: Effect of Oxygen Partial Pressure. *J. Biol. Chem.* **1992** *267*, 4658-4663.

Klein, B.P.; S. Grossman; G. King; B.S. Cohen; and A. Pinksky. Pigment Bleaching, Carbonyl Production, and Antioxidant Effect During the Anaerobic Lipoxygeanse Reaction. *Biochim. Biophys. Acta* **1984** *793*, 72-79.

Klein, B.P.; D. King; and S. Grossman. Co-oxidation Reactions of Lipoxygenase of plant Systems. *Adv. Free Rad. Biol. Med.* **1985** *1*, 309-343.

Kransovskii, A.A.; and I.I. Paramonova. Interaction of Singlet Oxygen with Carotenoids: Rate Constants of Physical and Chemical Quenching. *Biophysics* **1983** *28*, 769-774.

Krinsky, N.I. Biological Roles of Singlet Oxygen. In *Singlet Oxygen,* Wasserman, H.H.; and Murray, R.W. Eds,; Academic Press: New York, 1979, pp. 597-641.

Krinsky, N.I. Antioxidant Functions of Carotenoids. *Free Rad. Biol. Med.* **1989** *7*, 617-635.

Krinsky, N.L.; and S.M. Denke. The Interaction of Oxygen and Oxyradicals with Carotenoids. *J. Natl. Cancer Inst.* **1982** *69*, 205-210.

Kurashige, M.; Y. Okazoe; E. Okimasu; Y. Ando; M. Mori; W. Miki; M. Inoue; and K. Utsumi. Oxidative Injury of Biological Membranes Mediated by Free Radical and the Inhibition by Astaxanthin. *Cytoprotect. Biol.* **1989** *7*, 383-391.

Kurashige, M.; E. Okimasu; M. Inoue; and K. Utsumi. Inhibition of Oxidative Injury of Biological Membranes by Astaxanthin. *Physiol. Chem. Phys. Med. NMR* **1990** *22*, 27-38.

La Roe, E.G.; and P.A. Shipley. Whiskey Composition: Formation of Alpha- and Beta-Ionone by the Thermal Decomposition of Beta-Carotene. *J. Agric. Food Chem.* **1970** *18*, 174-175.

Land, E.J.; A. Sykes; and T.G. Truscott. Photochemistry pf Biological Molecules. II. The Triplet States of β-Carotene and Lycopene Excited by Pulse Radiolysis. *Photochem. Photobiol.* **1971** *13*, 311-320.

Lee, E.C.; and D.B. Min. Quenching Mechanisms of β-Carotene on the Chlorophyll-Sensitized Photooxidation of Soybean Oil. *J. Food Sci.* **1988** *53*, 1894-1895.

Lee, E.C.; and D.B. Min. Effects, Quenching Mechanisms and Kinetics of Carotenoids in Chlorophyll-Sensitized Photooxidation of Soybean Oil. *J. Agric. Food Chem.* **1990** *38*, 1630-1534.

Leibovitz, B.; M.L. Hu; and A.L. Tappel. Dietary Supplements of Vitamin E, Beta-Catotene, Coenzyme Q10, and Selenium Protect Tissues Against Lipid Peroxidation in Rat Tissue Slices. *J. Nutr.* **1990** *120*, 97-104.

Levin, G.; and S. Mokedy. Antioxidant Activity of 9-*cis* Compared to All-*trans*-β-Carotene in vitro. *Free Rad. Biol. Med.* **1994** *17*, 77-82.

Liebler, D.C.; and T.D. McClure. Antioxidant Reactions of β-Carotene: Identification of Carotenoid Radical Adducts. *Chem. Res. Toxicol.* **1996** *9*, 8-11.

Lim, B.P.; A. Nagao; J. Terao; K. Tanaka; T. Suzuki; and K. Takama. Antioxidant Activity of Xanthophylls on Peroxy Radical Mediated Phospholipid Peroxidation. *Biochim. Biophys. Acta* **1992** *1126*, 178-184.

Mader, I. Thermal Degradation of β-Carotene. *Science* **1964** *144*, 533-534.

Martin, H.D.; C. Ruck; M. Schmidt; S. Sell; S. Beutner; B. Mayer; and R. Walsh. Chemistry of Carotenoid Oxidation and Free Radical Reactions, *Pure Appl. Chem.* **1999** *71*, 2253-2262.

Marty, C.; and C. Berest. Degradation of *trans*-β-Carotene During Heating in Sealed Glass Tubes and Extrusion Cooking. *J. Food Sci.* **1986a** *51*, 698-702.

Marty, C.; and C. Berest. Degradation Products of *trans*-β-Carotene Produced During Extrusion Cooking. *J. Food Sci.* **1986b** *53*, 1880-1886.

Marty, C.; and C. Berest. Factors Affecting the Thermal Degradation of All-*trans*-β-Carotene. *J. Agric. Food Chem.* **1990** *38*, 1063-1067.

Mathews-Roth, M.M. Photoprotection by Carotenoids, *Fed. Proc.* **1987** *46*, 1890-1893.

Mathews-Roth, M.M. Recent Progress in the Medical Applications of Carotenoids. *Pure Appl. Chem.* **1991** *63*, 147-156.

Mathews-Roth, M.M.; T. Wilson; E. Fujimoriti; and N.I. Krinsky. Carotenoid Chromophore Length and Protection Against Photosensitization. *Photochem. Photobiol.* **1974** *19*, 217-222.

Mathis, P.; and J. Klero. The Triplet State of β-Carotene and Analog Polyenes of Different Length. *Photochem. Photobiol.* **1973** *18*, 343-346.

Mayo, F.R. Free Radical Autoxidation of Hydrocarbons. *Acc. Chem. Res.* **1968** *1*, 193-201.

McClure, T.D.; and D.C. Liebler. A Rapid Method for Profiling the Products of Antioxidant Reactions by Negative Ion Chemical Ionization Mass Spectrometry. *Chem. Res. Toxicol.* **1995** *8*, 128-135.

McKeown, G.G. Composition of Oil-Soluble Annatto Food Colors. III. Structure of the Yellow Pigments Formed by the Thermal Degradation of Bixin. *J. Assoc. Off. Anal. Chem.* **1965** *48*, 835-837.

Miki, W. Biological Functions and Activities of Animal Carotenoids. *Pure Appl. Chem.* **1991** *63*, 141-146.

Milon, A.; G. Wolff; G. Ourisson; and Y. Nakatani. Organization of Carotenoid-Phospholipid Bilayer Systems: Incorporation of Zeaxanthin, Astaxanthin, and Their C50 Homologues into Dimyristoylphosphatidylcholine Vesicles. *Helv. Chim. Acta* **1986** *69*, 12-24.

Mordi, R.C. Mechanisms of β-Carotene Degradation. *Biochem. J.* **1993** *292*, 310-312.

Mordi, R.C.; J.C. Walton; G.W. Burton; L. Hughes; K.U. Ingold; and D.A. Lindsay. Exploratory Study of β-Carotene Autoxidation. *Tetrahedron Lett.* **1991** *32*, 4203-4206.

Mordi, R.C.; J.C. Walton; G.W. Burton; L. Hughes; K.U. Ingold; D.A. Lindsay; and D.J. Moffatt. Oxidative Degradation of β-Carotene and β-Apo-8'-Carotenal, *Tetrahedron Lett.* **1993** *49*, 911-928.

Nielsen, B.R.; A. Mortensen; K. Jorgensen; and L.H. Skibsted. Singlet versus Triplet Reactivity in Photodegradation of C40 Carotenoids. *J. Agric. Food Chem.* **1996** *44*, 2106-2113.

Nishino, H. Cancer Chemoprevention by Natural Carotenoids and Their Related Compounds. *J. Cell Biochem.* **1995** *Suppl. 2*, 231-235.

Ogunlesi, A.T.; and C.Y. Lee. Effect of Thermal Processing on the Stereoisomerism of Major Carotenoids and Vitamin A Value of Carrots. *Lebensmittel Wissenschaft und Technologie* **1979** *4*, 311-318.

Olson, J.A.; and N.L. Krinsky. Introduction: The Colorful Fascinating World of the Carotenoids, Important Physiological Modulators. *FASEB J.* **1995** *9*, 1547-15550.

Onyewu, P.N.; C.-T. Ho; and H. Daun. Characterization of β-Carotene Thermal Oxidation Products in a Model Food System. *J. Am. Oil Chem. Soc.* **1986** *63*, 1437-1441.

Orlandi, G.; F. Zerbetto; and M.Z. Zgierski. Theoretical Analysis of Spectra of Short Polyenes. *Chem. Revs.* **1991** *91*, 867-891.

Oshima, S.; F. Ojima; H. Sakamoto; Y. Ishiguro; and J. Terao. Inhibitory Effect of β-Carotene and Astaxanthin on Photosensitized Oxidation of Phospholipid Bilayers. *J. Nutr. Sci. Vitaminol.* **1993** *39*, 607-615.

Ouyang, J.; H. Daun; S. Chang; and C.T. Ho. Formation of Carbonyl Compounds from β-Carotene During Palm Oil Deodorization. *J. Food Sci.* **1980** *43*, 1214-1217.

Ozhogina, O.A.; and O.T. Kasaikina. β-Carotene as an Interceptor of Free Radicals. *Free Rad. Biol. Med.* **1995** *19*, 575-581.

Palozza, P.; and N.L. Krinsky. The Inhibition of Radical-Initiated Peroxidation of Microsomal Lipids by Both α-Tocopherol and β-Carotene. *Free Rad. Biol. Med.* **1991** *11*, 407-414.

Palozza, P.; and N.I. Krinsky. β-Carotene and α-Tocopherol as Synergistic Antioxidants. *Arch. Biochem. Biophys.* **1992a** *297*, 184-187.

Palozza, P.; and N.I. Krinsky. Astaxanthin and Canthaxanthin are Potent Antioxidants in a Membrane Model. *Arch. Biochem. Biophys.* **1992b** *297*, 291-295.

Palozza, P.; S. Moualla; and N.I. Krinsky. Effects of β-Carotene and α-Tocopherol on Radical-Initiated Peroxidation of Microsomes, *Free Rad. Biol. Med.* **1992** *13*, 127-136.

Palozza, P.; C. Luberto; and G.M. Bartoli. The Effect of Fatty Acid Unsaturation on the Antioxidant Activity of Beta-Carotene and Alpha-Tocopherol in Hexane Solutions. *Free Rad.*

*Biol. Med.* **1995** *18*, 943-948.

Papadopolou, K.; and J.M. Ames. Kinetics of All-*trans*-β-Carotene Degradation on Heating With and Without Phenylalanine. *J. Am. Oil Chem. Soc.* **1994** *71*, 893-896.

Parker, M.E.; E.H. Havey; and E.S. Stateller. *Elements of Food Engineering*, Reinhold Publishing Co.: New York, 1952, 386 pages.

Paul, G.; R. Redtke; R. Heiss; and K. Becker. Influence of Light on the Oxidative Deterioration of Edible Oils, IV. Dependence of the Rate of Oxidation on the Wavelength of Incident Light. *Fette Seifen Anstrichmittel* **1972a** *74*, 359-366.

Paul, G.; R. Heiss; K. Becker; and R. Redtke. Influence of Light on the Oxidative Deterioration of Edible Oils, II. Influence of Oxygen Partial Pressure and Light Intensity on Rate of Oxidation. *Fette Seifen Anstrichmittel* **1972b** *74*, 120-126.

Paul, G.; K. Becker; R. Redtke; and R. Heiss. Influence of Light on the Oxidative Deterioration of Edible Oils, V. Reaction Kinetics. *Fette Seifen Anstrichmittel* **1972c** *74*, 484-491.

Pfander, H., Ed. *Key to Carotenoids, 2nd Ed.,* Basel: Birkhäser, Switzerland, 1987, 296 pages.

Prabhala, R.H.; H.S. Garewal; F.L. Meyskens; and R.R. Watson. Immunomodulation in Humans Caused by Beta-Carotene and Vitamin A, *Nutr. Res.* **1990** *10*, 1473-1486.

Prabhala, R.H.; M. Braune; H.S. Garewal; and R.R. Watson. Influence of Beta-Carotene on Immune Functions. *Ann. N.Y. Acad. Sci.* **1993** *691*, 262-263.

Pryor, W.A.; D.L. Fuller; and J.P. Stanely. Reactions of Radicals. 41. Reactivity of the Methyl Radical. *J. Am. Chem. Soc.* **1972** *94*, 1632-1638.

Pryor, W.A.; and C.K. Govindan. Decomposition of Triphenylphosphine Ozonide in the Presence of Spin Traps. *J. Org. Chem.* **1981** *45*, 4679-4682.

Pullman, B.; and A. Pullman. *Quantum Biochemistry,* Interscience Publ.: New York, 1963, p. 445.

Radtke, R.; P. Smits; and R. Heiss. Influence of Light of Varying Intensity and Wavelength on the Oxidative Deterioration of Edible Oils. II. Experimental Results and Discussion. *Fette Seifen Anstrichmittel* **1970** *72*, 497-504.

Ramakrishnan, T.V.; and F.J. Francis. Stability of Carotenoids in Model Aqueous Systems. *J. Food Qual.* **1979a** *2*, 177-189.

Ramakrishnan, T.V.; and F.J. Francis. Coupled Oxidation of Carotenoids in Fatty Acid Esters of Varying Unsaturation. *J. Food Qual.* **1979b** *2*, 277-287.

Rawls, H.R.; and P.J. van Santen. A Possible Role for Singlet Oxygen in the Initiation of Fatty Acid Autoxidation. *J. Am. Oil Chem. Soc.* **1970** *54*, 234-236.

Rodgers, M.A.J.; and A.L. Bates. Kinetic and Spectroscopic Features of Some Carotenoid Triplet States Sensitization by Singlet Oxygen. *Photochem. Photobiol.* **1980** *31*, 533-537.

Russell, G.A. Deuterium Isotope Effects in the Autoxidation of Aralkyl Hydrocarbons: Mechanism of the Interaction of Peroxy Radicals. *J. Am. Chem. Soc.* **1957** *79*, 3871-3877.

Saleh, M.H.; and B. Tan. Separation and Identification of *cis/trans* Carotenoid Isomers. *J. Agric. Food Chem.* **1991** *39*, 1438-1443.

Saran, M.; C. Michel; and W. Bors. The Bleaching of Crocin by Oxygen Radicals. In *Chemical and Biochemical Aspects of Superoxide Dismutase,* Bannister, J.V.; and Hill, H.O.A., Eds.; Elsevier: New York, 1980, pp. 38-44.

Satter, A.; J.M. DeMan; and J.C. Alexander. Effect of Wavelength on Light-Induced Quality Deterioration of Edible Oils and Fats. *Can. Inst. Food Sci. Technol. J.* **1976a** *9*, 108-113.

Satter, A.; J.M. DeMan; and J.C. Alexander. Light-Induced Oxidation of Edible Oils and Fats. *Lebensm. Wiss. Technol.* **1976b** *9*, 149-152.

Schenck, G.O.; and G. Schade. [Photosensitized $O_2$-Transfer in Presence of β-Carotene Under

Exclusion of Singlet Oxygen]. *Chimia* **1970** *24*, 13-16.

Schreier, P.; F. Drawert; and A. Junker. The Quantitative Composition of Natural and Technologically Changed Aromas of Plants. IV. Enzymic and Thermal-Reaction Products Formed During the Processing of Tomatoes. *Z. Lebensm. Unters. Forsch.* **1977** *165*, 23-27.

Schreier, P.; F. Drawert; and S. Bhiwapurkar. Volatile Compounds Formed by Thermal Degradation of β-Carotene. *Chem. Microbiol. Technol. Lebensm.* **1979** *6*, 90-91.

Scott, G. *Atmospheric Oxidation and Antioxidants*, Second edn., Elsevier Publishing Co.: New York, 1992, 215-219.

Scotter, M.J. Characterization of the Colored Thermal Degradation Products by Bixin from Annato and a Revised Mechanism for Their Formation. *Food Chem.* **1995** *53*, 177-185.

Sies, H.; W. Stahl; and A.R. Sundquist. Antioxidant Functions of Vitamins: Vitamin E and C, Beta-Carotene and Other Carotenoids. *Ann. N.Y. Acad. Sci.* **1992** *669*, 7-20.

Sieso, V.; and J.C. Crouzet. Tomato Volatile Components: Effects of Processing. *J. Am. Hotr. Sci.* **1977** *95*, 461-464.

Simpson K.L. Relative Value of Carotenoids as Precursors of Vitamin A. *Proc. Nutr. Soc.* **1983** *42*, 7-17.

Smith, G.N.; and J.B. Sumner. The Induced Reaction Between Methyl Linoleate and Bixin During Oxidation by Lipoxidase. *Arch. Biochem.* **1948** *17*, 75-80.

Stanescu, V.; and T. Eisenburger. Hydrolytic Changes in Lard During Rancidity Development: Prevention and Elimination of Rancidity. *Industri Alimentari* **1969** *20*, 369.

Stahl, W.; and H. Sies. Physical Quenching of Singlet Oxygen and *cis/trans* Isomerization of Carotenoids. *Ann. N.Y. Acad. Sci.* **1993** *691*, 10-19.

Stratton, S.P.; W.H. Schaefer; and D.C. Liebler. Isolation and Identification of Singlet Oxygen Oxidation Products of β-Carotene. *Chem. Res. Toxicol.* **1993** *6*, 542-547.

Strzalka, K.; and W.I. Gruszecki. Effect of Beta-Carotene on Structural and Dynamic Properties of Model Phosphatidylcholine Membranes. I. An ESR Spin Label Study. *Biochim. Biophys. Acta* **1994** *1194*, 138-142.

Sumner, R.J. Lipid Oxidase Studies. III. The Relation Between Carotene Oxidation and the Enzyme Peroxidation of Unsaturated Fats. *J. Biol. Chem.* **1942** *146*, 215-218.

Sundqvist, A.; K. Briviba; and H. Sies. Singlet Oxygen Quenching by Carotenoids. *Methods Enzymol.* **1994** *234*, 384-388.

Suzuki, H. *Electronic Absorption Spectra and Geometry of Organic Molecules*, Academic Press, New York, 1967, 568 pages.

Sweeney, J.P.; and A.C. Marsh. Vitamins and Other Nutrients: Separation of Cartene Stereoisomers in Vegetables. *J. Assoc. Off. Anal. Chem.* **1970** *53*, 937-940.

Takahashi, A.; N. Shibasaki-Kitakawa; and T. Yonemoto. Kinetic Model for Autoxidation of β-Carotene in Organic Solutions. *J. Am. Oil Chem. Soc.* **1999** *76*, 897-903.

Takahashi, A.; J. Suzuki; N. Shibasaki-Kitakawa; and T. Yonemoto. A Kinetic Model for Co-Oxidation of β-Carotene with Oleic Acid. *J. Am. Oil Chem. Soc.* **2001** *78*, 1203-1207.

Takahashi, A.; N. Shibasaki-Kitakawa; and T. Yonemoto. A Rigorous Kinetic Model for β-Carotene Oxidation in the Presence of an Antioxidant α-Tocopherol. *J. Am. Oil Chem. Soc.* **2003a** *80*, 1241-1247.

Takahashi, A.; N. Shibasaki-Kitakawa; and T. Yonemoto. Kinetic Analysis of β-Carotene Oxidation in a Lipid Solvent With or Without an Antioxidant. In *Lipid Oxidation Pathways*, vol. 1, Kamal-Eldin, A, Ed., AOCS Press: Champaign, Illinois, 2003b, pp. 111-137.

Terao, J. Antioxidant Activity of β-Carotene and Related Carotenoids in Solution. *Lipids* **1989** *24*, 659-661.

Terao, J.; and S. Matsushita. Products Formed by Photosensitized Oxidation of Unsaturated Fatty Acid Esters. *J. Am. Oil Chem. Soc.* **1977** *54*, 234-238.

Terao, J.; R. Yamauchi; H. Murkami; and S. Matsushita. Inhibitory Effects of Tocopherols and β-Carotene on Singlet Oxygen-Initiated Photooxidation of Methyl Linoleate and Soybean Oil. *J. Food Process Preserv.* **1980** *4*, 79-93.

Terao, J.; P.L. Boey; F. Ojima; A. Nagao; T. Suzuki; and K. Takama. Astaxanthin as Chain Breaking Antioxidant in Phospholipid Peroxidation. In *Oxygen Radicals,* Yagi, K.; Kondo, M.; Niki, E.; and Yoshikawa, T. Eds., Elsevier Sci. publ.: Amesterdam, 1992, pp. 657-660.

Tookey, H.L.; R.G. Wilson; R.L. Lohman; and H.J. Dutton. Coupled Oxidation of Carotene and Linoleate by Lipoxidase. *J. Biol. Chem.* **1958** *230*, 65-72.

Thrash, R.J.; H.L-B. Fang; and G.E. Leori. On the Role of Forbidden Low-Lying Excited States of Light Harvesting Carotenoids in Energy Transfer in Photosynthesis. *Photochem. Photobiol.* **1979** *29*, 1049-1050.

Truscott, T.G. The Photophysics and Photochemistry of the Carotenoids. *J. Photochem. Photobiol. B.* **1990** *6*, 359-371.

Tsuchihashi, H.; M. Kigoshi; and E. Niki. Action of β–Carotene as an Antioxidant Against Lipid Peroxidation. *Arch. Biochem. Biophys.* **1995** *323*, 137-147.

Tsuchiya, M.; G. Scita; H.-J. Freistleben; V. Kagan; and L. Packer. Antioxidant Radical-Scavenging Activity of Carotenoids and Retinoids as Compared to α-Tocopherol. *Methods Enzymol.* **1992** *213*, 460-472.

Tsukida, K.; and K. Saiki. Thermal Stereoisomerization of All-(*E*)-β-Carotene, (*Z*)-β-Carotene, and Electrocyclized β-Carotene. *J. Nutr. Sci. Vitaminol.* **1983** *29*, 111-122.

Vecchi, M.; G. Englert; R. Maurer; and V. Meduna. Separation and Characterization of β-Carotene Isomers. *Helv. Chim. Acta* **1981** *64*, 2746-2758.

Vile, G.F.; and C.C. Winterbourn. Inhibition of Adriamycin-Promoted Lipid Peroxidation by Beta-Carotene, Alpha-Tocopherol and Retinol at High and Low Oxygen Partial Pressure. *FEBS Lett.* **1988a** *238*, 353-356.

Vile, G.F.; and C.C. Winterbourn. Adriamycin Dependent Peroxidation of Rat Liver Microsomes Catalyzed by Iron Chelates and Ferritin: Maximum Peroxidation at Low Oxygen Partial Pressure. *Biochem. Pharmacol.* **1988b** *37*, 2893-2897.

Wasielewski, M.R.; D.G. Johnson; E.G. Bradford; and L.D. Kispert. Temperature Dependence of the Lowest-Excited Singlet State Lifetime of All-*trans*-β-Carotene and Fully Degraded All-*trans*-β-Carotene. *J. Chem. Phys.* **1989** *91*, 6691-6697.

Weber, F.; and W. Grosch. Co-Oxidation of a Carotenoid by the Enzyme Lipoxygenase: Influence on the Formation of Linoleic Acid Hydroperoxides, *Z. Lebensm.-Unters.-Forsch.* **1976** *161*, 223-230.

Wilkinson, F.; and J.G. Brummer. Rate Constants for the Decay and Reactions of the Lowest Excited State of Molecular Oxygen in Solution. *J. Phys. Chem. Ref. Data* **1981** *10*, 809-899.

Wolff, C.; and H.T. Witt. On Metastable States of Carotenoids in Primary Events of Photosynthesis. *Z. Naturforschg* **1969** *24B*, 1031-1037.

Woodall, A.A.; G. Britton; and M.J. Jackson. Antioxidant Activity of Carotenoids in Phosphatidylcholine Vesicles: Chemical and Structural Considerations. *Biochem. Soc. Trans.* **1995** *23*, 133S.

Yamauchi, R.; N. Miyake; H. Inoue; and K. Kato. Products Formed by Peroxy Radical Oxidation of β-Carotene. *J. Agric. Food Chem.* **1993** *41*, 708-713.

Zechmeister, L.; A.L. LeRosen; W.A. Schroeder; A. Polgar; and L. Pauling. Spectral Characteristics and Configuration of Some Stereoisomeric Carotenoids Including Prolycopene and Pro-γ-Carotene. *J. Am. Chem. Soc.* **1943** *65*, 1940-1951.

Zechmeister, L. *cis-trans* Isomeric Carotenoid Pigments. In *Progress in The Chemistry of Natural Products, vol. 18,* Zechmeister, L. Ed., Springer-Verlag: Vienna, Austria, 1960, pp. 223-239.

Zinsou, C. Degradation Enzymatique de β-Carotene. *Physiol. Veg.* **1971** *9*, 149-167.

# 8

# Co-oxidation of Proteins by Oxidizing Lipids

Karen M. Schaich
Department. of Food Science, Rutgers University, 65 Dudley Rd., New Brunswick, NJ, 08901-8520

## Introduction and Overview of Macromolecular Damage

Lipid oxidation is the chemical reaction that most limits shelf life of foods, and it is increasingly being recognized as a major contributor to oxidative damage in vivo. In foods, the most obvious indicators of lipid oxidation or "rancidity" are off-odors and off-flavors that arise directly from lipid oxidation products, particularly aldehydes. Less recognized, but perhaps even more important in both foods and biological systems, are the co-oxidations that broadcast oxidative damage from lipids to all kinds of molecules via reactions of lipid free radicals in early stages of oxidation and later reactions of product epoxides and carbonyls as oxidation progresses. The critical effect of co-oxidations is to redirect the damage processes and confuse the reaction picture, quenching propagation of lipid free radical chains and decreasing apparent lipid oxidation (as commonly measured), but leaving in its place footprints of lipid oxidation in damage to proteins, DNA, and other molecules.

Damage to proteins from oxidizing lipids has been recognized for decades, but perspectives have changed with improvements in analytical capabilities. Most early studies were rather global in approach and primarily reported changes in general behaviors (texture, crosslinking, scission, loss of nutritional value), molecular functionality (enzyme activity, browning, and color fading in foods), and alterations in molecular function such as cell signaling, gene response, and apoptosis in vivo (Fig. 8.1) with only cursory, if any, determination of mechanisms. Model system studies with isolated amino acids provided more information about potential damage processes, but these are not always perfect models for reactions in proteins. In the past ten years there has been a marked shift to more sensitive analysis of reaction products facilitated by advances in chromatography in tandem with mass spectrometry, development of immobilized enzymes that hydrolyze proteins without destroying individual modified amino acids, and increased use of immunological methods to track specific changes of intact proteins in tissues. In some ways, recent gains in understanding using these methods have only raised more questions about the exact role of lipid oxidation products in cellular cytotoxicity and the detailed mechanisms involved, both direct and from co-oxidations, and relative to other oxidants such as hydroxyl radicals.

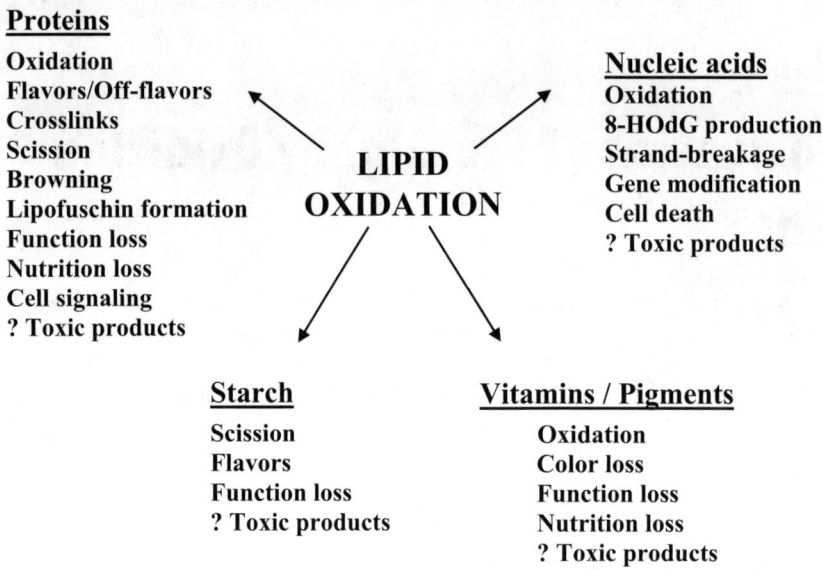

Fig. 8.1. Types of damage that occur when oxidizing lipids co-oxidize cellular molecules in foods and living tissues (plant and animal).

Although it is clear that co-oxidations by lipids do occur and contribute to protein deterioration, quantitative and qualitative cause-effect relationships remain poorly established except in some model systems. Until recently, lipid hydroperoxides and/or aldehydes were always measured when determining lipid oxidation involvement in food quality loss or in pathology, while their co-oxidation products have been largely ignored. The problem of identifying lipid oxidation damage is confounded further in that oxidizing lipids induce nearly the same protein oxidations and react with the same amino acids as hydroxyl radicals (Davies 1987a, 1987b). As a consequence, it is very likely that the true extent of lipid oxidation in both foods and tissues is consistently underestimated and macromolecular damage, especially in vivo, is routinely attributed to other oxidants (e.g., hydroxyl radical, HO•).

Reactions of oxidizing lipids with proteins are important to food science, chemistry, biochemistry, and medicine, and research approaches taken by each field have varied tremendously. Information gained by each of these fields needs to be integrated into a comprehensive picture of the chemistry, causes, and effects of protein co-oxidation by lipids. To begin the process and to encourage broader consideration and more accurate determination of the impact of oxidizing lipids in complex systems, this chapter links commonly observed macromolecular modifications in protein properties to the chemistry of four classes of lipid oxidation products—free radicals, hydroperoxides, epoxides, and carbonyl (mostly aldehyde) secondary products—differentiating the types of reactions and patterns of damage induced by each oxidant and showing where such reactions occur in some specific food applications and pathological conditions. Due to space limitations, most mechanistic details of lipid reactions with proteins, as well as specific reactions and products of individual amino acids, will be presented in a subsequent publication.

## Causative Agents in Molecular Damage to Proteins from Oxidizing Lipids

When analyzing lipid oxidation in either foods or physiological tissues, we clearly need to look beyond peroxide and TBARS values to include alternate lipid pathways (Schaich, 2005) and co-oxidation products (Schaich, 1980b; Borg and Schaich, 1984). As noted previously, oxidizing lipids generate multiple reactive species, all of which have potential to react with non-lipid molecules:

- Peroxyl and alkoxyl radicals transfer radicals to other molecules, leading to crosslinking and polymerization, molecular scissions, and a variety of co-oxidations.

- Hydroperoxides hydrogen bond to proteins and nucleic acids, leading to induced decomposition in situ and subsequent H abstraction from or lipid radical addition to protein sites.

- Epoxides bind to proteins, forming adducts.

- Carbonyl products, particularly aldehydes, participate in a variety of addition reactions leading to the formation of adducts, crosslinking of macromolecules, fluorescent products, and browning. Both saturated and unsaturated aldehydes also autoxidize and lead to free radical damage.

Because of the dynamic nature of lipid oxidation, these four classes of lipid oxidation products react independently, and sequentially. In rapidly oxidizing or extensively oxidized systems, they can react simultaneously. It is both dishonest and inaccurate to claim that any one of these oxidants is the major damaging agent. The apparent dominant oxidant changes with the reaction system, the conditions for oxidation, the timing of the reaction before analysis, the methodology used to detect both lipid and co-oxidation products, and the endogenous processes that may degrade or remove some products. Most attention has been given to aldehyde reactions, particularly in medical applications, because individual aldehydes can be isolated and reacted with proteins in targeted studies, and the adduct products tend to be more stable and less rapidly cleared in vivo. As a result, the reactions are easier to follow. Lipid radical and hydroperoxide reactions have not been studied with the same intensity or scrutiny because the reaction systems are less easy to define and control, the intermediates and products are constantly transforming, and they are more difficult to track and isolate. Pathways to clear minimally oxidized proteins in vivo are known, but the long-term effects of replacing hydrophilic oxidants, such as hydroxyl radicals (HO•), with hydrophobic oxidants from lipids has not been evaluated.

The discussion of lipid oxidants and associated damage presented in the rest of this chapter is intended to demonstrate the complexity of lipid co-oxidations of proteins and encourage research that investigates beyond individual lipid oxidants and considers the dependence of protein oxidation on reaction system and solvent, reaction time, concentration of lipid oxidation products, and other factors before assigning causality. In most cases, protein co-oxidations in foods and tissues involve multiple lipid oxidants that generate a wide range of protein products. The task for the future is to decipher the pathways that sequentially or simultaneously lead to each.

Although presented first, the damage reactions reviewed in this chapter were actually elucidated after observations of global effects as a means to explain why and how amino acid destruction, crosslinking, formation of fluorescent products, and loss of enzyme and other functional activities occur. Some connections of basic chemistry to specific proteins, foods, or pathological conditions are made along the way to provide orientation and practical grounding. However, the main purpose of this section is to detail the various reactions of lipid oxidants that lead to co-oxidation of proteins to illustrate their similarities and differences and to provide a conceptual framework for explaining protein changes. The global effects on proteins discussed in the section "Reactions Underlying Molecular Damage to Proteins" are all mediated by each of the four lipid oxidants. As will be shown, much of the detailed localized chemistry from this section is repeated in each effect—just the consequences of the chemistry differ depending on the protein and the reaction conditions.

## Transfer of Lipid Radicals to Proteins

### Evidence

The first reactive lipid species to form are free radicals, and as long as the lipids are in close proximity to susceptible protein sites, free radical transfer to proteins can occur rapidly, either by hydrogen abstraction or radical addition. Electron paramagnetic resonance (EPR or ESR, electron spin resonance) first provided direct evidence that oxidizing lipids transfer free radicals to proteins in pilot studies by Roubal and Tappel (Roubal and Tappel, 1966a; Roubal, 1970), then in more definitive studies by Schaich and Karel (1975,1976; Schaich, 1980b). Dry systems were used to stabilize protein radicals, eliminate interference of water, and facilitate radical detection. Starting with unoxidized lipid, EPR signal intensity increased just behind increase in peroxide values (Schaich and Karel, 1975). When pre-oxidized lipid was mixed with powdered proteins, radical transfer began essentially instantaneously and detectable signals developed in minutes (Schaich, unpublished data). EPR signals show a strong singlet center line with $g$ values of 2.0035–2.0060 (Figures 8.2 and 8.3). Lower g-values in this range are typical of N•; higher g-values are more likely NO• (nitrogens are immobilized so expected triplet signals are not observed). For non-sulfhydryl proteins, spectra show little structure and total spectrum widths are narrower (~50 G) (Fig. 8.2). Spectra from sulfur proteins are broader (~75 G), showing strong high field peaks with g ~ 2.015 and 2.023 from sulfinyl (RSO•) and/or sulfonyl (RSOO•) radicals, consistent with formation of various sulfur oxide products (Finley and Lundin, 1980). Shoulders (g ~ 2.001) from alkoxyl or peroxyl radicals also are evident in some spectra (Fig. 8.3).

What do EPR spectra reveal about lipid radical transfer sites? Clearly, thiol (SH) groups on cysteine are major targets for radical transfer from lipids. Cysteine reacted with oxidizing lipid forms the same sulfur radicals as in proteins, but the g ~ 2.015, 2.023 peaks are now dominant; some H abstraction from the α-amino group is also evident (Fig. 8.4). However, an obligate prerequisite for reaction is accessibility. Thiols on the protein surface react with lipids very rapidly at very low levels of oxidation; sulfur radical peaks are the first detected in protein EPR spectra, and they continue to

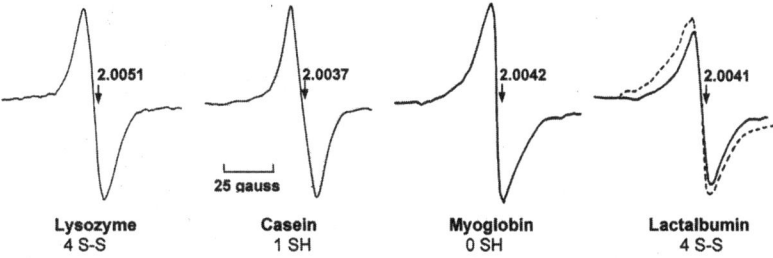

**Fig. 8.2.** EPR signals of free radicals induced in non-sulfhydryl proteins by reaction with oxidizing methyl linoleate in dry model systems. Lactalbumin: solid line, 2 mW power; dotted line, 20 mW power, showing minimal contributions from radicals with different saturation characteristics. Scale for all spectra as indicated for casein. Adapted from (Schaich and Karel, 1976).

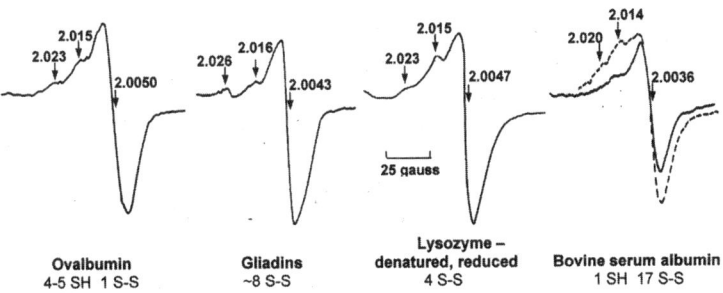

**Fig. 8.3.** EPR signals of free radicals induced in sulfhydryl proteins by reaction with oxidizing methyl linoleate in dry model systems. Scale for all spectra as indicated for denatured lysozyme. Bovine serum albumin: solid line, 2 mW power; dotted line, 20 mW power, enhancing peroxyl and sulfoxyl signal detection at higher power and providing clear evidence of multiple radical centers. Adapted from (Schaich and Karel, 1976).

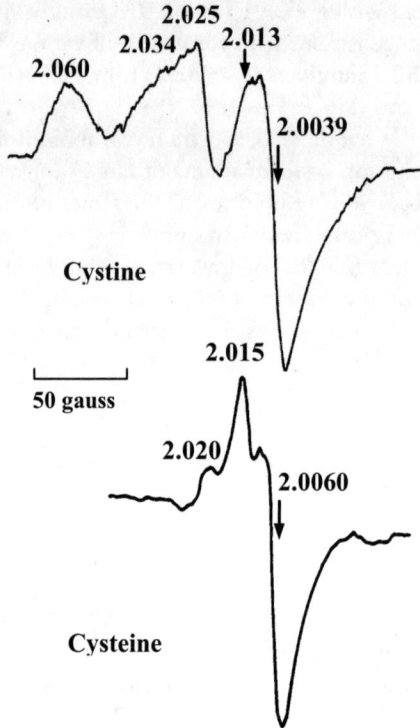

**Fig. 8.4.** EPR spectra of free radicals transferred from oxidizing methyl linoleate to cystine (top) and cysteine (bottom) in dry model systems. Cysteine reacted for 1 day; cystine reacted for 6 days at 37°C. Adapted from (Schaich and Karel, 1976). Similar spectra are produced in emulsions (Schaich, 1980b).

increase over time if incubation with the lipid denatures the protein and new cysteine residues become available (Schaich, 1980b). In contrast, oxidizing lipids do not react readily with buried –SH or with cystine double bonds. S–S radicals clearly present in irradiated proteins are not produced by oxidizing lipids, even after proteins are denatured; oxidizing lipids attack the S-S in cystine only after long incubation times (Fig. 8.4) (Schaich and Karel, 1976; Schaich 1980b). Although the possibility of electron migration to disulfides from other sites can not be ruled out, the main radical with g = 2.055, 2.024, and 1.99 is most likely a R-S-S$^•$ radical formed by abstraction of a hydrogen on the carbon α or β to one S, as has been shown for t-butoxyl radicals (Adams, 1970) and hydroxyl radicals (Elliot et al., 1981):

$$\text{Pept-CH}_2\text{-CH}_2\text{-S-S-CH}_2\text{-CH}_2\text{-Pept} \xrightarrow{\text{RO}^•} \text{Pept-CH}_2\text{-CH}_2\text{-S-S-}\overset{\bullet}{\text{C}}\text{H-CH}_2\text{-Pept} + \text{ROH}$$

$$\text{HO}^• \downarrow \qquad\qquad \downarrow$$

$$\text{H}_2\text{O} + \text{Pept-CH}_2\text{-CH}_2\text{-S-S-CH}_2\text{-}\overset{\bullet}{\text{C}}\text{H-Pept} \longrightarrow \text{Pept-CH}_2\text{-CH}_2\text{-S-S}^• + \text{CH}_2\text{=CH-Pept}$$

Other unidentified radical species are also present.

The broad central envelopes of proteins reacted with oxidizing lipids reflect overlapping unresolved hyperfine structures from multiple radical sites and delocalization of free electrons on the peptide backbone. Although steric hindrance limits accessibility of peptide α-carbons to lipid alkoxyl and peroxyl radicals, amines and thiols on amino acid side chains on the protein surface provide ready sources of hydrogen for abstraction. Histidine, tryptophan, arginine, lysine, and cysteine all produce stable radicals when incubated alone with oxidizing linoleic acid (Fig. 8.5); in proteins, their signals combine to produce the broad envelopes observed in EPR signals. Proteins that have more open structure and higher concentrations of reactive side chains on their surface show more hyperfine in their EPR signals, as can be seen with ovalbumin, serum albumin, and lactalbumin in Fig. 8.3, while highly structured proteins, such as casein, have narrower signals with little structure (Fig. 8.2). The signal wings also show more intensity and structure in proteins reacted after denaturation increases side chain accessibility (e.g., lysozyme in Fig. 8.3), consistent with increases in damage noted for denatured proteins.

Recent studies reported similar results for lysozyme, ovalbumin, arginine, ly-

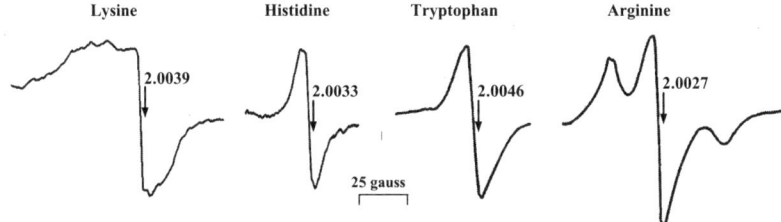

**Fig. 8.5.** EPR signals of amino acids reacted seven days with oxidizing methyl linoleate in dry lyophilized emulsions. g-values of center lines indicate N-centered radicals for lysine, histidine, and tryptophan; in arginine the radical is on a terminal amine or the carbon connected to it. Adapted from (Schaich and Karel, 1976).

sine, and histidine reacted with oxidizing lipids in lyophilized emulsions (Saeed et al., 1999). However, fish myosin signals had lower g-values (2.0021) consistent with carbon rather than nitrogen-centered radicals. This is somewhat puzzling considering myosin's high content of arginine and lysine (Connell and Howgate, 1959), which should generate N-centered radicals. Three speculative explanations may be offered for this difference in behavior from other proteins.

First, the normally reactive amino acid side chains may be inaccessible. Myosin is highly organized in a coiled coil double helix that does not denature readily; in addition, fish myosin aggregates and exhibits extensive loss of solubility with freezing (Connell, 1960; Buttkus, 1970) and lyophilization (Huss, 1995). Second, fish myosin has a high content of glutamic acid, leucine, and valine which may offer alternative radical attack sites that would produce C• radicals—glutamic acid by decarboxylation and aliphatic amino acids by H abstraction from side chains. Finally, myosin, like collagen and gelatin, is a structural protein with high glycine and proline content, providing α-C• sites for radical localization on the peptide backbone; thus it may be more prone to peptide scission than other proteins, especially in a dry reaction system. Scission would generate C• radicals by reactions that will be described later in this chapter. Electrophoresis and amino acid analyses conducted over time on the damaged myosin may shed some light on dominant reaction mechanisms with structural proteins.

It must be stressed that EPR only detects radicals that are sufficiently long-lived to maintain a steady-state radical concentration of about $10^{13}$ spins/sec. Consequently, there is the possibility that lipid radicals react with protein sites other than the ones noted previously, but the resulting radicals are too short lived to detect or they react or convert too fast to non-radical products. Thus, lack of EPR signals alone cannot be considered proof that radical reactions do not occur with other amino acids, especially serine, threonine, and the aliphatic amino acids that are also destroyed by oxidizing lipids.

*Consequences of Free Radical Transfer to Proteins*

Clearly, protein free radicals are red flags for oxidative damage. Regardless of the transfer site or mechanism, radical transfer from lipid oxyl radicals, LOO• and LO•, to proteins by H abstraction generates protein radicals (P•) with consequences that vary with the protein and the reaction system (Fig. 8.6). Protein hydroperoxides (Davies et al., 1995) and all of the protein radical species shown in Fig. 8.6 are reactive. They oxidize to secondary products that alter protein conformation and solubility and inhibit enzyme activity, recombine to form polymers, add oxygen to form peroxyl radicals on both peptide backbones and amino acid side chains, and they undergo

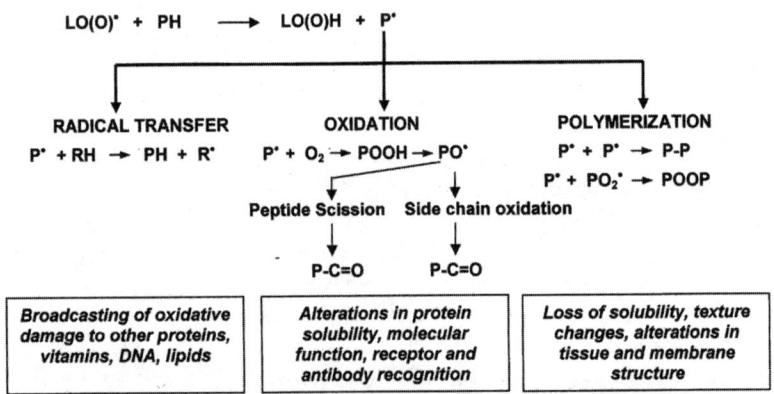

**Fig. 8.6.** Pathways and consequences of free radical production in proteins. P• is a protein radical on any α-carbon of the main peptide backbone or on an amino acid side chain, and RH is any molecule with abstractable hydrogens.

scission reactions with decarboxylation; under some conditions they also show deamination in patterns similar to irradiated proteins (Davies, 2005). These changes will be discussed in more detail in the Section "Reactions Underlying Molecular Damage to Proteins".

Protein radicals also transfer radicals to other proteins (Soszylqski et al., 1996), DNA (Gebicki and Gebicki, 1999; Luxford et al., 2000), lipids (Gardner and Weisleder, 1976; Gardner et al., 1977; Avdulov et al., 1997) and potentially other molecules to broadcast and perpetuate oxidative damage. One example of this is the addition of cysteine radicals to methyl linoleate hydroperoxide in the presence of $FeCl_3$ (Gardner and Weisleder, 1976; Gardner et al., 1977). The same H abstraction reaction that quenches a LO• also generates a thiyl radical RS• which adds to the double bond of the lipid to generate a new radical on the adjacent carbon. Continued H donation from cysteine maintains a constant supply of RS• for more addition reactions, thus perpetuating and expanding the oxidative damage.

$$RSH + Fe^{3+} \longrightarrow RS^\bullet + Fe^{2+} + H^+$$

$$CH_3(CH_2)_4\overset{OOH}{C}HCH=CHCH=CH(CH_2)_7COOH \xrightarrow[Fe^{2+}]{RS^\bullet} CH_3(CH_2)_4\overset{O^\bullet}{C}HCH=CHCH-\overset{\bullet}{C}H(CH_2)_7COOH$$

$$\text{New L'-RS radicals} \xleftarrow{L_2OOH} RS^\bullet + Fe^{2+} \xleftarrow{} RSH + Fe^{3+} \xrightarrow{} RS$$

$$CH_3(CH_2)_4\overset{OH}{C}HCH=CHCHCH-(CH_2)_7COOH \overset{RS}{\underset{OH}{|}} \xleftarrow{O_2} CH_3(CH_2)_4\overset{OH}{C}HCH-\overset{\bullet}{C}HCHCH-(CH_2)_7COOH \overset{}{\underset{RS}{|}}$$

$$\text{New L'-RS radicals} \xleftarrow{L_2OOH} 2(Fe^{2+} + RS^\bullet \quad RSH + Fe^{3+})$$

The downstream molecular and functional degradation that ensues has significant potential for dramatic effects on food quality and is increasingly being recognized as mediators of pathological processes in vivo. Similar addition of cysteine or glutathione to prostaglandin $A_1$ has been observed in model systems (Ham et al., 1975) and in red blood cells in vivo (Cagen et al., 1976), presumably by free radicals.

Radical transfer occurs early in lipid oxidation and in effect is an antioxidant process for lipids. Paradoxically, lipid oxidation may appear to be low when radical transfer to proteins is high. As a consequence, co-oxidation effects of lipid radicals too often are missed or misinterpreted.

## Reactions of Lipid Hydroperoxides

Lipid hydroperoxides are not reactive species per se, unlike lipid radicals, epoxides, and carbonyl oxidation products. Nevertheless, proteins incubated with lipid hydroperoxides are damaged within minutes, so possible roles for hydroperoxides beyond decomposition to free, diffusible alkoxyl radicals must be considered. Under some circumstances, kinetic analyses argue for the presence of concerted lipid hydroperoxide–protein reactions in which there is induced decomposition of LOOH and direct reaction of the resulting LO(O)• with amino acid targets within a reaction cage. The process may be metal mediated or metal independent.

Metal contaminants, particularly iron and copper that are always present in tissues and lab reagents, decompose lipid hydroperoxides in solution and release free LO(O)•. Metal reactions in solution are anticipated and even used in model system studies to enhance reaction rates. Cage reactions of metals, however, are more damaging (Fig. 8.7). The unusual sensitivity of metallo-proteins is due in part to binding and reduction of LOOH in a reaction cage, leading to oxidation of amino acids near the ligand site, particularly histidine (Kowalik-Jankowska et al., 2004). Most non-metalloproteins also have metal-binding sites, for example on histidine, glutamic acid, or aspartic acid side chains, that can serve as foci for metal-catalyzed reduction of LOOH in cage reactions on protein surfaces. In support of this concept, most of the iron in buffered solutions of β-lactoglobulin oxidized by methyl linoleate was bound to the protein (Yuan et al., 2007).

When metals are in solution rather than on the protein, reducing agents must be present to cycle metals; lipid peroxide reduction rates, radical lifetimes, migration,

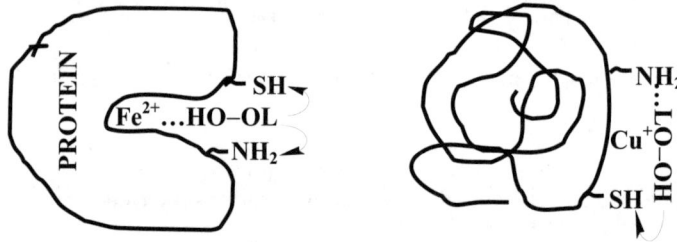

**Fig. 8.7.** LOOH-induced radical transfer to proteins: two forms of cage reactions in which lipid hydroperoxides are bound in close proximity to a metal and reduced in situ; the resulting LO• then abstracts a hydrogen from a susceptible amino acid nearby before release. Such facilitated LOOH reduction is evident in faster transfer of radicals to proteins and increased rates of lipid oxidation (chain propagation by LO• >>LOO•).

and contact with the protein also become limiting critical issues, and radical transfers to protein tend to slow down. Under some conditions, damage from lipid hydroperoxides to proteins occurs so rapidly and at such low concentrations, even in reactions with demetalled amino acids (Schaich, unpublished data), that metal-independent mechanisms must also be operating under many conditions. Molecule-assisted homolysis (MAH) of hydroperoxides is well known (Pryor, 1966). In MAH, hydroperoxides hydrogen bond to protein sites that induce LOOH decomposition, and released LO(O)• radicals abstract hydrogens from nearby amino acids in a reaction cage. Noting that the kinetics of radical transfer to lysozyme exceeded the apparent rates of oxidation of methyl linoleate, Schaich and Karel proposed protein-facilitated decomposition of LOOH and cage transfer of radicals that enhance the reactivity of LOOH (Karel et al., 1975; Schaich and Karel, 1976):

$$\text{LOOH} + \text{PH} \longrightarrow \text{LO-OH}\cdots\text{H-P]} \longrightarrow \text{LO}^{\bullet} + \text{P}^{\bullet} + \text{H}_2\text{O}$$

Solid state EPR signals showed N(O)• and S(O)• radicals, which are consistent with observations that amines (Harris and Olcott, 1966) and sulfhydryls (Little and O'Brien, 1968) both undergo concerted reactions with [lipid] hydroperoxides:

$$R_3N + R'OOH \longrightarrow [R_3N...HOOR'] \longrightarrow R_3NO + R'O^\bullet$$

$$2\ RSH + LOOH \longrightarrow RSSR + H_2O + LO^\bullet \xrightarrow{L'H} LOH + L'^\bullet$$

This kind of concerted reaction may contribute to the sensitivity of his, arg, lys, trp, cys, ser, and thr to LOOH, all of which contain hydrogen bonding amino, carboxylic acid, and hydroxyl groups on their side chains (Gardner 1979, Schaich 1980b). It also may be enhanced in lipid-bonding proteins, such as bovine serum albumin, where hydrophobic side chains facilitate associations with lipids and bring reactive residues into close proximity with LOOH.

Concerted reactions between LOOH and proteins have now been demonstrated for:

a. β-lactalbumin reacted with LOOH, LnOOH, and AnOOH, where one mol LOOH per 18,000 mw protein oxidized tryptophan and cysteine and generated extensive crosslinking within one hour (Hidalgo and Kinsella, 1989);

b. HDL apoA1 and A2 reacted with cholesterol and phospholipid hydroperoxides, where appearance of met oxidized to methionine sulfoxide occurred concurrently with LOOH reduction (Garner et al., 1998a).

c. LDL and cyt c incubated with LOOH or phospholipid (PE) hydroperoxides, where loss of LOOH led to reduction of cyt c, release of $O_2^{-\bullet}$, and generation of fluorescent products (Fruebis et al., 1992). The concerted reaction mechanisms involving hydroperoxide (peroxyl radical) addition proposed to explain the kinetics and products is shown in Fig. 8.8. Pathway A has parallels in radiation and HO• chemistry. Pathway B is sterically unlikely in linoleic or linolenic acid but could be feasible in arachidonic acid and ω-3 fatty acids. Both mechanisms merit further testing.

Although mechanisms were not identified, very rapid protein degradation suggests that concerted reactions were involved in substantial losses of trp, met, cyh, pro, val, leu, as well as fragmentation and crosslinking of lupine conglutins with MLOOH at pH 9 (Fruebis et al., 1992; Lqari et al., 2003), in reactions of butylamine with LOOH in $CHCl_3$ (Zamora and Hidalgo, 1995), and in the very rapid formation of protein carbonyls and loss of lysine without lipid aldehydes when MLOOH, LnOOH, and AnOOH were incubated with BSA (Refsgaard et al., 2000). The latter study is also an excellent example of the broadcasting action of lipid oxidation; chain initiation was not affected, but yields of downstream lipid oxidation products were substantially depressed when oxidation was transferred to BSA. Again, this shows that if only normally measured lipid oxidation products are monitored in complex systems

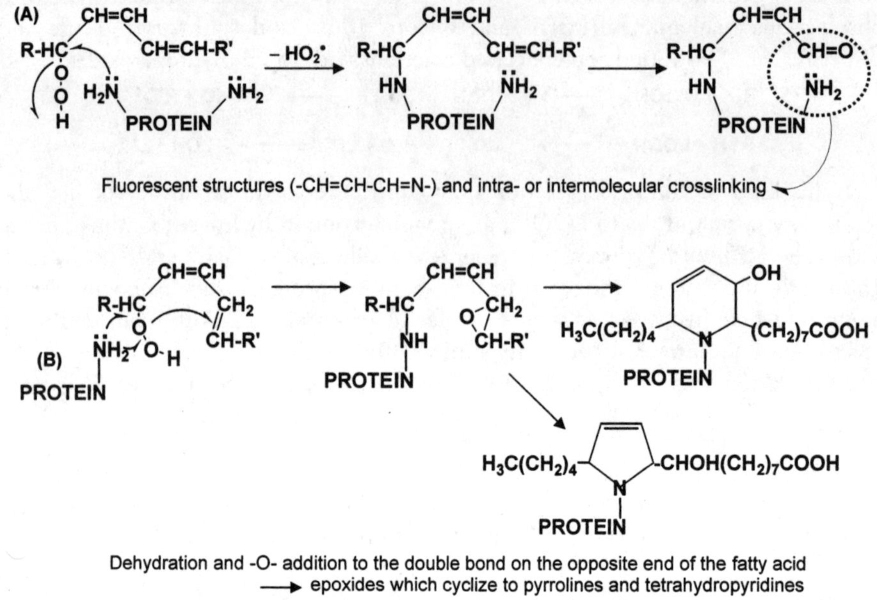

**Fig. 8.8.** Two reaction pathways proposed to explain concerted reactions of lipid hydroperoxides with proteins (shown for reaction of 13-HOO-linoleic acid, LOOH, with protein-bound lysines). Modified from (Fruebis et al., 1992): original reaction scheme assumed that LOOH (generated from lipoxygenase and purified before incubation with proteins) was converted to LOO• before reaction with amino groups on proteins. However, EDTA-complexed iron and small amounts of unoxidized linoleic acid could not account for the large concentrations of LOO• needed to drive the rapid reaction. Replacing LOO• with LOOH gives reactions that are consistent with radiation and HO• chemistry, $O_2^{-\bullet}$ equilibrium (at neutral pH, $O_2^{-\bullet} \rightleftarrows HO_2^{\bullet}$), and still account for observed products.

and footprints of co-oxidations are ignored, the true extent of oxidative degradation can be greatly underestimated.

## Reactions of Lipid Epoxides

It is interesting that lipid epoxides are carcinogenic (Chung et al., 1993; Blair, 2001; Lee et al., 2001), mutagenic (Lee et al., 2002), and strongly cytotoxic; yet their reactions with proteins have received relatively little attention. Epoxides (also called oxiranes) are cyclic products generated by internal reactions of lipid hydroperoxides (Hamberg and Gotthammar, 1973), peroxyl addition products or alkoxyl radicals (Gardner, 1989; Schaich, 2005), or reaction between hydroxynonenal and lipid hydroperoxides or hydrogen peroxide (Chen and Chung, 1996):

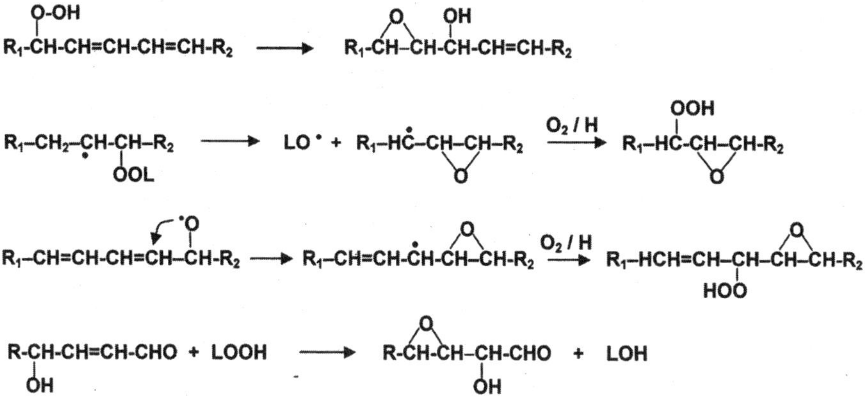

$$R_1-\overset{\overset{O-OH}{|}}{CH}-CH=CH-CH=CH-R_2 \longrightarrow R_1-\overset{\overset{O}{\diagup \diagdown}}{CH-CH}-\overset{\overset{OH}{|}}{CH}-CH=CH-R_2$$

$$R_1-CH_2-\overset{\overset{\bullet}{C}H}{}-CH-R_2 \longrightarrow LO^{\bullet} + R_1-\overset{\bullet}{HC}-\underset{\underset{O}{\diagdown \diagup}}{CH-CH}-R_2 \xrightarrow{O_2/H} R_1-\overset{\overset{OOH}{|}}{HC}-\underset{\underset{O}{\diagdown \diagup}}{CH-CH}-R_2$$
$$\phantom{R_1-CH_2-}\overset{|}{OOL}$$

$$R_1-CH=CH-CH=\overset{\curvearrowleft\overset{\bullet O}{|}}{CH}-CH-R_2 \longrightarrow R_1-CH=CH-\overset{\bullet}{CH}-\underset{\underset{O}{\diagdown \diagup}}{CH-CH}-R_2 \xrightarrow{O_2/H} R_1-\overset{}{HCH}=CH-\underset{\underset{HOO}{|}}{CH}-\underset{\underset{O}{\diagdown \diagup}}{CH-CH}-R_2$$

$$R-\overset{\overset{}{CH}-CH=CH-CHO}{|}+LOOH \longrightarrow R-\underset{\underset{O}{\diagdown \diagup}}{CH-CH}-\overset{\overset{}{CH-CHO}}{|}+LOH$$
$$\phantom{R-}\overset{|}{OH}\phantom{abcdefghijklmn}\overset{|}{OH}$$

4,5-epoxy-2-decenal has been reported as a common secondary product of ω-6 fatty acids (e.g., initial oxidation at C-13 of linoleic acid, cyclization of alkoxyl radical to form epoxide, secondary oxidation at C-9, α-scission of alkoxyl radical to form aldehyde); 4,5-epoxy-2-heptenal is an epoxide from ω-3 fatty acids (e.g., initial oxidation at C-16 of α-linolenic acid, cyclization of alkoxyl radical to form epoxide, secondary oxidation at C-12, α-scission of alkoxyl radical to form aldehyde) (Gardner, 1989).

It is interesting that these reactions not only generate a new oxidant species but also shift the inherent macromolecular damage capability of oxidizing lipids as the product mix changes. Internal cyclizations or rearrangements of LO(O)• to epoxides are always present, competing with H abstraction during lipid oxidation. Cyclization is favored when abstractable protons are limited (i.e., at low lipid concentrations), in aprotic solvents, in neat unsaturated lipids oriented on surfaces, and at lower temperatures (Schaich, 2005). Thus, rather than being intriguing reaction artifacts or trace side products, epoxides are common components of most complex reaction systems, especially with multiple phases. As very reactive compounds, epoxides are probably the most underrated and understudied lipid-derived oxidant.

What accounts for the marked reactivity of epoxides, especially with proteins and nucleic acids? Epoxy functions vicinal to olefin double bonds are particularly susceptible to hydrolysis and nucleophilic attack (Lederer, 1996). When association of a nucleophile with the epoxide generates a partially charged transition state (Ingold, 1969), the oxirane ring then opens easily because the three-membered ring is strained and at a higher energy level. The oxygen stays on the less highly substituted carbon and the nucleophile adds to the opposite carbon (the allylic carbon when there is an adjacent double bond) from the backside (Ege, 1999; McMurray, 2000). This basic process is shown below for reaction of lysine, histidine, and cysteine (left to right) with a hypothetical isolated epoxide from oxidation of linoleic acid.

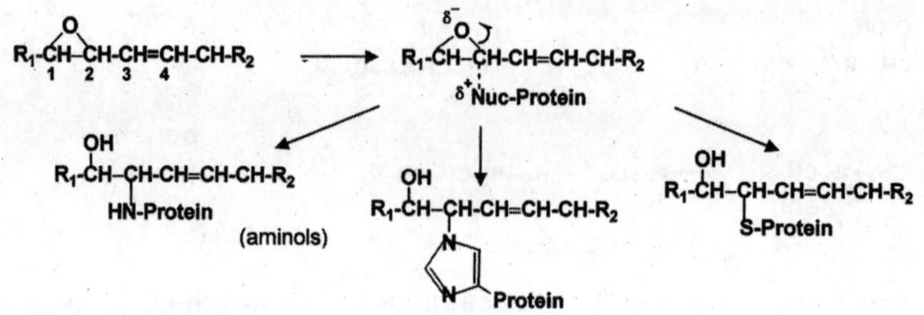

Model system studies of lipid epoxides reactions have revealed some important characteristics. Lederer used epoxyhexenol (trans-4,5-epoxy-trans-2-hexen-1-ol) as a simple model for the reactive region of oxidized fatty acids and propylamine as a model for lysine, eliminating competing reactions with carboxylic acid and amine groups (Lederer, 1996). When the reaction was run in an aprotic solvent (THF) to avoid nucleophilic attack of solvent molecules, the amine added exclusively at C-2 (the allylic position) by an $S_N2$ mechanism to form an aminol as shown previously. In aqueous solutions, aminols formed in >50% yields at pH 9 and production decreased dramatically as pH decreased: only 3% aminols were formed after 24 hours at physiological pH, and there was no reaction at pH ≤ 6. Acid reduces the nucleophilicity of the lysine ε-amino function and protonates the epoxide, facilitating C-O bond scission and release of –OH from the oxirane ring; the amine can then add to C-2 in classical nucleophilic addition (A) or to C-4 in the double bond in an $S_N1$-like reaction (B), as shown below (Ege, 1999; McMurray, 2000). At the same time, protonation of the epoxide also increases epoxide hydrolysis (C) to rates competitive with amine addition (Lederer, 1996):

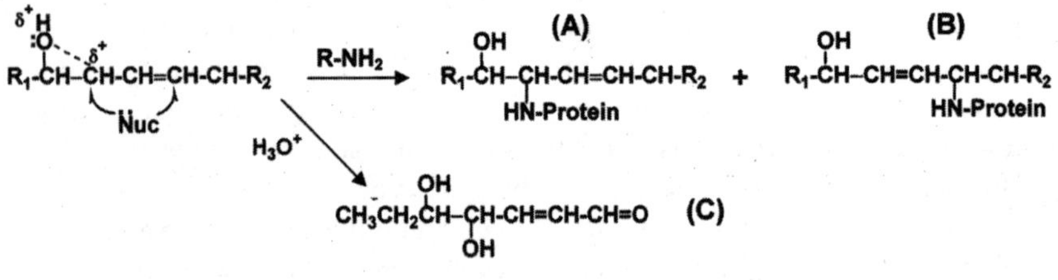

What factors direct the balance between these three reactions and products? Obviously, solvent influences are critical. Lysine with blocked amino and carboxyl groups showed classical nucleophilic addition (A) to 9,10-epoxy-13-hydroxy-11 octadecenoic acid t-butyl ester (Lederer et al., 1998). The reaction was slow in methyl pyrrolidone, due to limited availability of solvent protons, but the rate increased dramatically with addition of up to 20% water (the solubility limit). As expected water also increased hydrolysis rates (C) competitively, but it did not change the addition mechanism; in all cases, the exclusive reaction was backside nucleophilic attack at the allylic position yielding the aminol adduct (A) with four diastereomers (Lederer, 1996). These results suggest that epoxide reactions with proteins are most important under anhydrous conditions, for example in dry foods and in hydrophobic interior regions of biomembranes and blood lipoproteins.

The nature of the nucleophile and its conformational accessibility also affect reactivity and pathways. When reacted with butadiene moloxide in phosphate buffer (pH 7.4), the α-amino group of valine added at both epoxide carbons in 2:1 ratio, C2:C1 (terminal C) (Moll, 1999):

(C-1 adduct)    (C-2 adduct)

In intact proteins, such as mouse hemoglobin, reaction of N-terminal valines occurred almost totally at C-1 due to steric hindrance. Epoxide adducts also formed with lysine, serine, histidine, and methionine, although the regioisomer could not be distinguished by ESI-MS (Moll et al., 2000). In all systems, higher temperatures (37°C) increased hydrolysis at the expense of amine adducts. Similarly, LC-MS-MS analyses of human hemoglobin reacted with styrene oxide, ethylene oxide, and butadiene dioxide revealed major epoxide addition sites were the N-terminal valines of both α and β Hb chains, plus cysteine and histidine of specific sequences (Badghisi and Liebler, 2002); adduct structures were not specified.

Epoxide addition by this classical mechanism is probably responsible for uncharacterized binding of 9,10-epoxy stearic acid to albumin reported thirty years ago (Pokorny et al., 1966); C-2 addition to epoxides has been demonstrated with sulfhydryl compounds (Buttkus, 1972) and cysteine (Gardner et al., 1977; Gardner and Jursinic, 1981).

When the epoxide is in an alkenal rather than a fatty acid, the reaction changes due to competition with Schiff base formation at the carbonyl (Zamora and Hidalgo, 1995, 2003b, 2005; Zamora et al., 1999; Hidalgo and Zamora, 2000). In chloroform, 70% acetonitrile, or aqueous methanol, histidine reacted with 4,5-epoxy-2-alkenals by classical epoxide addition (A, above), accompanied by a parallel increase in protein carbonyls. However, with primary amines, such as lysine and ethyl amine, Schiff base adducts preferentially formed with the aldehyde and the epoxide group remained intact. The imine was added to the epoxides via backside attack, forming a pyrrole ring with oxyl side chain. However, without a solvent or amine proton source, the epoxide oxygen remained reactive as an anion (-O$^-$) rather than forming the more stable alcohol (-OH). This intermediate then transforms to two sets of products: (1) the hydroxide anion protonates, giving a hydroxyalkyl pyrrole, and (2) the side chain is released as an aldehyde, for example formaldehyde or acetaldehyde (dominant reaction with α-amino acids), leaving an N-substituted pyrrole. Hydroxyalkyl pyrroles polymerize over time (see Section "Crosslinking" for reactions).

[Reaction scheme showing RCH(O)-CH-CH=CH₂-CH=O reacting with R'-NH₂ to form intermediates leading to pyrrole products (1) and (2), plus R-CHO, with (1) leading to Polymers]

Depending on the amine and temperature, two other reaction pathways have been observed with epoxy-alkenals (2:1 acetonitrile–water or buffer as solvent) (Zamora and Hidalgo, 2005; Zamora et al., 2006):

a. direct amine addition to the epoxides → alkyl-substituted pyrroles and furans. The rate of this reaction increases 10-fold when reaction temperature is changed from 37 to 60°C.

[Reaction scheme showing epoxyalkenal + R'-NH₂ → hydroxyamine intermediate → pyrrole + furan + H-CHO]

b. Schiff base imine formation followed by elimination of the amine R' and cyclization of remaining product to alkylpyridines

[Reaction scheme showing epoxyalkenal + R'-NH₂ → imine → hydroxy intermediate → (via H₂O and -H₂O) → alkylpyridine + R"C=O]

An alternate mechanism for formation of 1,2-dihydropyridines is C-4 addition of imines to vinyl epoxides in presence of a Lewis acid (e.g., water) (Brunner et al., 2006). Analogous products should be expected in reactions of epoxyalkenals with amino acids:

$$\text{RCH-CH-CH=CH-CH=O} + \text{R}_2\text{-CH=NH-R}_3 \longrightarrow \text{R-C-CH=CH-CH-CHO, CH-R}_2, \text{HN-R}_3 \longrightarrow \text{(pyridine ring with CHO, R, R}_2, \text{R}_3\text{)}$$

Aminols from epoxide-amino acid reactions have distinctive structures that can differentiate the oxidation from free radicals, hydroperoxides, and carbonyl products. However, variously substituted pyridines and pyrroles are also products of extensively studied aldehydes, as will be detailed in the next section. Thus, it can be exceedingly difficult to distinguish epoxides from aldehyde reactions, except perhaps by kinetics, and it is quite likely that reactions of epoxides with proteins have been overlooked and misinterpreted as aldehyde-mediated damage. New analytical methods will be necessary to track the lipid oxidants responsible for common protein products under various reaction conditions and determine the full role of epoxides in protein degradation.

## Addition Reactions of Secondary Products from Lipid Oxidation

Secondary lipid oxidation products are responsible for the off-odors and flavors associated with rancidity and they also appear to be long lived in biological tissues, so they have received considerable attention in co-oxidation studies. Most research has focused on aldehyde reactions, even though scission reactions of lipid alkoxyl radicals generate a wide variety of reactive secondary oxidation products.

All aldehydes react with nucleophilic groups on proteins to form adducts, with three general outcomes:

a. Linear adduct formation (Schiff base, Michael addition, or a combination of both) that changes surface chemistry and protein recognition. This is the initial step for all aldehydes.

b. Cyclic products, especially dihydropyridines and pyrroles, as adducts or in crosslinks.

c. Protein crosslinks, both intra- and inter-molecular, via Schiff base, combinations of Schiff base and Michael additions, Schiff base-Michael addition-ring links, or a complex combination of any of these.

The reaction pathways that occur or dominate in a given system are influenced by the nature of the protein, relative protein-aldehyde concentrations, pH, phase or solvent, oxygen tension, and other factors.

*Monofunctional alkanals*
Monofunctional alkanals (Fig. 8.9) have low reactivity and high selectivity; they re-

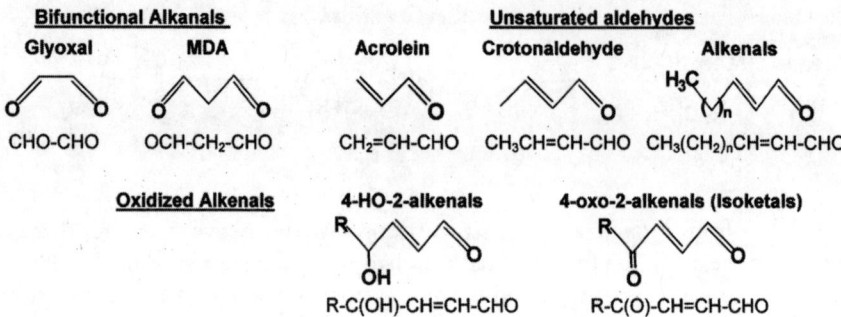

**Fig. 8.9.** Structures of aldehydes that play important roles in oxidative degradation of proteins.

act with amines exclusively by Schiff base formation with preference for N-terminal residues, and at low aldehyde concentrations and low $pO_2$ the reaction often goes no farther. For example, hexanal reacts with insulin B chain only at N-terminal phenylalanine residues and a single lysine (lys29) near the end of the peptide chain (Fenaille et al., 2003). Nonanal and its oxidation product 8-oxononanoic acid form Schiff bases with both the amine and thiol groups of cysteine to yield thiazolidine dicarboxylic acid derivatives (Gardner et al., 1977; Gardner and Jursinic, 1981):

Gardner and Jursinic proposed this as a generalizable model for reaction of low levels of lipid aldehydes with N-terminal amino acids of proteins (Gardner and Jursinic, 1981). When the N-terminal residue has side chains containing amines in place of the thiol, various analogous heterocyclic products could form, with the ring size determined by the specific side chain.

In the presence of excess aldehyde, three molecules of alkanal form pyridines on reaction with amino acids (Suyama and Adachi, 1979). Reaction details are presented in the "Bifunctional Saturated Aldehydes" section.

In the literature, Schiff base adducts are cited indiscriminately and often incorrectly both as the imine (–N=CH-CH$_2$-CHO) and the enamine (–NH-CH=CH-

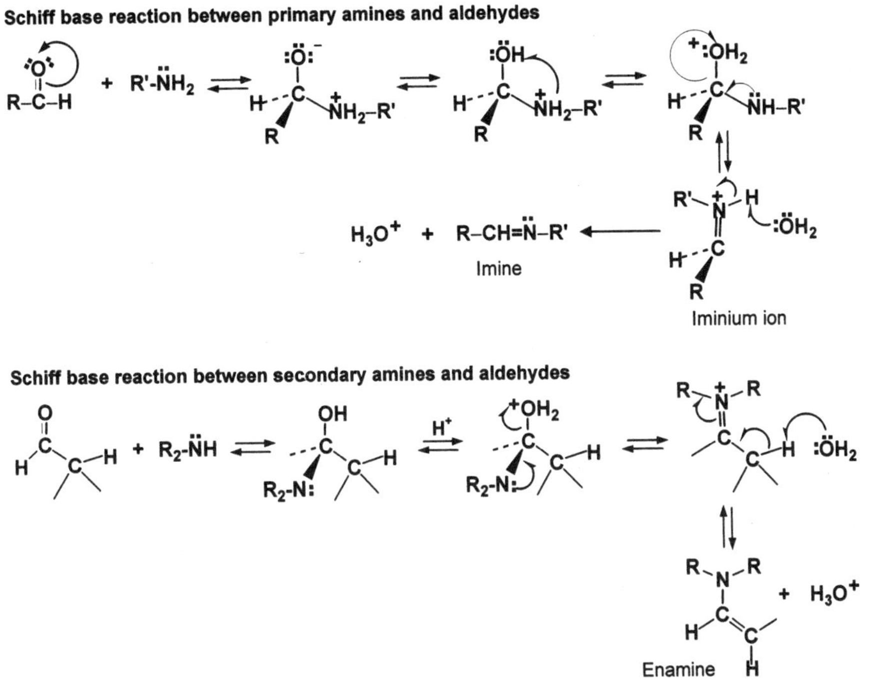

**Schiff base reaction between primary amines and aldehydes**

$$R\text{-}C(=O)\text{-}H + R'\text{-}NH_2 \rightleftharpoons \cdots \rightleftharpoons \cdots \rightleftharpoons \text{Iminium ion} \rightarrow H_3O^+ + R\text{-}CH=N\text{-}R' \text{ (Imine)}$$

**Schiff base reaction between secondary amines and aldehydes**

$$\cdots \rightleftharpoons \cdots \rightleftharpoons \cdots \rightleftharpoons \text{Enamine} + H_3O^+$$

**Fig. 8.10.** Formation of imines and enamines from schiff base additions of primary and secondary amines with carbonyls (McMurray, 2000).

CHO). As shown in Fig. 8.10, the imine forms first with primary amines, such as lysine, and in acidic solutions rearranges to the enamine, whereas only the enamine forms with secondary amines, such as histidine. Unless otherwise specified, Schiff base structures written in reactions of this chapter assume formation from primary amines at neutral pH and thus will be shown as the imine.

*Bifunctional Saturated Aldehydes*
Bifunctional saturated aldehydes (e.g., glyoxal and malonaldehyde (MDA), Fig. 8.9) are more reactive due to the second carbonyl and keto-enol tautomerism (Esterbauer et al., 1991):

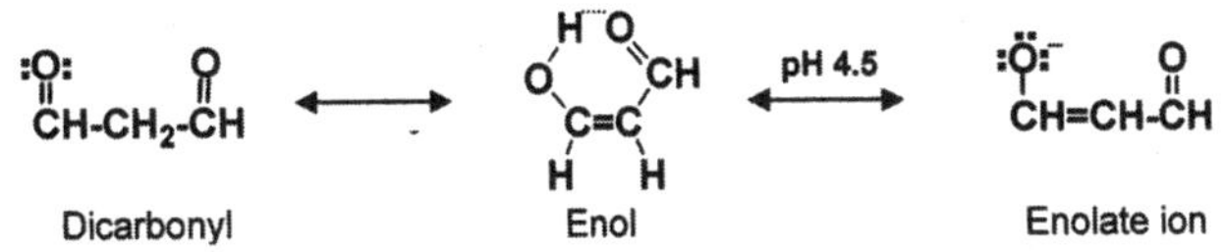

$$\underset{\text{Dicarbonyl}}{CH(=O)\text{-}CH_2\text{-}CH(=O)} \longleftrightarrow \underset{\text{Enol}}{\cdots} \underset{pH\,4.5}{\longleftrightarrow} \underset{\text{Enolate ion}}{CH=CH\text{-}CH(=O)}$$

Nucleophilic double bonds facilitate Schiff base, aldol, and other additions to enol and enolate forms of MDA, and the enol –OH is very susceptible to dehydration as well (Nair et al., 1981; Ege, 1999). MDA is most reactive in mildly acidic aqueous solutions where its enol form (β-hydroxy acrolein) dominates. Under these conditions active formation of hydrogen-bonded dimers

$$2 \; \underset{\underset{H}{|}}{CH}=CH-\underset{\underset{}{\|}}{CH} \quad \longleftrightarrow \quad \text{(cyclic hydrogen-bonded dimer)}$$

and a strong tendency to self-polymerize by aldol condensation, especially at high concentrations, compete with MDA addition to proteins (Esterbauer et al., 1991) and reduce efficiency of MDA as a protein oxidant. This is one reason why MDA is on the low end of the scale in damage to proteins when reactivities of various aldehydes are compared.

MDA and other bifunctional aldehydes react by three main mechanisms:

*Schiff base addition to nucleophilic groups on amino acids and proteins.* As with saturated aldehydes, the major reaction of MDA with proteins is via Schiff base formation. The main protein target is lysine ε-amino groups; 50-60% of the lysine ε-amino are destroyed when myosin is reacted with MDA (Buttkus, 1967). Cysteine–SH, and histidine imidazole groups on side chains are also important targets, but Schiff base formation with these groups must compete with facile hydrogen abstractions and rapid Michael addition reactions, respectively. Tyrosine, arginine, and tryptophan side chains form Schiff bases more slowly except at pH < 4.2, where their α-amino groups become the exclusive reaction sites for many amino acids (Nair et al., 1981).

Schiff base formation proceeds as shown in Fig. 8.10, but with the complication of two carbonyls and tautomeric forms. Reaction of proteins with one aldehyde group of MDA yields Schiff base adducts; reaction with both carbonyls generates enamine or iminopropene crosslinks. In acid, amines add both to dicarbonyl and enol forms of MDA and yield different products:

$$\underset{CH-CH_2-CH}{\overset{:O:}{\|}} \;\; \overset{O}{\|} + R-NH_2 \;\; \overset{+/- \; H_2O}{\longleftrightarrow} \;\; R-N=CH-CH_2-CH=O \; \text{(imine)}$$

$$\underset{HC-HC=CH}{\overset{O}{|}} + R-NH_2 \;\; \overset{+/- \; H_2O}{\longleftrightarrow} \;\; \text{(cyclic intermediate)} \;\; \longleftrightarrow \;\; R-NH-CH=CH-CH=O \; \text{(enamine)}$$

In a given system, the equilibrium distribution between imine versus enamine tautomers is directed by the pH and hydrogen bonding capacity of the solvent (Yildiz et al., 1998; Nazir et al., 2000). Acid increases MDA protonation and drives the enol reaction forward, while aqueous solutions inhibit elimination of water and drive the equilibrium towards dissociation. In general, imines are the dominant tautomer at equilibrium under physiological conditions (neutral pH) (Burcham and Kuhan, 1996) where hydrogen bonding with the polar solvent interferes with proton transfer within the aldehyde-amine complex (Yildiz et al., 1998; Nazir et al., 2000). Acid and organic solvents increase intramolecular proton transfer capabilities and facilitate conversion to enamines.

The ease with which imine versus enamine tautomers can be formed and stabilized in an aldehyde-amine complex is determined by the component aldehyde; the amine has no effect. For some aldehydes starting with cyclic 4-HO-2-enals hydroxyl imines were the only products in any solvent, while for other aldehydes reacting with the same amine, enamines (ketoamines) formed in varying proportions in acidified organic solvents (chloroform and benzene) but did not appear in hydrogen-bonding solvents (Yildiz et al., 1998; Nazir et al., 2000). Comparable studies documenting MDA tautomer distributions and associated products under different conditions will contribute greatly to understanding the reactivity of this aldehyde, particularly in relation to other aldehydes.

The existence of MDA tautomeric forms with their corresponding reactivity and Schiff base products is emphasized here because the literature is totally inconsistent about the structures of MDA and other lipid aldehyde Schiff base structures cited. Review of the hundreds of papers reporting lipid-protein Schiff base products reveals imine and enamine structures, and saturated congeners as well, used interchangeably and indiscriminately, usually with little or no consideration of the reaction conditions. In many cases the structures reported were inconsistent with reaction conditions and solvents employed. Researchers need to be actively cognizant of how their reaction conditions affect the forms of MDA available for reaction and to use this information in interpreting and publishing results. New studies are increasingly using NMR and LC-MS/MS to identify structures precisely. In the absence of specific structural information, studies should report either the tautomer equilibrium or the dominant imine plus specific reaction conditions.

*Formation of cyclic products (dihydropyridines) with amines:* Schiff base formation is the first step in most MDA reactions with proteins. However, when a molar excess of MDA is present, dihydropyridines form via secondary condensation of the Schiff bases. MDA is a short molecule, so two are needed to complete the pyridine ring; related products are generated by two molecules of MDA plus one monofunctional aldehyde, such as acetaldehyde or formaldehyde. (Kikugawa and Ido, 1984; Nair et al., 1988; Freeman et al., 2005):

$$2\, OHC\text{-}CH_2\text{-}CHO + R_1CHO + R_2NH_2 \xrightarrow{pH\ 7} \underset{R_2}{\underset{|}{N}}\!\!\!\!\!\!\!\!\!\!\!\!\!\!\!\!\!\! \begin{array}{c} R_1 \\ OHC\diagup\!\!\!\!\diagdown CHO \end{array}$$

(1,4-dihydropyridine-3,5-dicarbaldehydes)

When $R_2$ is a protein, the dihydropyridine adds a reactive group with two carbonyls to the protein surface and becomes a site primed for further reaction. Dihydropyridines are flavor precursors in foods (Buttery et al., 1977; Suyama and Adachi, 1980; Maga, 1981). Unfortunately, when bound to proteins they are not absorbed in the gut or hydrolyzed in tissues, which translates to loss of nutritional quality (Giron-

Calle et al., 2003). Dihydropyridines also form in vivo and have been identified as adducts using immunochemical techniques (Yamada et al., 2001) and as crosslinks in tissues (Slatter et al., 1998).

*Michael addition reactions with amines:* The double bond in the enol and enolate forms of MDA provides an electrophilic site with enhanced susceptibility to nucleophilic reaction via Michael-type addition of the amine to the β-carbon of the double bond. Michael-type is the most accurate term since classical Michael additions are between carbon compounds. However, the term "Michael addition" is commonly used with amine additions to α,β-unsaturated carbonyls, so it will be used in this manner throughout this chapter. Details of Michael addition reactions are in the following section on unsaturated aldehydes.

## Unsaturated Aldehydes

Unsaturated aldehydes, primarily 2-enals, are extraordinarily reactive compounds. α,β-unsaturation makes three tautomers possible, two of which have carbocations activated towards nucleophilic addition. Thus, 2-alkenals and their oxidized derivatives have three potential reaction sites: Schiff base formation at the carbonyl and Michael-type 1,2 and 1,4 addition at the carbocations (Esterbauer et al., 1991; Ege, 1999; McMurray, 2000).

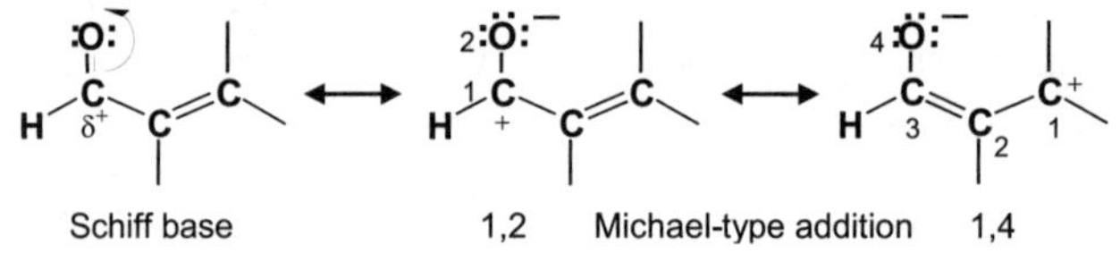

Michael additions to the double bond are preferred and generate the most products, both immediately and in subsequent transformations. The nucleophilic thiol of cysteine, ε-amine of lysine, and imidazole nitrogen of histidine are the main targets of unsaturated aldehydes (Esterbauer et al., 1991; Petersen and Doorn, 2004). With direct (1,2) addition, the amine (or thiol) adds to the carbonyl carbon with α,β unsaturation, generating a carbinolamine intermediate that rearranges and dehydrates to a Schiff's base (only the reactive groups of the 2-alkenals are shown below) (Ege, 1999; McMurray, 2000):

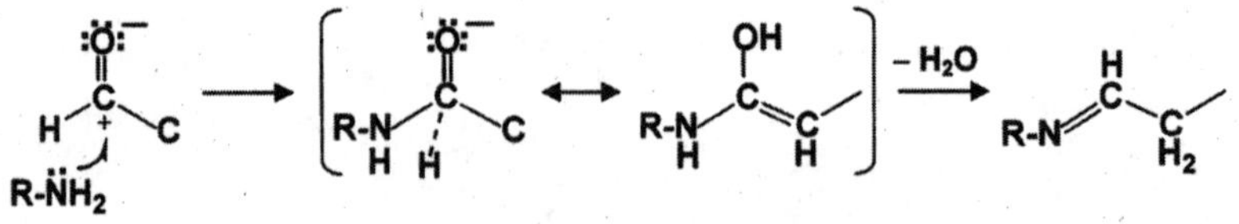

The reaction is catalyzed by acid and facilitated in organic solvents and hydrophobic microenvironments; it is also reversible (Esterbauer et al., 1991). With most lipid aldehydes and amino acids, it is a precursor for some minor products. An important side-effect is the reduction of reactive carbonyls to alcohols.

Conjugated (1,4) addition is the dominant initial process in aqueous phase reaction of lipid alkenals with amino acids. The amine adds to the β-carbon of the double bond in conjugation with the carbonyl:

$$\underset{H}{\overset{H}{C}}=\underset{H}{\overset{:\ddot{O}:^-}{C}}-\overset{H}{\underset{}{C}}{}^+\underset{}{\overset{}{\diagdown}} \;\; \overset{R-NH_2}{\longrightarrow} \;\; \left[ \underset{H}{\overset{H}{C}}=\underset{H}{\overset{:\ddot{O}:^-}{C}}-\underset{H}{\overset{H_2\overset{+}{N}\text{-}R}{C}}\diagdown \;\; \longleftrightarrow \;\; \underset{H}{\overset{:\ddot{O}:}{C}}-\underset{H}{\overset{}{\overset{\shortparallel}{C}}}-\underset{H}{\overset{H_2\overset{+}{N}\text{-}R}{C}}\diagdown \right] \;\; \overset{H_3O^+}{\longrightarrow} \;\; H\overset{O}{\overset{\shortparallel}{C}}-\underset{H_2}{\overset{}{C}}-\underset{H}{\overset{HN-R}{C}}\diagdown$$

Importantly, the carbonyl remains intact and can contribute to carbonyls detected in oxidized proteins or react further with an additional amine to form a Schiff base. Michael addition followed by Schiff base formation is the initial sequence found repeatedly in the formation of complex adducts (shown in following sections). Like Schiff base formation, Michael additions are reversible in an aqueous buffer without excess amine to prevent hydrolysis and aldol condensation of C=O (Nadkarni and Sayre, 1995). Aqueous solvents favor 1:1 amine:C=O adduct formation, while organic solvents favor 2:1 complexes.

Adding to the complexity of alkenal reactions with amines is the marked tendency of the resulting adducts to cyclize and undergo further reaction. Multiple pathways compete, yielding complex mixtures of many different products. Michael and Schiff base additions, alone and in combination, plus various cyclizations and some crosslinking probably all occur simultaneously with at least low yields in most systems. A full accounting of all products has never been reported. Accessibility of specific amino acids in proteins as well as reaction conditions, such as pH, solvent, oxygenation, degree of lipid oxidation, and timing of analyses, all influence which pathways and products will dominate in a given system. Thus, the discussion of reactions that follows does not attempt to portray any one product as the dominant lipid oxidation product causing damage to proteins, but instead describes the broad range of reactions that probably occur simultaneously although in different proportions under various conditions.

The importance of recognizing multiple competing pathways when analyzing damaged proteins or amino acids cannot be overstressed. Too often studies have focused on a single product, and when it was found in proteins, claimed that product to be the major oxidant. LDLox, which has been the target of innumerable studies of protein damage mechanisms, is an excellent example for putting the problem of simultaneous multiple oxidation pathways in perspective. Just about every oxidant reacted with LDL has generated oxidation products that could be identified in LDLox and atherosclerotic plaques by both chemical and immunological techniques. The major oxidized protein constituent of LDLox is $N^\varepsilon$-(2-propenal) lysine formed by direct addition of MDA to lysine, but variable levels of pyrroles, hydroxynonenal, and oxononenal adducts are also present in the same samples (Uchida, 2000). Which is the most toxic modification? Perhaps all are important in independently inducing different cell responses that together generate full-blown pathology in atherosclerosis and other diseases of oxidative stress. Clearly more quantitative differentiation of multiple pathways and products is needed to elucidate factors controlling reaction routing and product stability under different conditions.

### Acrolein and Crotonaldehyde

The simplest alkenals are the ubiquitous air pollutants acrolein and crotonaldehyde (Fig. 8.9). Although acrolein is most often associated with thermal degradation of frying oils, it is also a metabolite in the transformation of allyl compounds and has been identified among secondary lipid oxidation products of arachidonic acid and ω-3 PUFA in vivo (Uchida et al., 1998a, 1998b; Uchida, 1999; Kehrer and Biswal, 2000). Crotonaldehyde is a carcinogenic oil component found in wood smoke and car exhausts (ATSDR, 2002) and is claimed to be produced in vivo by $Cu^+$ and $Fe^{3+}$/ascorbate-catalyzed oxidation of linoleic and linolenic acids (Ichihashi et al., 2001). Both aldehydes are highly electrophilic, so they have very strong reactivity with nucleophiles, such as thiols, the imidazole groups of histidine, and the ε-amino group of lysine, via Schiff base and Michael additions (Esterbauer et al., 1991; Uchida et al., 1998a).

The dominant reaction varies with the protein and amino acid accessibility, but Michael addition is usually faster and the products are more stable. Histidine prefers Michael addition almost exclusively, with reaction at the τ nitrogen (Esterbauer et al., 1991; Uchida and Stadtman, 1992; Uchida et al., 1998a, 1998b)[1]. The products are β-substituted propanals.

Lysine forms Michael addition products when aldehyde concentrations are low, but when aldehydes are present in excess of the amines, multiple additions generate cyclic pyrrole derivatives.

---

1 ⸿⸺ in the reactions represents his and lys, respectively, attached to a protein, with only the reactive group shown.

$$\text{R-NH}_2 + \text{CH}_3\text{CH=CH-CHO} \longrightarrow \text{R-NH-CH(CH}_3\text{)CH}_2\text{CH=O}$$

$$\text{R-NH}_2 + 2\,\text{CH}_2\text{=CH-CHO} \longrightarrow \text{R-(CH}_2)_4\text{-N-ring-CH=O}$$

As noted previously, Michael addition of amino acids to alkenals has two important consequences: 1) the products introduce into proteins carbonyl groups that are detected as part of the standard protein carbonyl assay; and 2) the aldehyde carbonyls provide new reactive sites for further reactions, so they act as nuclei for crosslinking and cyclization. In contrast, Schiff base adducts with lysine and cysteine remove carbonyls by condensation with amines:

$$\text{CH}_3(\text{CH}_2)_n\text{CHO} + \text{R-NH}_2 \longrightarrow \text{CH}_3(\text{CH}_2)_{n\text{-}1}\text{CH=CH-NH-R} \longleftrightarrow \text{CH}_3(\text{CH}_2)_n\text{CH=N-R}$$

Since Schiff bases activate proteases (Davies, 1987b), this modification may play an important protective role in clearing oxidized proteins from biological systems.

With extended incubation time or high aldehyde concentrations, accumulated lysine Schiff base and Michael adducts of acrolein and crotonaldehyde cyclize to yield EMP [$N^\varepsilon$-(5-ethyl-2-methylpyridinium)] and FDP [$N^\varepsilon$-(2,5-dimethyl-3-formyl-3,4-dehydropiperidino)] structures (Fig. 8.11). Although two aldehydes are required for each amine, the amino group appears to direct the condensations. EMP structures are formed via initial 1,2 addition of lysine to the carbonyl carbon, followed by a 1,4-addition of the imine to a second aldehyde, carbonyl condensation to close the ring, and a final dehydration and dehydrogenation. In the process, the second carbonyl is reduced. While EMP-lysine is a minor product for crotonaldehyde and not formed by acrolein, it is the dominant pathway for 2-hexenal and 2-octenal (Ichihashi et al., 2001). A significant consequence is fixing a permanent positive charge on the ε-amino groups involved.

In contrast, FDP adducts are formed via successive 1,4 additions of lysine, followed by carbonyl condensation to close the ring, then dehydration. Unlike EMP adducts, FDP adducts retain a carbonyl function and thus will be detected in carbonyl assays of oxidized proteins. Acrolein and crotonaldehyde both form FDP-lysine adducts (Uchida et al., 1995, 1998b; Ichihashi et al., 2001), and analogous adducts are formed by 2-hexenal (dipropyl-FDP-lys) and 2-pentenal (diethyl-FDP-lys) (Ichihashi et al., 2001). Antibodies to FDP and EMP functions have detected adducts of acrolein and crotonaldehyde in LDL oxidized by copper (II) (Uchida et al., 1998a, 1998b; Ichihashi et al., 2001) and in BSA co-oxidized with methyl arachidonate (Uchida et al., 1998b).

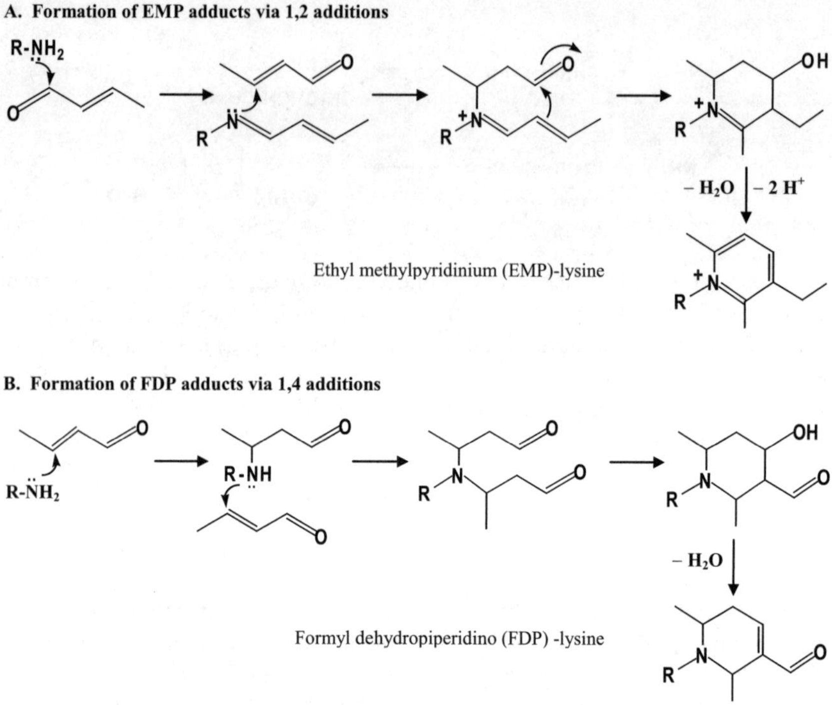

**A. Formation of EMP adducts via 1,2 additions**

Ethyl methylpyridinium (EMP)-lysine

**B. Formation of FDP adducts via 1,4 additions**

Formyl dehydropiperidino (FDP) -lysine

**Fig. 8.11.** Formation of cyclic products from additions of lysine to unsaturated aldehydes such as crotonaldehyde: A. EMP (ethyl methylpyridinium)-lysine via 1,2-addition (Ichihashi et al., 2001); B. FDP (formyl dehydropiperidino)-lysine via 1,4-addition (Uchida et al., 1995, 1998b; Ichihashi et al., 2001).

*4-Hydroxy-2-Alkenals and 4-Oxo-2-Alkenals*

Oxidized alkenals (Fig. 8.9) are the aldehydes most reactive with proteins. Perhaps the best known and most studied aldehydes in this class are 4-hydroxy-2-nonenal (HNE) and 4-oxo-2-nonenal (ONE). Reactions and chemistry of HNE have been reviewed in elegant detail by Esterbauer et al., (1991); other useful reviews covering various aspects of HNE reactions with proteins and subsequent physiological effects are also available (Uchida, 2000; Schaur, 2003; Davies et al., 2004; Petersen and Doorn, 2004). 4-Oxo-2-alkenals, also called γ-ketoaldehydes, may be formed independently but are actually tautomers of 4-HO-2-alkenals:

$$\underset{\text{R-C-CH=CHCH=O}}{\overset{\text{OH}}{|}} \longleftrightarrow \underset{\text{R-C-CH}_2\text{CH}_2\text{CH=O}}{\overset{\text{O}}{\|}}$$

Thus, under oxidizing conditions, the reactions of these two oxidants are often difficult to distinguish, except by kinetics (ONE reactions are orders of magnitude faster) and balance between pathways. The major reaction for both HNE and ONE is (1,4) Michael addition to nucleophilic amino acid side chains (thiol, imidazole, lysine) (Schauenstein et al., 1971; Esterbauer et al., 1975, 1976; Uchida and Stadtman,

1992, Bruenner et al., 1995; Schaur, 2003). For HNE, the free carbonyl can then undergo secondary Schiff base formation with another amine. The second carbonyl in ONE makes it markedly more electrophilic and provides a site for Schiff base formation as well as Michael addition in both initial and subsequent reactions. ONE also reacts with arginine, but HNE does not (Lin et al., 2005).

To provide a quantitative frame of reference, ONE reacted up to 30% faster than 4-HNE with RNAse A and β-lactoglobulin (Lin et al., 2005). With [amine] > 10x [aldehyde], the following $t_{1/2}$ for the reactions were observed: for RNAse, 95 h and 6 min for HNE and ONE, respectively; and for β-lactoglobulin, 14 h and 3 min for HNE and ONE, respectively. The order of Michael addition reactivity for the two aldehydes with amino acids is as follows (relative rate constants are in parentheses) (Doorn and Petersen, 2002, 2003):

HNE:   cys (1) >> his (0.002) > lys (0.001)
ONE:   cys (186) >>> his (0.02) > lys (0.007) > arg (0.0006)

Michael additions of cysteine on proteins to HNE and ONE occur almost spontaneously, proffering a staggering potential for biological damage. In foods, lipid oxidation products accumulate, so thiol reactivity of HNE and ONE may be an important contributor to quality deterioration during long-term storage, limited only by the availability of cysteine residues on proteins. Sulfhydryl proteins, such as protein disulfide isomerase, show extensive loss of cysteine in the presence of HNE in model systems (Carbone et al., 2005). However, extensive protective mechanisms counteract this reactivity in vivo. GSH-S-transferase redirects the reaction to intracellular glutathione (GSH) in a catalyzed reaction that is 600x faster than chemical conjugation, and aldehyde dehydrogenases rapidly reduce carbonyl groups to alcohols, effectively limiting cellular concentrations of these reactive aldehydes, even under conditions of oxidative stress (Siems and Grune, 2003).

Despite nearly instantaneous reaction rates with thiols, cysteine contents of most proteins are very low, so histidine and lysine become the major targets of HNE and ONE. However, these aldehydes differ in amino acid selectivity. HNE prefers Michael addition reactions with histidine and lysine, but the amino acid most damaged depends on protein primary structure, configuration, and residue availability. Histidine is the main target in β-lactoglobulin B and human hemoglobin (Bruenner et al., 1995), myoglobin (Alderton et al., 2003) and apomyoglobin (Bolgar and Gaskell, 1996), insulin B chain (Fenaille et al., 2003), bovine heart cytochrome c oxidase (Musatov et al., 2002), and RNAse A (Liu et al., 2003; Lin et al., 2005), while lysine is the dominant reaction site in glucose-6-dehydrogenase (Uchida and Stadtman, 1992; Friguet et al., 1994a), human LDL (Refsgaard et al., 2000), bovine serum albumin (Zamora and Hidalgo, 2003a), and a secondary site in RNAse A (Liu et al., 2003; Lin et al., 2005). In contrast, the fastest reaction between ONE and RNAse or β-lactoglobulin was Schiff base formation with specific lysines, leading to ONE-lys pyrrolidone on K16 and K145, and ONE-his-lys pyrrole crosslink K16—H24 (Lin et al., 2005).

Schiff base and Michael additions of hydroxy and oxo alkenals to proteins generate linear adducts that remove peptide thiols and amines and add carbonyls. The order of efficiency in generating protein carbonyls is acrolein > oxononenal > hydroxynonenal > dodecadienal > MDA (Yuan et al., 2007). However, these adducts are not always detected because they dissociate reversibly (Esterbauer et al., 1991; Nadkarni and Sayre, 1995), especially under conditions used to hydrolyze proteins (Lin et al., 2005), and the aldehyde also cyclizes rapidly to form a hemiacetal (Esterbauer et al., 1975, 1976; Uchida and Stadtman, 1992; Lin et al., 2005) (shown below for histidine, cysteine, and lysine, respectively):

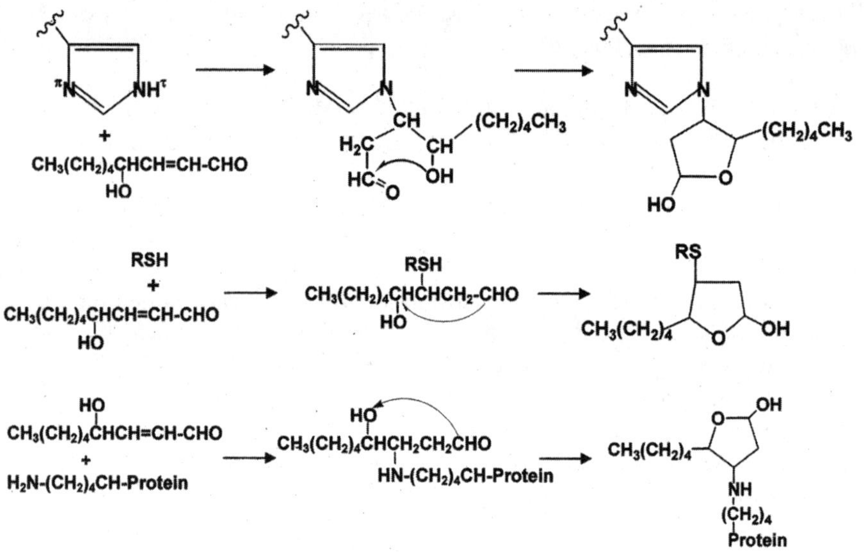

When amines are present in molar excess over the aldehydes, pyrrole derivatives are formed, as shown below. The ring structure originates from reaction of two amines (usually lysine) with HNE and ONE in a complex sequence involving Schiff base addition, (oxidation for HNE), Michael addition, oxidation, and cyclization, (Sayre et al., 1993, 1997; Xu et al., 1999b; Schaur, 2003; Zhang et al., 2003). HNE and ONE both form stable fluorescent 2-pentyl-2-hydroxy-1,2-dihydropyrrol-3-one iminium complexes involving addition of two lysines, as shown in the reaction sequence below (Xu et al., 1999a, 1999b; Zhang et al., 2003). The net result is formation of a peptide crosslink. In addition, if the pyrrole–OH in the iminium complex is not removed in dehydration, it can react further, for example with another lysine, to form cyclic or acyclic mixed aminals.

$$\text{P-lys-NH}_2 + \text{RCH(OH)CH=CH-CHO} \xrightarrow{\text{HNE}} \underset{\text{Schiff base}}{\text{RCH(OH)-CH=CH-CH=NH-lys-P}} \longleftrightarrow \text{RCH=CH-CH=CH-NH-lys-P} \downarrow [\text{O}]$$

$$\underset{\text{P-lys-NH}}{\text{RCH=CH-CH-CH=NH-lys-P}} \longleftarrow \underset{\underset{\text{Michael adduct}}{\text{P-lys-NH}}}{\text{RCHCH-CH-CH=NH-lys-P}} \xleftarrow{\text{P-lys-NH}_2} \text{RCHCH=CH-CH=NH-lys-P}$$

$$[\text{O}] \downarrow \qquad\qquad\qquad\qquad\qquad\qquad\qquad \text{P-lys-NH}_2 + \text{RCH-CH=CH-CHO (ONE)}$$

[Cyclization scheme showing 2-pentyl-2-hydroxy-1,2-dihydropyrrol-3-one iminium link intermediates leading to HN-lys-P pyrrole product with loss of $-H_2O$]

In addition to pyrroles, HNE and ONE generate a wide range of secondary cyclic products, including epoxides, pyrroles, pyrrolidones, and thiazolidines as adducts and crosslinks under oxidizing conditions. As with pyrroles, these products arise from various combinations of Schiff base and Michael additions, oxidations, cyclizations, and dehydrations. Amine and aldehyde concentrations are major determinants of product pathways. Intermolecular cyclization is facilitated at high aldehyde concentrations, and crosslinking occurs when amines are present in molar excess over aldehydes. Secondary cyclic products are responsible for many of the observed changes in properties of proteins reacted with HNE and ONE. The reactions generating various cyclic products will be discussed in the sections on Crosslinking and Formation of Fluorescent Products.

Detailed structures are now being identified by electrospray ionization (ESI) (Bruenner et al., 1995; Bolgar and Gaskell, 1996; Brame et al., 1999; Fenaille et al., 2002, 2003; Liu et al., 2003; Schöneich and Sharov, 2006) and MALDI-TOF (Doorn and Petersen, 2002; Kapphahn et al., 2006) mass spectrometry of modified peptides, with and without proteolysis (Fenaille et al., 2004). The presence of individual adduct structures in individual proteins can be tracked by specific antibodies (Uchida et al., 1993; Xu et al., 2000), although consideration must be given to cross-reactivity when interpreting data.

## Physiological γ-Ketoaldehydes–Levuglandins

Isoketals form on free fatty acids in transport and also on fatty acids still esterified in phospholipids where they generate novel PL-protein adducts that profoundly affect protein function in membranes, completely blocking $K^+$ ion channels (Brame et al., 2004). A special class of isoketals are isoprostanes or levuglandins formed by rearrangement of endoperoxides in arachidonic acid (Brame et al., 1999; Davies et al., 2004). The reaction below shows only the 5-series of hydroperoxides, but analogous isoketals are also formed with the precursor hydroperoxide at C-8, 12, and 15, with 5-OOH and 15-OOH being dominant.

IsoLevuglandin D    IsoLevuglandin E

All isoketals show remarkable proclivity for polymerization (Xu et al., 1999a; Brame et al., 2004; Davies et al., 2004). Levuglandin E2 is orders of magnitude more effective in crosslinking proteins than any other oxidation product of arachidonic acid (Iyer et al., 1989). In aqueous systems, isoketals react with proteins faster than all other secondary products (first order $k=10^6-10^8$ s$^{-1}$); in organic solvents the rate of pyrrole formation is $10^3-10^4$ times slower (second order $k=10^2-10^4$ M$^{-1}$ sec$^{-1}$) because the Schiff base product dominates and cyclization does not occur until water is added (Amarnath et al., 1995). This observation provides compelling evidence that phase localization of the reaction will determine the dominant pathway. The reason for the exceptional crosslinking reactivity of isoketals is that, like other 4-HO-enals and 1,4-dicarbonyls, they almost instantaneously cyclize to form pyrroles when reacted with amines (Amarnath et al., 1995). When a 3x molar excess of levuglandin E2 was reacted with BSA, >50% of the isoketal had reacted within 20 seconds, accompanied by production of BSA crosslinks (Brame et al., 1999). In buffered systems a small portion of the products also arise from classical carbonyl-amine condensation to form imine Schiff base adducts, but the major products are lactams, hydroxylactams, and polymers from oxidation of pyrroles (Fig. 8.12). The first step is a Paal-Knorr condensation of dicarbonyls with protein amines (primarily lysine) to form monoalkylpyrroles (Amarnath et al., 1995; Xu et al., 1999a). In the presence of oxidizing lipids, the pyrroles then oxidize further to lactams and hydroxylactams (Xu et al., 1999a; Brame et al., 2004; Davies et al., 2004) which retain available carbonyls.

### Addition Reactions from Free Radical Oxidation of Unsaturated Aldehydes
Although Michael addition to α,β-unsaturated aldehydes is generally presumed, at least two studies have shown that secondary autoxidation of unsaturated aldehydes can yield radicals that induce crosslinking in proteins (Funes and Karel, 1981) and also react with nucleic acids (Yang, 1993). Although routinely ignored, autoxidation of saturated aldehydes to carboxylic acids and other products is very rapid and involves intermediate radicals (Pokorny et al., 1985); secondary oxidation of unsaturated aldehydes should be even more facile. Free radical oxidation of aldehydes presents still another pathway for lipid oxidation damage to proteins, and whether these radicals act conventionally by hydrogen abstraction or have a greater tendency to add to proteins needs to be investigated further.

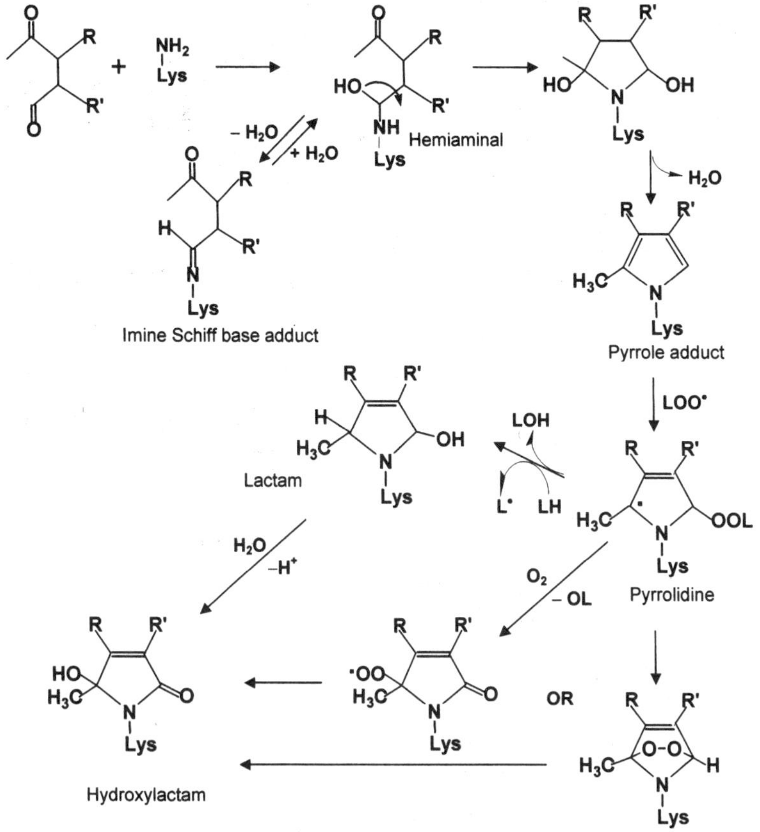

**Fig. 8.12.** Competing pathways for reaction of isoketals with ε-amino groups of proteins. The pyrrolidine pathway is dominant. For one, R and R' = H. Reaction sequence adapted from (Brame et al., 1999, 2004; Xu et al., 1999a; Davies et al., 2004).

## Reactions Underlying Molecular Damage to Proteins

The reactions of lipid radicals, hydroperoxides, epoxides, and aldehydes discussed previously are of theoretical value only until they are translated into observed behaviors in foods and biological systems. Indeed, these damage mechanisms were presented first to provide a basis for understanding the long list of macromolecular properties changed when oxidizing lipids are reacted with proteins. This section now connects specific lipid co-oxidation chemistry to commonly observed changes in protein physical and chemical properties.

## Amino Acid Losses

The potential for oxidizing lipids to damage proteins has been documented since the early 1960s (Wills, 1961a,b; Lewis and Wills, 1962; Tappel, 1965, 1973; Roubal and Tappel, 1966a), with one of the earliest observations being amino acid losses in proteins reacted with oxidizing lipids (Desai and Tappel, 1963; Roubal and Tappel, 1966b; Chio and Tappel, 1969a; Gamage et al., 1973; Horigome and Miura, 1974; Horigome et al., 1974; Kanazawa et al., 1975; Matsushita, 1975; Gardner, 1979; Matoba et al., 1984b; Nielsen et al., 1985b). Over a large number of studies, the following amino acids have consistently been shown to be most sensitive to damage by peroxidizing lipids:

| | | |
|---|---|---|
| Cysteine | Serine | Glycine |
| Tryptophan | Threonine | Alanine |
| Histidine | | Valine |
| Lysine | | Proline |
| Arginine | | Leucine |
| Tyrosine | | Isoleucine |
| Methionine | | |

The amino acids in the first column (on left) all are located primarily on protein surfaces (Fig. 8.13); with the exception of methionine, they have readily abstractable hydrogens, and cys, his, lys, trp, and arg form stable radicals when reacted with oxidizing lipids. The phenoxyl hydrogen of tyrosine is probably also easily abstracted, but without protecting ortho groups the resulting radical is not stable enough for detection. The susceptibility of these amino acids to damage from oxidizing lipids

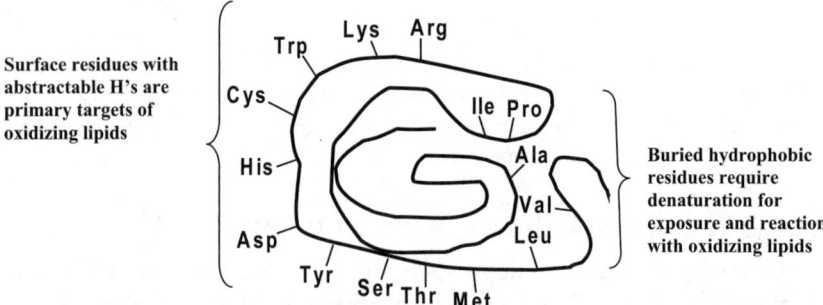

**Fig. 8.13.** Diagrammatic representation of amino acids most susceptible to attack by oxidizing lipids. The model protein shown is a typical globular or structural protein in an aqueous environment: surface residues are primarily hydrogen-bonding or charged amino acids and aliphatic amino acids are mostly buried. Hydrophobic proteins or proteins embedded in membranes have higher proportions of aliphatic and aromatic amino acids exposed for reaction.

is thus predictable and parallels damage from irradiation, photolysis, and hydroxyl radicals (Schaich, 1980b). Side chain amine thiol and amine groups of these amino acids also react rapidly with carbonyl products of lipid oxidation to form Schiff bases, Michael adducts, and their cyclic products. Thus, these amino acids are involved in early stages of oxidation and remain reactive through secondary and even terminal stages, although perhaps changing their pathways, as proteins denature and expose new amino acid residues.

Damage to the other amino acids is less easily explained. Serine and threonine (center column) have –OH groups, but the pKs are >>14 and the O-H bond energies are high (464 kJ/mol, 111 kcal/mol), so these –OH groups do not donate protons as easily and thus should not be favorable targets for oxidizing lipids. Reaction with other oxidants also shows low reactivity for the side chain hydroxyls. Oxidation by silver (Shi et al., 2007), $KMnO_4$ (Halligudi et al., 2000), and electrical current (Huerta et al., 1997) all documented initial attack at the –COOH group with accompanying decarboxylation, followed by deamination; the side chain –OH groups were never modified. However, serine and threonine side chain hydroxyls do mediate rather strong hydrogen bonding. Thus, it is interesting to speculate that these amino acids hydrogen bond to lipid hydroperoxides and induce LOOH decomposition (molecule-assisted homolysis), accompanied by radical transfer in cage reactions to generate side chain radicals and subsequent hydroperoxides and breakdown products:

$$\begin{array}{c} H_2N\text{-}CH\text{-}COOH \\ | \\ CH_2 \\ | \\ OH\cdots HOOL \end{array} \longrightarrow \begin{array}{c} H_2N\text{-}CH\text{-}COOH \\ | \\ CH_2 \\ | \\ [OH\cdots H^{\bullet} + {}^{\bullet}OOL] \end{array} \longrightarrow \begin{array}{c} H_2N\text{-}CH\text{-}COOH \\ | \\ {}^{\bullet}CH_2 \\ + H_2O + {}^{\bullet}OOL \end{array} \longrightarrow \begin{array}{c} H_2N\text{-}CH\text{-}COOH \\ | \\ {}^{\bullet}OOCH_2 \end{array}$$

In contrast, the hydrophobic amino acids (last column) have no readily abstractable hydrogens, do not participate in hydrogen bonding, and are buried in the interior of native protein; they do not become exposed without denaturation. Indeed, damage to these amino acids has been reported primarily in proteins incubated with oxidizing lipids at higher temperatures (55–60°C) where denaturation is facilitated (Horigome and Miura, 1974; Horigome et al., 1974). Thus, their sensitivity to damage by oxidizing lipids is more difficult to explain, and indeed, little has yet been elucidated regarding mechanisms of lipid reactions with these amino acids. Nevertheless, three possible explanations for observed damage may be offered.

   a. When free hydrophobic amino acids are reacted with a variety of strong oxidants, the main sites of reaction are the carboxylic acid group, followed by the amine group; decarboxylation and deamination frequently occur (Schaich, 1980a; Bobrowski and Schöneich, 1996; Huerta et al., 1997; Guitton et al., 1998). The only oxidant that has shown ability to react with valine, leucine, and isoleucine side chains is the highly electrophilic hydroxyl radical, HO•, which abstracts hydrogens rather nonspecifically from hydrocarbons (Hasegawa and Patterson, 1978; Patterson and Hasegawa, 1978), and attacks the side chains and terminal methyl groups of these amino acids (Fu et al., 1995). It may seem counterintuitive, but when HO• generated cleanly by radiation were reacted with individual amino acids, it was proline, valine, leucine, isoleucine that showed the highest yields of protein radicals and hydroperoxides (Simpson,1992). Analyses revealed that a tertiary carbon atom or a segment containing at least two methylene groups was required to stabilize the initial radical so that $O_2$ could add to form peroxyl radicals. By analogy, HO• are produced when lipid hydroperoxides decompose under UV light or at elevated temperatures (60°C is sufficient), and they are generated when ferrous iron autoxidizes. Thus, it is reasonable to expect that the adventitiously produced HO•, rather than the expected LOO• or LO• radicals, may mediate oxidation of aliphatic amino acids.

b. It is well documented in irradiated proteins that free radicals generated anywhere along the main peptide or amino acid side chains tend to migrate along the protein backbone to glycine residues, where they are stabilized (Schaich, 1980a). The hydrocarbon side chains of alanine, valine, leucine, isoleucine, and proline may similarly provide electron sink sites where radicals are stabilized (Fig. 8.14), leading to oxidation and production of hydroperoxides on side chains of these amino acids.

c. Iron is often present adventitiously or deliberately added as a catalyst in lipid oxidation studies. LOOH can complex to $Fe^{2+}$ to form ferryl iron ($Fe^{4+}$ =O) or equivalent complexes that are highly electrophilic and abstract hydrogens even more rapidly and non-specifically than HO• (k > 109 L $M^{-1}s^{-1}$). Ferryl iron complexes also catalyze oxygen insertion into C-H bonds to yield epoxides, ketones, and alcohols (Schaich, 2005).

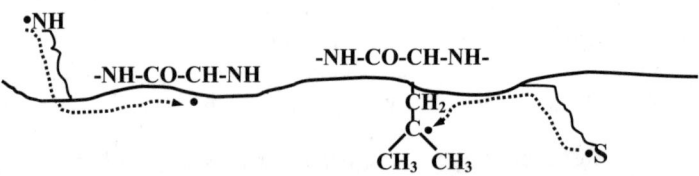

**Fig. 8.14.** Radicals initially formed at amino or sulfhydryl sites on amino acid side chains migrate along peptide backbone and tend to localize at glycine residues (left). Aliphatic side chains may similarly serve as electron sinks (right), leading to free radical oxidation of hydrophobic residues.

$$\text{LOOH-Fe}^{2+} \longrightarrow \text{Fe}^{4+}\begin{Bmatrix} {}^-\text{OH} \\ {}^-\text{OL} \end{Bmatrix} \xrightarrow{\text{RH}} \text{Fe}^{4+}\begin{Bmatrix} {}^-\text{OH} \\ \text{RH} \end{Bmatrix} + \text{LOH} \longrightarrow \text{Fe}^{2+} + \text{R}^{\bullet} + \text{H}_2\text{O}$$

$$\xrightarrow{\text{RH}} \text{Fe}^{3+} + {}^-\text{OH} + \text{R}^{\bullet} + \text{LOH}$$

$Fe^{3+}$-LOOH yields the $Fe^{3+}$-$^{\bullet}$OH complex which is functionally equivalent to $Fe^{4+}$ (Bossman et al., 1998). Detailed studies are needed to determine which of these explanations, or others, accounts for lipid oxidation damage to aliphatic amino acids.

Specific amino acid breakdown products and the lipid oxidant inducing them are listed in Table 8.1. Mechanisms and reactions leading to these products will be presented in a separate paper.

The most obvious consequence of amino acid destruction by oxidizing lipids is diminution of nutritional quality of foods, as has been shown in a variety of food systems, for example intermediate moisture meats (Obanu et al., 1980), freeze-dried beef (Dvorak, 1968), casein (Horigome and Miura, 1974; Horigome et al., 1974; Yanagita and Sugano, 1975; Matoba et al., 1984b; Kanazawa et al., 1987), whey protein (Nielsen et al., 1985a, 1985b) chickpea legumin (Sánchez-Vioque et al., 1999), ovalbumin (Horigome and Miura, 1974; Horigome et al., 1974, Yanagita and Sugano, 1974), and lupine conglutins (Lqari et al., 2003). Loss of nutritional value increases markedly when lipid-protein reactions occur at high temperatures (50–60°C), as might be expected.

Interestingly, the problem seems to involve more than mere destruction of amino acids or reduction in digestibility or absorption, since simultaneous supplementation of diets with amino acids found to be affected does not always restore normal nutritional value (Horigome and Miura, 1974; Horigome et al., 1974; Yanagita et al., 1976). Effects of damaged ovalbumin could be repaired by adding back lys and met (Yanagita and Sugano, 1974), but amino acids could not reverse damage from casein reacted with oxidizing ethyl linoleate (Yanagita and Sugano, 1975; Yanagita et al., 1976). Absorption of damaged proteins or proteolysis of damaged proteins with recycling of component amino acids in vivo has a cascade of effects, starting with decreased liver nitrogen and lipids (Yanagita and Sugano, 1975; Yanagita et al., 1976). This, in turn, causes imbalances in the amino acid pool that lead to impaired protein synthesis as well as other as yet unidentified metabolic and functional impairments. Amino acid imbalances may also be exacerbated by the conversion of polar essential fatty acids to non-polar non-essential fatty acids, such as alanine and leucine, histidine degradation to aspartic acid and asparagine, and proline transformation to glutamic acid (Table 8.1).

These and other indirect effects on protein nutrition are exceedingly difficult to track but may well contribute to aging, oxidative stress, and a wide range of pathologies by creating an environment in which the direct molecular damage to key proteins from lipids or oxidants cannot be counteracted.

Another critical impact of amino acid modification is on protein structure and associated extensive effects on protein functionality in foods and biological activity in

**Table 8.1.** Examples of Degradation Products Formed in Reactions of Amino Acids with Lipid Oxidation Intermediates and Products.

| Amino Acid | Lipid Oxidant | Reaction[a] | Products | References |
|---|---|---|---|---|
| Cysteine | LO(O)• | radical recombination | cystine | (Little & O'Brien, 1967, 1968) |
| | | R-S scission | alanine, H$_2$S, cystine | (Roubal & Tappel, 1966b) |
| | | oxidation | alanine sulfinic acid, cysteic acid, | (Lewis & Wills, 1962; Little & O'Brien, 1968) |
| | | | cystine disulfoxide, cystine thiosulfinate, cystinethiosulfonate, cysteine sulfoxides, | (Finley & Lundin, 1980) |
| | | | cysteine sulfinic and sulfonic acids | |
| | LOOH | addition | lipid epoxides, cysteine oxides | (Gardner & Weisleder, 1976; Gardner & Kleiman, 1981) |
| | | | 9-S-cysteine-13-HO-10-ethoxy-*trans*-11-octadecenoic acid | (Gardner et al. 1977) |
| | | | 9-S-cysteine-10,13-diHO-*trans*-11-octadecenoic acid | |
| | epoxides | addition | hydroxylated adducts | (Buttkus, 1972) |
| | alkanals | 1:1 addition | Schiff base adducts | (Doorn & Petersen, 2002; Petersen & Doorn, 2004; Lin et al., 2005) |
| | alkenals | 1:1 addition | Michael addition products | |
| | | Schiff base addition/cyclization | thiazolidinecarboxylic acid derivatives | (Schmolka & Spoerri, 1957; Buttkus, 1968) |
| | MDA (excess) | Michael + Schiff base addition | 3:2 complex condensation products | (Buttkus, 1968) |
| | HO-alkenals | 1:1 Michael addition | cyclic hemiacetals | (Esterbauer et al., 1975; Esterbauer et al., 1991) |
| | oxoalkenals | Schiff base addition/cyclization | thiazolidinecarboxylic acid derivatives | (Gardner & Jursinic 1981) |

Table 8.1., cont. Examples of Degradation Products Formed in Reactions of Amino Acids with Lipid Oxidation Intermediates and Products.

| Amino Acid | Lipid Oxidant | Reaction[a] | Products | References |
|---|---|---|---|---|
| Methionine | LOOH | oxidation | methionine sulfoxide | (Garner et al., 1998a; Brock et al., 2007) |
| | | strong oxidation | methionine sulfone | (Wainwright et al., 1972) |
| | | deamination, decarboxylation | methional, methane thiol, acrolein | (Wainwright et al., 1972) |
| | MDA | Schiff base addition | HOOC-CH(R)-NH-CH=CH-CHO | (Buttkus, 1968) |
| Histidine | | deamination | imidazole lactic acid, imidazole acetic acid | (Yong & Karel, 1978, 1979) |
| | LO(O)• | decarboxylation, scission | histamine, ethylamine | (Roy & Karel, 1973) |
| | | ring scission | aspartic acid, asparagine, valine | (Roy & Karel, 1973) |
| | HNE | Michael addition | β-substituted propanal adducts, hemiacetals (?) | (Uchida & Stadtman, 1992) |
| | epoxy alkenals | Michael addition | β-substituted adducts, epoxide intact | (Zamora et al., 1999) |
| | crotonaldehyde | nucleophilic addition to C=C | β-substituted butanal adducts | (Ichihashi et al., 2001) |
| Lysine | | decarboxylation | diaminopentane | (Karel et al., 1975) |
| | | α-deamination/cyclization | pipecolic acid | (Karel et al., 1975) |
| | | ε-deamination/rad recombination | 1,10-diamino-1,10-docarboxydecane | (Karel et al., 1975) |
| | LO(O)• | ε-deamination/oxidation/decarboxylation | α-amino adipic acid, aspartic acid, alanine glycine | (Karel et al., 1975) |
| | MLOOH | concerted scission/addition | $N^\varepsilon$- (hexanoyl)lysine, $N^\varepsilon$- (azelayl)lysine | (Kato et al., 1999; Kawai et al., 2003) |
| | pentanal | Schiff base addition | 2-propyl-2-heptenal-lysine | (Dalsgaard et al., 2006) |
| | hexanal | 1:2 Schiff base, fragment, cyclize | α-amino-ε-caprolactam, immonium-lysine, diacetyl lysine, pipecolic acid | (Fenaille et al., 2004b) |

*Cont. on p. 218.*

Table 8.1., cont. Examples of Degradation Products Formed in Reactions of Amino Acids with Lipid Oxidation Intermediates and Products.

| Amino Acid | Lipid Oxidant | Reaction[a] | Products | References |
|---|---|---|---|---|
| Lysine, cont. | 2 MDA + 1 alkanal | Schiff base, cyclization | 1,4-dihydropyridine-3,5-dicarbaldehydes | (Freeman et al., 2005) |
| | crotonaldehyde | Michael addition 1:1 | β-substituted adducts | (Ichihashi et al., 2001) |
| | crotonaldehyde, acrolein | Michael addition 2:1 | FDP-lysine (3-formyl-3,4-dehydropiperidino adducts) | (Uchida, 2000) |
| | crotonaldehyde (excess) | Schiff base 2:1 | EMP-lysine (pyridinium adduct) | (Uchida, 2000) |
| | 2-octenal | Michael addition 1:1 | β-substituted adduct | (Alaiz & Girón, 1994) |
| | epoxy alkenals | Schiff base, cyclic addition | amino carboxypentylpyrroles | (Zamora & Hidalgo, 1995; Zamora et al., 2000) |
| | epoxy alkenals | Schiff base, scission, addition | N-pyrrolylnorleucine | (Zamora et al., 1995) |
| | epoxy heptenal | Michael addition, cyclization | N-substituted pyrroles and 2-(1-hydroxyalkyl)pyrroles | (Zamora et al., 2000) |
| | HO-ω-oxoalkenoic acids | Schiff base, cyclization | 2-(ω-carboxyalkyl)pyrroles | (Gu et al., 2003) |
| | HNE | Michael addition 1:1 | β-substituted adducts, cyclic hemiacetals | (Szweda et al., 1993; Nadkarni & Sayre, 1995) |
| | HNE | 1:2 Schiff base + Michael addition | amine-hemiacetal-amine crosslinks | (Nadkarni & Sayre, 1995) |
| | HNE | 1:2 Michael + Schiff base addition | Hydroxy-imino-dihydropyrroles | (Cohn et al., 1996; Itakura et al., 1998; Tsai et al., 1998) |
| Arginine | LO(O)• | H abstraction, oxidation, hydrolysis | 5-oxoproline | (Spiteller, 2006) |
| | radicals | H abstraction, peroxidation | γ-glutamyl semialdehyde | (Pietzsch, 2000; Requena et al., 2001) |
| | MLOOH | oxidation | NO• release | (Schaich, 2002c) |

Table 8.1., cont. Examples of Degradation Products Formed in Reactions of Amino Acids with Lipid Oxidation Intermediates and Products.

| Amino Acid | Lipid Oxidant | Reaction[a] | Products | References |
|---|---|---|---|---|
| Tryptophan | LO(O)• | H abstraction, peroxidation | kynurenine, N-formyl kynurenine | (Yong et al., 1980; Krogull & Fennema, 1987) |
| | | addition | dioxindole-3-alanine | (Yong et al., 1980) |
| | | ring scission | indoleacetic acid and indolelactic acid | (Yong et al., 1980) |
| | LO(O)• | H abstraction, radical dimerization | dityrosine | (Hunter et al., 1989; Kikugawa et al., 1991; Giulivi et al., 2003) |
| | | H abstraction, oxidation | tyr hydroquinone, tyr quinone | (Giulivi et al., 2003) |
| Tyrosine | | addition, oxidation, scission | DOPA, dopamine, 5,6-diHO-indole | (Giulivi & Davies, 1993; Giulivi et al., 2003) |
| | HO• | addition, decarboxylation | dopamine, dopamine quinone | (Giulivi & Davies, 1993; Giulivi et al., 2003) |
| Phenylalanine | HNE | Schiff base, cyclization | N-substituted-2-pentyl pyrrole, phenylacetaldehyde | (Hidalgo et al., 2005) |
| Glycine | LO(O)• | H abstraction, deamination | formic acid | (Berger et al., 1999) |
| | MDA | Schiff base | N-prop-2-enal aminoacetic acid | (Crawford et al., 1966). |
| Valine | HO• | H abstraction, peroxidation | 3 valine hydroperoxides | (Fu et al., 1995) |
| | LO(O)• | H abstraction, peroxidation | 5-oxoproline | (Spiteller, 2006) |
| Proline | HO•, t-Bu-O• | addition, ring scission | glutamic semialdehyde | (Requena, 2001) |
| | HO•, radiation | addition, oxidation, scission | γ-aminobutyric acid, glutamic acid, and 2-pyrrolidone | (Schuessler & Schilling, 1984; Kato et al., 1992; Matysik et al., 2002) |

[a] Reactant proportions cited are aldehyde:amine

vivo. Structural alterations are the subjects of Sections "Transfer of Free Radicals from Oxidizing Lipids to Proteins" through "Alteration or Loss of Biological Function."

## Transfer of Free Radicals from Oxidizing Lipids to Proteins

Most dry foods with lipids and proteins show stable EPR signals from radical transfer. EPR signal intensities correlate with the extent of lipid and protein oxidation and provide clear footprints of lipid-mediated damage (Schaich and Karel, 1975; Schaich, 1980b; Saeed et al., 1999, 2006) (Fig. 8.15). Thus, use of protein EPR signals in conjunction with lipid oxidation analyses can reveal the extent to which oxidation has been broadcast beyond lipids and improve predictions of quality deterioration during processing and storage.

Some cautions in analysis and interpretation of free radicals in foods and biological systems are necessary. Radical quantitation is complicated by moisture beyond interference with EPR measurements. Maintaining low moisture slows radical recombinations and prolongs radical lifetimes, hence the common use of dry models to study protein radical reactions. Nevertheless, much spectral information is lost in these systems due to molecular immobilization. In addition, it can be argued that such reaction conditions are relevant only to dehydrated foods and do not accurately reflect reaction conditions in biological systems. On the flip side, even low moisture provides protons to quench protein free radicals and in effect repairs limited damage, while higher moisture increases molecular mobility and enhances recombination of side chain radicals (Zirlin and Karel, 1969).

Thus, in solutions or emulsions, tracking free radical transfer to proteins requires special techniques to detect short-lived radical species. Radicals may be detected directly by low temperature (liquid nitrogen) measurements or by continuous flow-mix methods, which will also be able to provide reaction rates, (Schaich 2002a, 2002b, 2002c). Radical transfer may be followed indirectly by spin trapping (Lion et al., 1981; Culbertson et al., 2003; Headlam and Davies, 2003; Augusto et al., 2004; Davies and Hawkins, 2004; Mason, 2004) or immuno-spin trapping (Mason, 2004) as has been demonstrated with HO· and NO· reactions with proteins, or by chemical modifications to proteins (Reubsaet et al., 1998; Davies et al., 1999; Akagawa et al., 2006). Spin traps offer the advantage of being able to detect and distinguish both short-lived lipid radicals and longer lived protein radicals in the same sample and in multiple solvents. Since nitroxide radical line shapes and splittings are very sensitive to local molecular motion and to solvent polarity (Schaich 2002a, 2002d), they can also uniquely reflect alteration in protein conformation, for example unwinding in denaturation (Saeed et al., 2006) or collapsing in aggregation or crosslinking, that accompanies radical transfer.

## Changes in Physical Properties: Solubility, Hydrophobicity, Conformation, Aggregation

Any modification of amino acids through oxidation, scissions, or adduct formation must be reflected in protein properties. Protein oxidation, whether by HO· or oxidizing lipids, can induce marked changes in conformation or structure of proteins

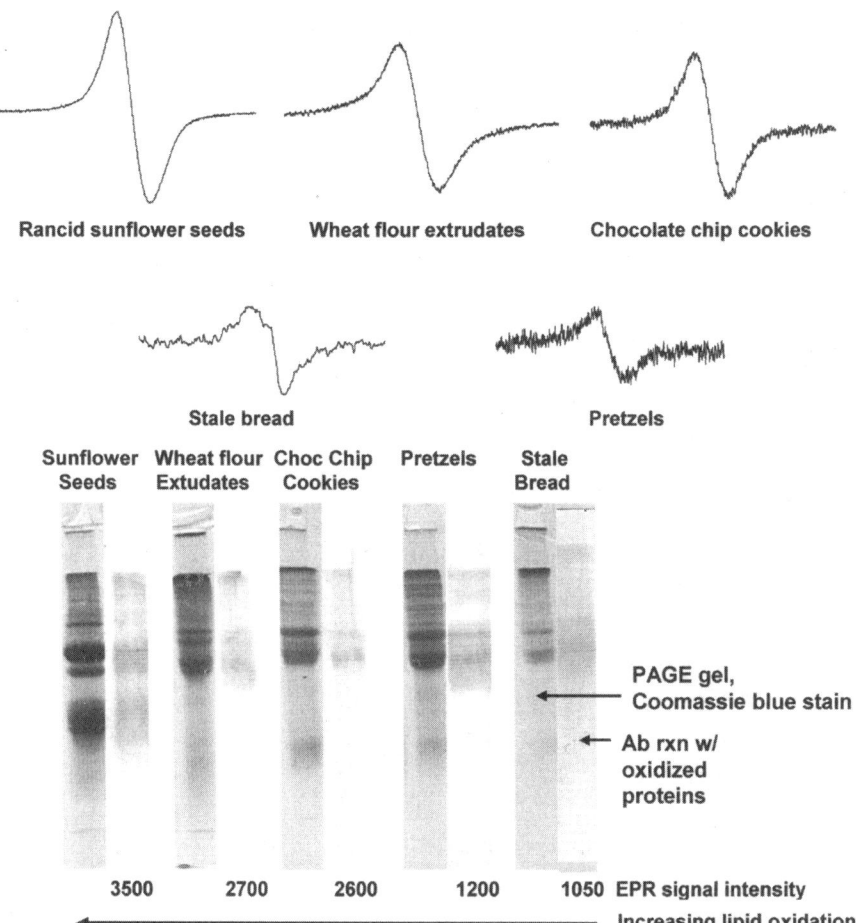

**Fig. 8.15.** Correlation of lipid oxidation, protein oxidation (antibody reaction), and protein free radicals (EPR signal intensity) in common foods (Tanczos et al., 2002). (Top) Room temperature EPR signals of several common foods. Sulfur radicals are only detected at liquid nitrogen temperatures or lower (-196°C). (Bottom) Left lane for each food: protein bands on 12% polyacrylamide electrophoresis gels (PAGE). Right lane: Same proteins transferred to Western blots and reacted with antibodies raised against carbonyl dinitrophenylhydrazones in oxidized proteins.

by altering surface charges (Haberland 1982, 1984) or by increasing hydrophobicity, inducing denaturation, or complexing lipids to the protein (Lea, 1957; Davies and Delsignore, 1987). By reaction with free amines, MDA denatured rabbit myosin (King and Li, 1999), reduced its α-helix content, increased random structure, and eliminated β-strand structure (Li and King, 1999). MDA binding to his, tyr, arg, and met residues of myosin altered the isoelectric point and reduced solubility (Buttkus, 1967). High concentrations of MDA caused conformational changes in the globular heads and destroyed active sites of myosin $Ca^{2+}$-ATPase (King and Li, 1999). Some of these changes lead to aggregation of proteins without crosslinking, as has been observed with plasma β-lipoproteins incubated with methyl linoleate hydroperoxide (MLOOH) (Nishida and Kummerow, 1960; Suyama and Adachi, 1979) and with chicken myofibrils lyophilized with methyl linoleate (Smith et al., 1990).

All of these changes reduce protein extractability and solubility in a wide range of systems. Mechanisms contributing to loss of solubility vary with the protein, the extent of lipid oxidation or specific lipid oxidation product reacted, and the reaction system. At low levels of lipid oxidation, decreased solubility in egg albumin complexed with oxidized fatty acids was attributed to hydrogen bonding between LOOH and proteins; no covalent crosslinking or loss of functional groups was observed (Narayan and Kummerow, 1958). Covalent binding of actively oxidizing lipids with a concurrent increase in hydrophobicity was observed with γ-globulin and albumin (Nielsen, 1978), cytochrome c (Desai and Tappel, 1963), casein (Leaver et al., 1999a), β-lactoglobulin (Leaver et al., 1999b), and soy protein (Huang et al., 2006). Incubation of soy protein with soy phospholipids reduced solubility by nearly one-half via disulfide crosslinking; it also oxidized sulfhydryls to products not reducible by mercaptoethanol and generated protein carbonyls that also altered protein solubility (Boatright and Hettiarachchy, 1995). Non-covalent aggregation and covalent crosslinking via carbonyls or free radicals also decrease solubility (Huang et al., 2006).

Changes in protein solubility, conformation, and intermolecular associations mediated by oxidized lipids translate into deterioration of protein function, such as enzyme inactivation, impaired baking properties of store flours (Lea, 1957), decreased gel strength and water-holding capacity of chicken myofibrils (Smith et al., 1990) and gelatin (Matoba et al., 1984a), membrane leakiness and impaired $Ca^{2+}$-accumulation in muscle mitochondria (Player and Hultin, 1978), and loss of functional properties of soy proteins in foods (Boatright and Hettiarachchy, 1995), to cite a few examples. The same changes taking place in vivo contribute to loss of normal physiological function and development of abnormal function and pathology.

## Crosslinking

Crosslinking of proteins is a major effect of reaction with lipids at all stages of oxidation—oxyl radicals, epoxides, and secondary carbonyl products alike. Crosslinking, which leads to decreased solubility, changes in food textures, loss of functionality, and browning, has been recognized for decades. Indeed, the association of LOOH-induced protein polymerization products with aging (Hendley et al., 1963a, 1963b) and with toughening of meats (El-Gharbawi and Dugan, 1965) initially attracted attention to the importance of protein reactions with oxidizing lipids. Despite long-term recognition of this phenomenon, understanding of the responsible mechanisms is still evolving. A major obstacle is that lipid oxidation is a dynamic and constantly changing process, yet most studies focus on one stage or one intermediate or product. Thus, trying to fit individual pieces of the puzzle together for oxidation processes to determine damage causation in intact foods or living tissues under oxidative stress can be a distinct challenge.

*Free Radical Crosslinking*

As with physical properties, the susceptibility to crosslinking and the mechanisms involved are strongly dependent on the nature of the protein, the unsaturation of the peroxidizing lipid and its products, the reaction system (particularly moisture), and the reaction time. In systems with actively peroxidizing polyunsaturated fatty acids or esters, lipid radicals and hydroperoxides provide a major early source of protein radicals which recombine to generate dimers, trimers, and higher polymers within short incubation times (Fig. 8.16). Oxygen bridges may or may not be present.

The general reaction for free radical crosslinking (polymerization) of intact peptides is

$$2 \text{ LOOH or } LO(O)^{\bullet} + PH \longrightarrow 2 \text{ LO(O)H} + P^{\bullet} \longrightarrow P\text{-}P, P\text{-}P\text{-}P, \ldots (P)_n$$

where $P^{\bullet}$ may be $C^{\bullet}$ radicals on a peptide backbone or $N^{\bullet}$, thiyl -$S^{\bullet}$, or tyrosyl phenoxyl radicals ⓞ-$O^{\bullet}$ on amino acid side chains. $P^{\bullet}$ may remain localized on the side chain or migrate along the peptide backbone and become stabilized on glycine or alanine residues (Schaich, 1980b). Crosslinks form when any of these peptide or side-chain radicals recombine (Roubal and Tappel, 1966c; Schaich and Karel, 1975), as shown in the reactions below (P is any peptide chain with parts of side chains up to the specified reactive functional group). Disulfide and dityrosine crosslinks are well-documented in proteins. Carbon and nitrogen radical polymerizations are expected from electrophoresis data but have not yet been documented in oxidized proteins. Recombination of α-C radicals is expected but sterically hindered. Side-chain $C^{\bullet}$ are common with $HO^{\bullet}$ oxidations but less with lipids. Azides resulting from side chain $N^{\bullet}$ recombinations are energetically less likely and probably unstable. However, recombination of side chain $C^{\bullet}$ and $N^{\bullet}$ is certainly feasible and would generate an isopeptide with links very similar to Schiff bases. Advances in enzyme technology and LC-MS detection should facilitate verification of such crosslinks and identification of the amino acids involved.

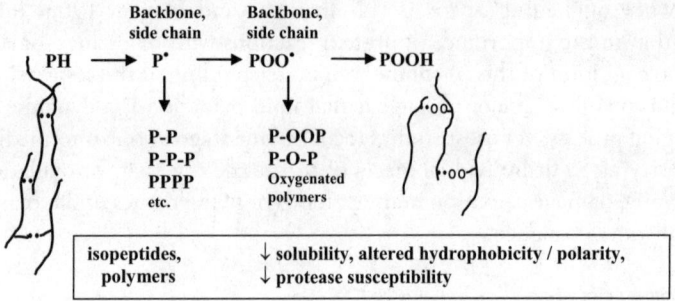

**Fig. 8.16.** Free radical crosslinking induced by oxidizing lipids produces mostly polymers of intact protein monomers, with and without oxygen bridges. The accompanying decrease in solubility and changes in physical properties have profound effects on protein function both in foods and physiologically.

$$2\ LOOH\ or\ LO(O)^{\bullet} + 2\ P\text{-}SH \longrightarrow 2\ LO(O)H + 2\ PS^{\bullet} \longrightarrow P\text{-}S\text{-}S\text{-}P$$

$$2\ LOOH\ or\ LO(O)^{\bullet} + 2\ P\text{-}\bigcirc\text{-}OH \longrightarrow 2\ LO(O)H + 2\ P\text{-}\bigcirc\text{-}O^{\bullet} \longrightarrow P\text{-}\bigcirc\text{-}OH$$
$$P\text{-}\bigcirc\text{-}OH$$

$$2\ LOOH\ or\ LO(O)^{\bullet} + 2\ P\text{-}\alpha C(R)\text{-}P \longrightarrow 2\ LO(O)H + 2\ P\text{-}\alpha C^{\bullet}(R)P \longrightarrow P\text{-}\alpha C(R)P$$
$$P\text{-}\alpha C(R)\text{-}P$$

$$2\ LOOH\ or\ LO(O)^{\bullet} + 2\ P\text{-}NH_2 \longrightarrow 2\ LO(O)H + 2\ P\text{-}N(H)^{\bullet} \longrightarrow P\text{-}N=N\text{-}P$$

$$2\ P\text{-}C^{\bullet} + 2\ P\text{-}N(H)^{\bullet} \longrightarrow 2\ P\text{-}CH_2\text{-}NH\text{-}P,\ P\text{-}CH_2\text{-}N=CH\text{-}P,\ or\ P\text{-}CH=CH\text{-}NH\text{-}P$$

Additions of LOO$^{\bullet}$ (or LO$^{\bullet}$) to proteins also generate protein radicals. Subsequent oxidation and recombination results in protein crosslinks with lipid peroxo bridges (Desai and Tappel, 1963).

$$LOO^{\bullet}\ (LO^{\bullet}) + P \longrightarrow {}^{\bullet}LOOP\ ({}^{\bullet}LOP)$$

$$^{\bullet}LOOP\ ({}^{\bullet}LOP) + O_2 \longrightarrow {}^{\bullet}OOLOOP\ ({}^{\bullet}OOLOP)$$

$$^{\bullet}OOLOOP\ ({}^{\bullet}OOLOP) + P \longrightarrow POOLOOP\ (POOLOP)$$

Unspecified mixed crosslinking involving recombination of radicals on intact proteins and protein scission fragments has also been reported, particularly when the reacting lipids were the more unsaturated linolenic (Ln), arachidonic (An), or docosahexaenoic (DHA) acids. Protein complexes with increased molecular weights but not full polymerization and not involving disulfide bonds in soy protein (Huang et al., 2006) or aldehydes in bovine serum albumin (Liu and Wang, 2005) suggest the following sequence mediated by lipid radicals, where $P_f$ is a protein fragment:

$$LO(O)^{\bullet} + PH \longrightarrow LO(O)H + P^{\bullet} \longrightarrow P_f^{\bullet} + P_f\text{-}C=O$$

$$P^{\bullet} + P_f^{\bullet} \longrightarrow P\text{-}P_f$$

It is interesting and somewhat surprising considering current analytical capabilities that even though mechanisms and structures of crosslinks involving aldehydes have been elucidated in considerable detail, no specific structures or locations of lipid-induced radical crosslinks have yet been identified.

Polymeric crosslinking from free radicals has been shown with RNAse (Roubal and Tappel, 1966c; Gamage et al., 1973; Januszewski et al., 2005), lysozyme (Schaich and Karel, 1975; Funes and Karel, 1981; Funes et al., 1982), β-lactoglobulin (Hidalgo and Kinsella, 1989), and heavy chain myosin in fish myofibrillar proteins (Ooizumi, 1999). In contrast, cod sarcoplasmic reticulum proteins (Soyer and Hultin, 2000) and bovine serum albumin (Liu and Wang, 2005) appear to polymerize in fragment-monomer combinations rather than $(P)_n$ polymers, while pepsin and trypsin do not polymerize yet still develop fluorescence from reaction with carbonyl lipid oxidation products (Gamage et al., 1973).

### Carbonyl-Mediated Crosslinking

Over time, hydroperoxides decompose to secondary products which mediate crosslinking. The relative crosslinking activity of different aldehydes in neutral solutions with Fe or Cu catalysts is oxononenal >> acrolein ≅ decadienal > hydroxynonenal ≅ MDA (Yuan et al., 2007); trienals from high PUFA are even more active (Ran, 2004). Four major crosslinking mechanisms are now recognized for aldehydes; the mechanism that dominates in a given reaction system will be influenced by pH (functional group dissociation and charge), relative concentrations of amines and carbonyls, amino acid availability and orientation on the protein surface, and solvent environment.

*1. Schiff Base Condensation of Amines with Saturated Bifunctional Aldehydes or Other Dicarbonyls that Provide Bridges Between Protein Chains:*

$$O=CH\text{-}CH_2\text{-}CH=O + 2 H_2N\text{-}R\text{-}Protein \longrightarrow Protein\text{-}R\text{-}N=CH\text{-}CH=CH\text{-}NH\text{-}R\text{-}Protein$$

This type of crosslink also forms between amines on one protein chain and carbonyl oxidation sites on a second protein:

$$P_1\text{-}CHO + P_2\text{-}lys\text{-}NH_2 \longrightarrow P_1\text{-}CH=N\text{-}lys\text{-}P_2$$

Nucleophilic amino acids provide the linkage points and vary with the protein, for example lysines in RNAse (Chio and Tappel, 1969a, Gerrard et al., 2002), sulfhydryls in lens proteins (Riley and Harding, 1993), and mixed histidine and cysteine links with lysine in β-lactoglobulin (Yuan et al., 2007). Schiff base crosslinks with conjugated -N=CH-CH=CH- structure are fluorescent and are often are accompanied by a yellow to brown color.

*2. Michael Addition of Thiols and Amines (primarily lysine and histidine) to C-3 of α,β-unsaturated Aldehydes and Hydroxyl Alkenals:*

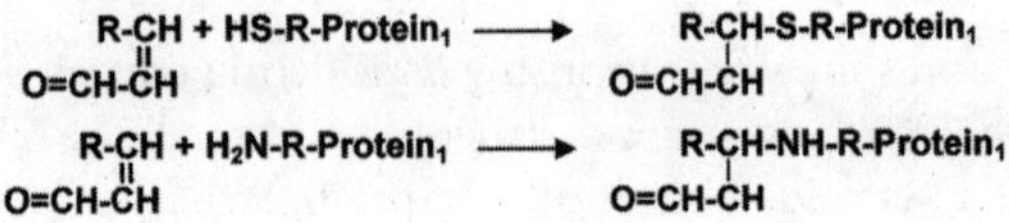

$$\begin{array}{c} \text{R-CH} + \text{HS-R-Protein}_1 \longrightarrow \text{R-CH-S-R-Protein}_1 \\ \text{O=CH-CH} \qquad\qquad\qquad\qquad \text{O=CH-CH} \end{array}$$

$$\begin{array}{c} \text{R-CH} + \text{H}_2\text{N-R-Protein}_1 \longrightarrow \text{R-CH-NH-R-Protein}_1 \\ \text{O=CH-CH} \qquad\qquad\qquad\qquad \text{O=CH-CH} \end{array}$$

Michael addition products may or may not be fluorescent (see Section "Formation of Fluorescent Adducts and Age Pigments"). Note that the direct Michael addition only generates lipid-protein adducts. Crosslinking occurs in a subsequent step when the free aldehyde forms a Schiff base product with a free amine, imidazole, or thiol on another segment of the same protein or on a neighboring protein:

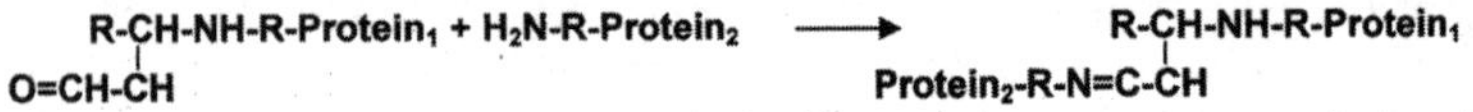

$$\begin{array}{c} \text{R-CH-NH-R-Protein}_1 + \text{H}_2\text{N-R-Protein}_2 \longrightarrow \text{R-CH-NH-R-Protein}_1 \\ \text{O=CH-CH} \qquad\qquad\qquad\qquad\qquad\qquad \text{Protein}_2\text{-R-N=C-CH} \end{array}$$

For histidine, Michael addition is the dominant reaction so the sequence shown above holds. However, for lysine or cysteine, the crosslinking may also occur in the reverse order, forming the Schiff base first, followed by Michael addition (Schaur, 2003). The crosslinks occur between single amino acids or between two different amino acids.

Michael addition-Schiff base crosslinking may be either (or both) intra- and intermolecular (Uchida and Stadtman, 1993), as has been observed with 4-hydroxynonenal and 4-oxononenal crosslinking of glucose-6-phosphate dehydrogenase (Friguet et al., 1994a, 1994b), β-lactoglobulin B and human hemoglobin (Bruenner et al., 1995), RNAse and BSA (Xu et al., 1999a), glyceraldehyde-6-phosphate dehydrogenase (Uchida and Stadtman, 1993), cytoskeletal proteins in P19 neuroglial cultures (Montine et al., 1996), and membrane proteins in erythrocyte ghosts (Hochstein and Jain, 1981; Beppu, 1986). Trienals, which are particularly strong crosslinkers (much greater than dienals and HNE), probably follow this reaction sequence with addition at multiple sites, although mechanism and product structures have not yet been determined (Ran, 2004).

Histidine is a strong copper binder, which contributes to its pro-oxidant activity. One interesting hybrid crosslink occurs when initial $Cu^{2+}$-mediated free radical oxidation of histidine yields an imidazol-2-one product that contains a reactive 4-oxo-2-ene region susceptible to Michael additions. Addition of the ε-amino group of lysine to C-2 generates a novel crosslink (Liu et al., 2004):

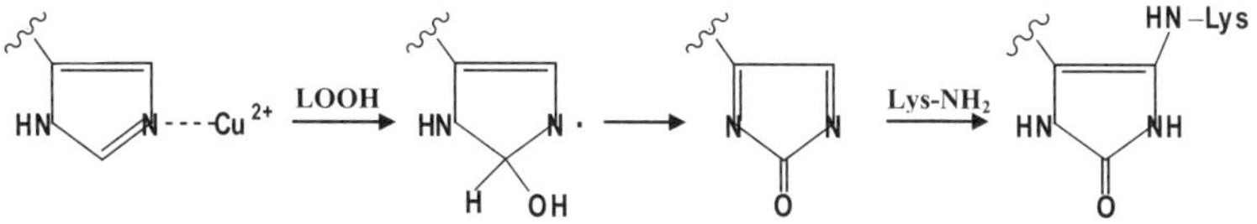

There is evidence that Schiff base formation and Michael addition of lysine are reversible, and that the presence of water or acid favors dissociation of both Schiff base and Michael adducts (Xu et al., 1999a), hence the crosslinks just described may be transient and not contribute to permanent crosslinks in aqueous media; they also are not detectable after acid hydrolysis (Kikugawa and Beppu, 1987; Lin et al., 2005), which may explain why Tappel's Schiff base products were only found in chloroform extracts (Chio and Tappel, 1969b). Both Schiff base and Michael addition products can react further via rearrangement, oxidation, reduction, and secondary additions (Schaur, 2003), so proposals have been made that stable crosslinks result from variations in secondary oxidations and rearrangements rather than initial adducts, as shown in the following mechanisms. Final crosslinks are presented here; more details of preceding reactions may be found in Section "Addition Reactions of Carbonyl Products from Lipid Oxidation".

*3. Pyrrole Links Formed when Aldehydes (usually hydroxy or keto derivatives of 2-alkenals) Cyclize Between Two Protein Nucleophiles (lys, his, cys):*

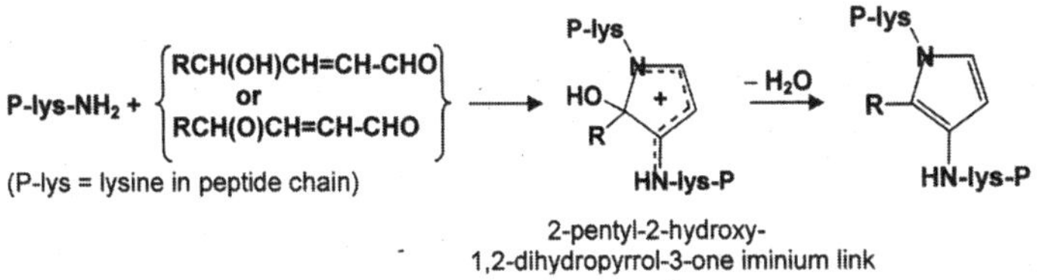

Pyrrole crosslinks require protein nucleophiles in at least 2:1 molar excess over the aldehydes. Monoalkylpyrroles form without the second amino acid (1:1 complexes), but do not contribute to crosslinking (Xu et al., 1999a). The basic process combines sequential Schiff base and Michael addition of amines to 2-alkenals, as described in the Michael addition in 2) above, but with additional oxidation (a critical requirement) the aldehyde cyclizes into an alkyl pyrrole ring that contains a nucleophilic adduct from a second amino acid. Dehydration of the iminium yields the pyrrole (Amarnath et al., 1994; Xu et al., 1998; Schaur, 2003).

This form of crosslinking has been identified in HNE and ONE reactions with RNAse and β-lactoglobulin (Zhang et al., 2003) and with RNAse and BSA (Xu et al., 1999a) in model systems. In vivo, 2-pentylpyrrole crosslinks from HNE have been found in oxidized LDL and atherosclerotic plaques (Salomon et al., 2000), and also are thought to be involved in aggregation of β-amyloid protein in Alzheimer's disease (Sayre et al., 1997).

Subsequent reactions of monoalkyl or amine-linked pyrroles lead to many variations in crosslinking: A) by lysine ε-$NH_2$ condensation with the pyrrole –OH to form cyclic or acyclic mixed aminals (Jirousek et al., 1990), B) by addition of a third protein nucleophilic link to a double bond in the pyrrole ring (Amarnath et al., 1994),

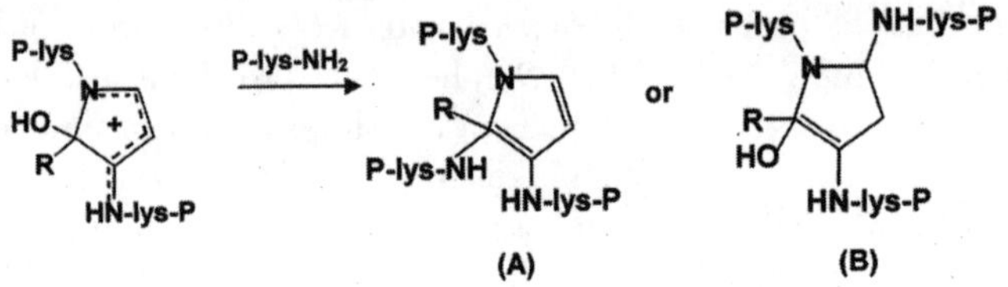

or C) by oxidation to pyrrole-pyrrole links. Nucleophiles, such as lysine, histidine, and cysteine, can add to pyrroles to extend crosslinking only when the ring is protonated or attached to a strong electron-withdrawing group. Oxygen activates the ring, imparting the requisite charge and creating a molecular precursor for crosslinking at C-2 or C-3 of the double bond (Amarnath et al., 1994), as shown in the following generalized process starting from an initial pyrrole-lysine adduct:

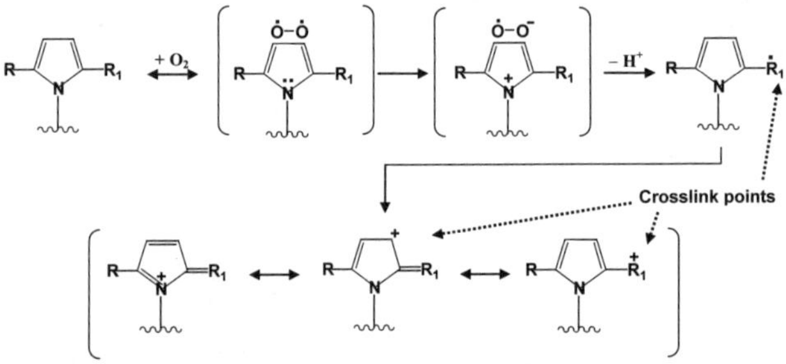

Several types of pyrrole dimers or polymers involving this type of addition have been identified (Amarnath et al., 1994, 1998; Zhu et al., 1994; Xu and Sayre, 1998). R = $R^1CH(OH)$- for HNE and ONE in the example below.

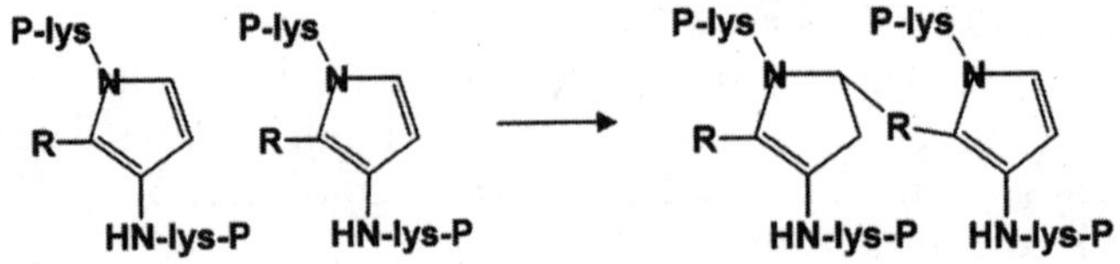

Epoxy alkenals form pyrroles with a hydroxylated side chain that provides an electrophilic site for crosslinking (Zamora and Hidalgo, 1995; Hidalgo and Zamora, 2000):

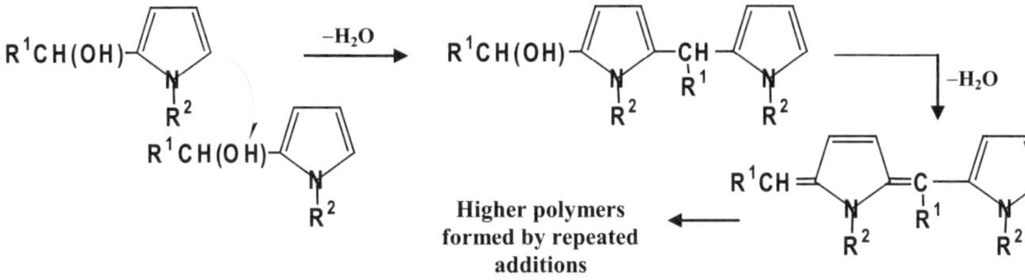

Consistent with this pyrrole activation also is a 2:2 pyrrole complex proposed to explain the very rapid crosslinking of proteins by 4-ketoaldehydes (Xu et al., 1999b; Xu and Sayre, 1999). Protein bound lysines form Schiff base adducts at each of the carbonyls; these then undergo end-to-end aldol condensation, cyclization of the aldehydes to linked pyrroles, and finally intramolecular pyrrole electrophilic substitution to generate a bicyclic compound:

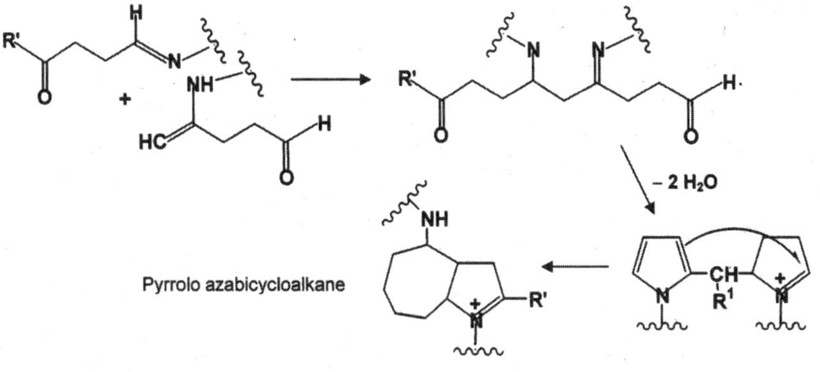

4. *Pyridinium Crosslinks Formed by Sequential Michael Addition and Schiff Base Reactions Under Conditions of Excess Aldehydes.* The initial ring structure results from condensation of two MDA and a saturated aldehyde with an amine (Esterbauer et al., 1991); the crosslinks form by subsequent Schiff base reactions with the free aldehydes (Kikugawa et al., 1985). The specific structure of the pyridine is dictated by the degree of aldehyde excess and nature of alkanal condensing with MDA (Beppu, 1986).

Pyridine crosslinks between lysines crosslinked erythrocyte proteins in red blood cell ghosts incubated with MDA (Kikugawa et al., 1985).

Considering all these potential pathways, it is clear that any attempt to attribute protein crosslinking to a single lipid oxidation product or a single type of crosslinking is an oversimplification at best. Every type of lipid oxidation intermediate or product induces protein crosslinking. In addition, crosslinking will start with free radical processes in actively oxidizing lipid, but if the reaction is followed long enough, carbonyl-mediated crosslinking eventually occurs as well, superimposed on previous oxidation and crosslinking processes. Therefore, multimodal protein crosslinking by lipids, with different kinetics, different products, and different effects should definitely be expected in most proteins. The dominant mode of crosslinking occurring will vary with the amino acid composition and the configuration of the protein, as well as the stage of lipid oxidation and the variety of products present. With actively oxidizing lipids (not isolated products), the dominant mode of crosslinking detected will vary with the length of reaction and time point at which proteins are analyzed.

For example, cytochrome c embedded in model membranes with oxidizing cardiolipin first crosslinked and aggregated into intramembranous particles by free radical reactions; with extended reaction, more extensive crosslinking by lipid aldehyde adducts led to the breakdown of membrane vesicles and formation of globular lipoprotein complexes in particles (Borovyagin et al., 1984). β-Lactoglobulin reacted with 13-LOOH in emulsion showed similar multimodal crosslinking, with rapid initial -S-S- crosslinking (dominant product) and slower C-C crosslinking; the S-S dimers also underwent further additive C-C crosslinking over time while retaining the S-S links (Hidalgo and Kinsella, 1989). Globular proteins with reactive side chains crosslink via surface links involving both radical and aldehyde reactions, while crosslinking of structural proteins, such as collagen and its derivatives, apparently involves the peptide chain itself. Inaccessibility of backbone sites to carbonyls severely limits the types of crosslinking possible, and may contribute to the greater tendency of collagen to degrade rather than crosslink when oxidized.

Multimodal crosslinking is more damaging than any single mode alone and probably explains why in most studies kinetics of appearance or disappearance of single lipid oxidation products do not exactly match crosslinking processes. The likelihood of multimodal crosslinking also means that measuring only one type of polymerization mechanism will, in most cases, underestimate and misrepresent the reactions actually occurring.

## Fragmentation

The flip side of radical recombination to crosslink proteins is radical scission to generate peptide fragments. The possibility of protein fragmentation by oxidizing lipids is noted frequently in literature introductions and discussions, but occurrence appears to be limited mostly to collagen and related structural proteins. Lipid-mediated oxidation of LDL also leads to extensive scission of apoB-100 into smaller peptides (Fong et al., 1987). Lyophilized gelatin-methyl linoleate mixtures undergo scission rapidly when incubated dry at 50°C, but as moisture content increases, fragmentation changes progressively to crosslinking (Zirlin and Karel, 1969; Matoba et al., 1984a). Connective tissue proteins (collagen) are degraded when lipids oxidize in cultured chondrocytes (Tiku et al., 2000). In α and γ-conglutinins from lupines reacted with 13-MLOOH at pH 9, protein scission occurs and the resulting fragments recombine with monomers to give a continuous range of protein complexes with increased molecular weights (Lqari et al., 2003). These observations suggest that fragmentation occurs only under extreme conditions and primarily with select proteins that have sensitive amino acid sequences.

To explain this sensitivity, Zirlin and Karel proposed that scission occurs in a process that parallels lipid oxidation, that is α-carbon radical → peroxyl radical → hydroperoxide → HOO- scission to alkoxyl radical → β scission to break peptide chain and generate protein carbonyls (Zirlin and Karel, 1969). Most aspects of this sequence have since been proven correct. Spin trapping of protein radicals has verified the presence of α-carbon radicals generated both directly and by migration of radicals initially formed on alkyl side chains (Headlam et al., 2000). Although Matoba et al. argued that only α-scission of the alkoxyl radical was consistent with a dramatic increase in amide and a much lower increase in carbonyl products in proteins reacted with oxidizing lipids (Matoba et al., 1984a), both α and β scission pathways for protein alkoxyl radicals have been documented for radiation and HO·, leading to different degradation patterns (Stadtman and Berlett, 1997; Stadtman and Levine, 2003; Kowalik-Jankowska et al., 2004; Davies, 2005).

Detailed reaction mechanisms for the diamide and α-amidation protein fragmentation pathways have been developed (Garrison, 1987; Davies et al., 1995; Stadtman and Berlett, 1997; Stadtman and Levine, 2003). A simplified version is presented in Fig. 8.17. C-C or β-scission is an oxidative process that decarboxylates the target amino acid (Garrison, 1987). Although it is the dominant process in oxygenated systems, β-scission is not a spontaneous process. It occurs under conditions of mild hydrolysis, which explains Zirlin and Karel's early observations that scission rarely occurred in dry proteins but increased with moisture content (Zirlin and Karel, 1969). Relatively minor differences in sample treatment or preparation for electrophoresis leading to hydrolysis may also account for unexpected globular protein fragmentation (Hunt et al., 1988; Soyer and Hultin, 2000; Liu and Wang, 2005). In contrast, N-C or α-scission is a reductive process that deaminates the target amino acid (Garrison, 1987). Both scission processes generate amides (marked with a star), the carbonyl products detected in standard assays. However, mild hydrolysis converts diamides to

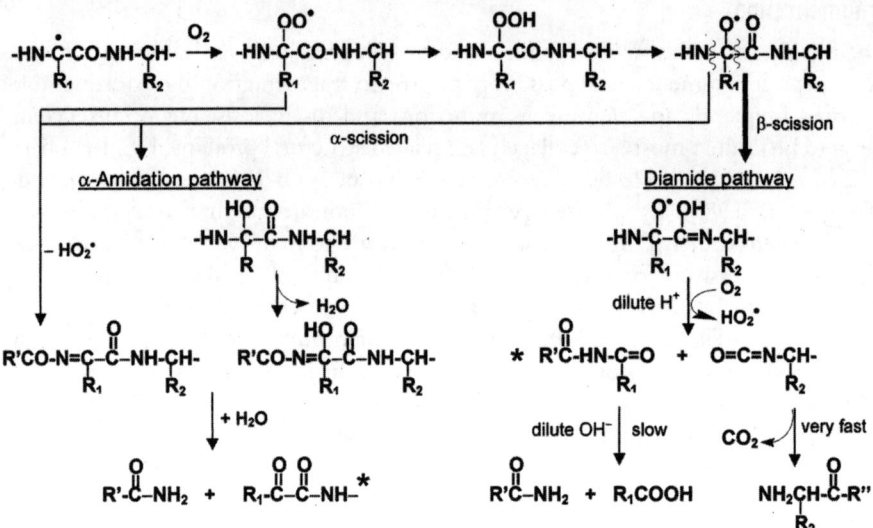

**Fig. 8.17.** Protein oxidation and fragmentation processes resulting from transfer of free radicals from oxidizing lipids to α-carbon sites on proteins. Adapted from (Garrison, 1987; Davies et al., 1995; Stadtman and Berlett, 1997; Stadtman and Levine, 2003). R and $R_2$ are amino acid side chains; R' and R" are continuations of the peptide chain. * denotes carbonyl compounds detected in oxidation assays.

amines, removes carbonyls used to diagnose and quantitate protein oxidation, and alters product distributions. This sensitivity to hydrolysis underscores the necessity to control reaction conditions scrupulously and to consider all the chemistry involved when interpreting results.

Collagen, structural proteins, and proteins with extensive α-helix regions may be especially susceptible to scission because they have more glycines, prolines, and aliphatic amino acids that are the sites of α-radical production. The proline residues of collagen oxidize readily when exposed to HO• radicals (Kato et al., 1992). Similarly, in gelatin reacted with LO(O)• radicals, proline, hydroxyproline, and glycine show the greatest loss (Matoba et al., 1984a). The α-carbon of proline is in the ring; oxidation at that point yields a 2-pyrrolidone compound upon scission (Kato et al., 1992).

Peptide chain cleavage via oxidation of glutamyl and aspartyl residues (side chain decarboxylation and deaminative scission of the peptide chain) has been shown with HO• (Garrison, 1987; Stadtman and Levine, 2003), but comparable degradation by lipid oxidants has not yet been reported. One study claimed that LO• generated from phospholipid hydroperoxides by Cu⁺ induced fragmentation of bovine serum albumen (Hunt et al., 1988). This system could be unusually damaging since BSA binds Cu⁺ rather extensively; multiple sites for cage reactions on the BSA surface then would generate reactive alkoxyl radicals near backbone sites that may ordinarily not be accessible. However, Cu⁺ autoxidizes and can damage proteins directly, yet controls accounting for this action were not run. Thus, whether oxidizing lipids can fragment globular proteins in solution remains to be demonstrated.

## Formation of Fluorescent Adducts and Age Pigments

Tappel first demonstrated that fluorescent ceroid age pigments in animal tissues and lipid-protein browning products in foods were co-oxidation products of polyunsaturated lipids and protein (Tappel, 1955), and research from his group was largely responsible for identification of the N,N'-disubstituted 1-amino-3-iminopropene (Schiff base) structures produced by reaction of carbonyl lipid oxidation products, particularly MDA, with side chain and terminal amino groups on proteins as the source of fluorescence (Chio and Tappel, 1969b; Fletcher and Tappel, 1970; Dillard and Tappel, 1971, 1973, 1984; Fletcher et al., 1973; Malshet and Tappel, 1973):

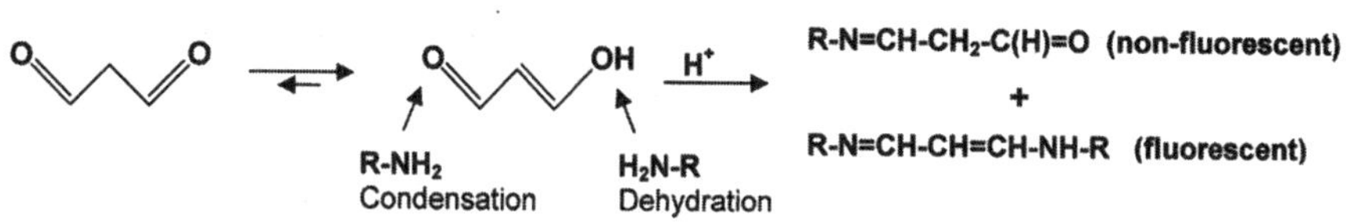

Lysine is by far the most reactive side chain, followed by histidine, tryptophan, and arginine. However, despite a common and mistaken expectation that all aldehydes generate fluorescent Schiff base products with amino acids, adducts are fluorescent only when an electron-donating group is present in conjugation with the imine (Malshet and Tappel, 1973). This explains, in part, why many of the MDA– and other aldehyde–protein products are not fluorescent; it points out the need to determine product structures and consider alternate pathways when comparing "reactivity" of different carbonyl products of lipid oxidation by measuring fluorescence.

Since Tappel's pioneering work, appearance of fluorescence has become a hallmark of protein oxidation reported in hundreds of articles, routinely attributed to and associated with generation of MDA. However, over time questions have been raised about the aldehyde sources of fluorescent lipid-protein adducts and the reaction mechanism and final structure of fluorescent adducts. Some observations that have led to disputes include:

- MDA is only produced in PUFA with three or more double bonds, yet fluorescence develops in some reactions with oxidizing linoleic acid, which does not generate MDA.

- Many lipid-derived aldehydes with varying chain lengths, both saturated and unsaturated, produce fluorescent products with proteins and amino acids.
- Fluorescent structures other than Schiff-base have been identified.
- Fluorescent iminopropene structures can be generated by reactions other than with aldehydes.
- Production of fluorescence does not always directly parallel loss of amino groups.
- Fluorescence characteristics (excitation and emission maxima as well as intensity) change over time as lipid oxidation progresses.
- In model systems, concentrations of aldehydes required to generate fluorescent products with proteins or amino acids are usually several orders of magnitude higher than aldehyde levels formed by lipids autoxidizing in foods or in vivo under conditions of oxidative stress, yet fluorescence develops rapidly in a wide range of "real" systems.

It is now clear that the chemistry underlying production of fluorescent products is much more complicated than originally proposed. Fluorescence arises both from some adducts and specific crosslinks formed between proteins and a family of lipid oxidation products—hydroperoxides as well as a range of secondary products (Kikugawa and Beppu, 1987; Hidalgo et al., 1999; Zamora and Hidalgo, 2003a). Simple Schiff base adducts (-N=CHCH$_2$-CHO or -NH-CH=CH-CHO) contribute to browning (Zamora et al., 2000) but are not fluorescent (Itakura and Uchida, 2001). The "classic" linear iminopropenes (RNH-CH=CH-CH=NR) to which most fluorescence is attributed are the initial products formed via Schiff base and Michael additions of 2-alkenals (Nadkarni and Sayre, 1995), but they form primarily under acidic conditions (where fluorescence is lowest), require high concentrations of reactants (Kikugawa and Beppu, 1987), and remain relatively minor products in most systems because they rapidly rearrange to other products. NMR and LC-MS/MS have identified pyridinium (Kikugawa and Ido, 1984; Kikugawa et al., 1984; Amarnath et al., 1998; Liu and Sayre, 2003) and pyrrole (Zhu et al., 1994; Hidalgo et al., 1999; Zamora et al., 2000; Liu et al., 2003; Zamora and Hidalgo, 2003a; Zhang et al., 2003; Zamora and Hidalgo, 2005) lipid-protein products in both monomer forms and crosslinks as the main sources of strong fluorescence that develops rapidly at physiological pH and at elevated temperatures in foods. Notably, these molecular structures contain the imino-ene conjugation required for fluorescence, as specified by Malshet and Tappel (1973). Other less dominant fluorescent products have been detected, but their structures remain to be identified.

Production of fluorescent co-oxidation products requires elevated temperatures. As will be shown later, little to no fluorescence develops in reactions at refrigerated or room temperatures, but a physiological temperature of 37°C is sufficient to drive rapid production of fluorescent protein adducts that are markers of oxidative damage; at accelerated shelf life testing of 60°C and higher food-processing temperatures fluorescent compounds become major products that contribute to characteristic browning and flavor production (Hidalgo et al., 1999, 2005).

Spectral characteristics of fluorescent products in various lipid oxidation systems are presented in Table 8.2 to show how fluorescence characteristics change with lipid oxidation product and amino acid or protein target. It also illustrates how assignment of structures based solely on excitation and emission maxima, although a common practice, is problematic and can lead to erroneous conclusions. Multiple products usually form in the same system. Averages of multiple products and differences in product distributions under different reaction conditions contribute to the variation in excitation and emission maxima that have been reported for different amino acids, proteins, and oxidants (Table 8.2). Solvents alter spectral characteristics: ex/em maxima and emission intensity both usually increase with solvent polarity (Kikugawa and Beppu, 1987; Chen et al., 1996). Sample handling is also critical. Traditional chloroform-methanol extracts of tissues develop artificial blue fluorescence that is distinctly different from the yellow fluorescence of age pigments when exposed to light or passed through silica gel columns (Kikugawa, 1994; Kikugawa et al., 1994). The blue fluorescence clearly arises from secondary oxidation processes, probably photosensitization by traces of heme compounds extracted in methanol and decomposition of preformed hydroperoxides by silica interactions and exposure to UV radiation from lab lights. Hence, great care must be exercised to obtain fluorescence data that accurately reflects system chemistry.

*Monofunctional Aldehydes. Alkanals*
Saturated aldehydes produce Schiff base adducts for which the fluorescence depends on amine source, temperature, and amine-aldehyde concentrations (both absolute and relative), and in some cases on the presence of oxygen (Kikugawa and Beppu, 1987). Glycine does not produce fluorescent products with hexanal, decanal, or acetaldehyde when incubated at refrigeration temperatures (4°C). Under the same conditions, products with lysine were only weakly fluorescent (Stapelfeldt and Skibsted, 1996; Veberg et al., 2006); even though emission intensities increased at elevated temperatures (30 or 37°C), absolute levels of fluorescence remained low (Kikugawa et al., 1985; Yamaki et al., 1992). However, when oxygen or a strong oxidizer, such as $H_2O_2$, was present, strong fluorescence traced to 2-HO-1,2-dihydropyrrol-3-ones developed (D, Fig. 8.18); a mechanism was proposed in which the aldehydes dimerize through Aldol condensation and then oxidize to tricarbonyls. Amines add to this compound rather than parent alkanals, and the complex cyclizes (Chen et al., 1996). Oxidation steps are also critical to develop fluorescent pyrroles after alkenal reaction with amines. Thus, it is clear that secondary oxidation of intermediate products significantly affect lipid-protein co-oxidations by shifting pathways and altering yields of fluorescent versus non-fluorescent products. Such differences become critically im-

**Table 8.2.** Variations in Excitation and Emission Maxima for Schiff Base Fluorescence from Different Amino Acid and Protein Substrates.[a] Unless otherwise noted, the oxidant is oxidizing lipids rather than an isolated product.

| Amine | Oxidant | Solvent | Ex (nm) | Em (nm) | Reference |
|---|---|---|---|---|---|
| Lysine | MDA | CHCl$_3$ | 395 | 470 | (Chio and Tappel, 1969a) |
| Lysine | hexanal | 10% ethanol | 345 | 415 | (Dalsgaard et al., 2006) |
| Lysine | heptadienal | | 382 | 434 | (Yamaki et al., 1992) |
| Lysine | alkanals | | 382 | 434 | (Yamaki et al., 1992) |
| Polylysine | LOOH, AnOOH | pH 7 buffer | 330 | 425 | (Fruebis et al., 1992) |
| Polylysine | MDA | pH 7 buffer | 398 | 470 | (Fruebis et al., 1992) |
| Glycine | LOOH | 0.5% SDS, pH 7 buffer | 360–370 | 435–450 | (Chio and Tappel, 1969b; Shimasaki et al., 1982) |
| Glycine | MDA | water | 370 | 450 | (Chio and Tappel, 1969b) |
| Glycine | alkanals+ H$_2$O$_2$ | | 345–375 | 415–440 | (Chen et al., 1996) |
| Glycine | heptadienal | | 397 | 470 | (Yamaki et al., 1992) |
| Ethylamine | HO-butanal | pH 7 buffer | 327 | 390 | (Liu and Sayre, 2003) |
| Valine | MDA | water | 370 | 450 | (Chio and Tappel, 1969b) |
| Leucine | MDA | water | 370 | 450 | (Chio and Tappel, 1969b) |
| Leucine | β-oxyacrolein | neat | 390 | 550 | (Buttkus, 1975) |
| Leucine | β-oxyacrolein | n-butanol | 385 | 460 | (Buttkus, 1975) |
| Glutathione | LOOH | pH 7 buffer | 350 | 440 | (Zamora et al., 1989) |
| Lysozyme | oxidizing ML | dry emulsion | 355 | 425 | (Leake and Karel, 1985) |
| BSA | LOOH, SP | pH 7 buffer | 350 | 425 | (Hidalgo et al., 1999) |
| BSA | LOOH, AnOOH | pH 7 buffer | 330 | 425 | (Fruebis et al., 1992) |
| BSA | LOOH | pH 7 buffer | 360 | 427 | (Shimasaki et al., 1982) |
| BSA | HNE | pH 7 buffer/MeOH | 360 | 430 | (Xu and Sayre, 1998; Xu et al., 1999b) |
| RNAse | oxidizing EtAn or MeLn | pH 7 buffer | 390, 395 | 470 | (Chio and Tappel, 1969b) |
| RNAse | 2-HO-heptanal | pH 7 buffer | 325 | 409 | (Liu and Sayre, 2003) |
| Porcine Myofibrils | oxidizing fibrils | pH 6 buffer | 395 | 485 | (Chelh et al., 2007) |
| Minced meat | 5 aldehydes | (solid) | 382 | 450–550 | (Veberg et al., 2006) |
| Frozen sardines | oxidizing lipids | CHCl$_3$-MeOH, | 327 | 415 | (Aubourg et al., 1998) |
| | | MeOH-water | 393 | 463 | (Aubourg et al., 1998) |

**Table 8.2., cont.** Variations in Excitation and Emission Maxima for Schiff Base Fluorescence from Different Amino Acid and Protein Substrates.[a] Unless otherwise noted, the oxidant is oxidizing lipids rather than an isolated product.

| Amine | Oxidant | Solvent | Ex (nm) | Em(nm) | Reference |
|---|---|---|---|---|---|
| Liver homogenate | oxidizing lipids | | 397 | 480 | (Li et al., 2006) |
| LDLox | atherosclerotic plaques | pH 7 buffer | 360 | 430 | (Xu et al., 2000) |
| LDLox | HNE or ONE | pH 7 buffer/ MeOH | 366 | 445 | (Xu et al., 2000) |
| Yellow lipofuschin | in vivo | 0.5% SDS, pH 7 buffer | 400 | 620 | (Kikugawa et al., 1994) |
| Blue lipofuschin | in vivo | $CHCl_3$-MeOH | 350 | 430 | (Kikugawa, 1994) |
| Amine pyridinium | MDA | pH 7 buffer | 403 | 462 | (Kikugawa and Beppu, 1987) |
| Amine pyridinium | MDA | $CHCl_3$ | 390 | 446 | (Kikugawa and Beppu, 1987) |

[a] Abbreviations: MDA, malonaldehyde; LOOH, linoleic acid hydroperoxide; BSA, bovine serum albumin; SP, secondary products of lipid oxidation; AnOOH, arachidonic acid hydroperoxide; HNE, hydroxynonenal; EtAn or MeLn, ethyl arachidonate or methyl linolenate; LDLox, oxidized low density lipoproteins; ONE, oxononenal; SDS, sodium dodecyl sulfate.

portant when they are used to interpret mechanism and determine molecular causes of protein damage, and argue for careful control of oxygen concentrations in reaction media and determination of oxygen dependence of reaction pathways in all experiments.

Excitation wavelengths and solvent also impact detected fluorescence: in 50% ethanol with excitation at 382 nm, emissions were absent or only barely detectable (Veberg et al., 2006), while in other studies emissions were detected with ex 327/em 415 nm or ex 393/em 463 nm (Aubourg et al., 1998), ex 348/em 416 nm (Yamaki et al., 1992), ex 357/em 430 mm (Kikugawa et al., 1985). Both excitation and emission wavelengths tend to increase with solvent polarity (Kikugawa and Beppu, 1987); this must be considered when comparing results between different systems.

Perhaps most importantly, amines appear to stimulate reactivity of saturated aldehydes, inducing increased aldol self-condensation and generation of 2-alkenals which cyclize to fluorescent products in the presence of oxygen (Suyama et al., 1981). Thus, alkanals may generate the same spectrum of products as alkenals, but at lower yields and slower rates.

*Bifunctional Aldehydes*

As shown previously, at neutral pH the monofunctional iminopropene MDA–lysine adduct (R-NH=CH-CH$_2$-CHO) does not fluoresce (Itakura and Uchida, 2001) because it lacks an electron-donating group in conjugation with the imine. The bi-

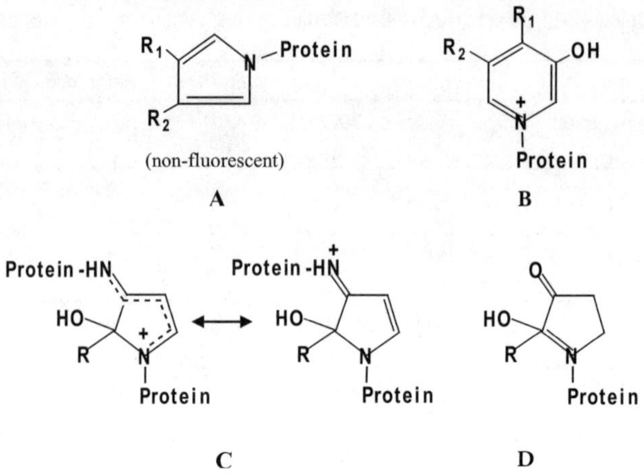

Fig. 8.18. Examples of fluorescent structures formed when lipid aldehydes react with lysine ε-amino groups of proteins.

functional iminopropene MDA–lysine adduct, however, does contain the necessary conjugation so the crosslink R-NHCH=CHCH=N-R is indeed fluorescent and may be the major source of fluorescence very early in reactions of actively oxidizing lipids (not isolated aldehydes) with proteins (Nadkarni and Sayre, 1995). As the reaction progresses, however, stronger fluorescence arises from pyridinium products (dihydropyridine dicarbaldehydes) formed by the condensation of two bifunctional aldehydes with primary amines, particularly the ε-NH$_2$ of lysine (Kikugawa and Beppu, 1987; Itakura and Uchida, 2001). Structure A shown below is formed when MDA reacts with an amine group either on a side chain linked to a protein or a peptide terminal amine. Similar fluorescent pyridine structures form in a ternary complex where monofunctional aldehydes condense with two molecules of MDA and an amine, introducing R-CHO at C-2 of MDA (Kikugawa et al., 1984; Freeman et al., 2005). In structure B, the third aldehyde is acetaldehyde, a decomposition product of MDA.

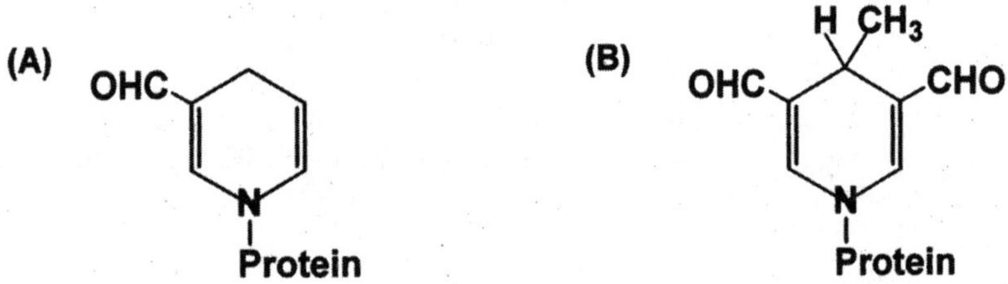

## Alkenals

Alkenals are significant precursors of fluorescence when reacted with proteins because they form Schiff base adducts, then undergo Michael addition at the double bond, and the amine-aldehyde-amine complex then rearranges to fluorescent pyrroles and (under some circumstances) pyrimidines. Hydroxy, epoxy, and oxo alkenals are the most reactive alkenals. When reacted with lysine, the major amino acid target, these aldehydes produce a variety of cyclic products, including non-fluorescent 4-alkyl-imidazolium crosslinks through Amadori rearrangements of initial Schiff base adducts (Fig. 8.18a) (Liu and Sayre, 2003), fluorescent 4,5-dialkyl-3-HO-pyridiniums (Fig. 8.18b) (Liu and Sayre, 2003), and 2-hydroxyalkyl-3-imino-1,2-dihydropyrrol derivatives (Fig. 8.18c) by a series of Schiff base and Michael additions plus oxidations (Tsai et al., 1998; Xu and Sayre, 1998; Hidalgo et al., 1999; Zamora and Hidalgo, 2003a). 2-HO-2-pentyl-1,2-dihydropyrrol-3-one iminium (D, Fig. 8.18) is a key structure on LDLox contributing to fluorescence in atherosclerotic plaques (Xu et al., 2000). Note that the bond sequence (-NH-CH=CH-CH=N-) initially cited by Tappel as being required for fluorescence is contained within each of these structures, and it is missing in the non-fluorescent imidazolium.

## Lipid Hydroperoxides

It is particularly important to recognize potential contributions of radicals and hydroperoxides since these products are generally ignored in global interpretations of fluorescence origins. In model systems with oxidized phospholipids and lysozyme, the precursor for fluorescence was a monomeric product of hydroperoxides formed without fragmentation (structure not identified); hydroxyl, keto, and epoxy derivatives as well as monofunctional aldehydes, such as hexanal and 2,4-decadienal, did not yield fluorescent products (Iio and Yoden, 1988). When hydroperoxides and various oxidation products (e.g., propanal, butanal, pentanal, and hexanal) of methyl linoleate were reacted with β-lactoglobulin in pH 7.4 phosphate buffer at 37°C, the aldehydes reacted quite rapidly with free amines (~20% in 2 h) but produced only weak or no fluorescence (Hidalgo and Kinsella, 1989). In contrast, the slow reaction of isolated 13-MLOOH with amines (~10% over 25 h) was accompanied by generation of strong fluorescence associated with a C-C or C-N dimer. Binding of 1 mole MLOOH per 18,000 mw protein did not correlate with fluorescence; thus, radicals formed rapidly on tryptophan and more slowly on other amines or their recombination products were judged to be the source of fluorescence. In many other reports protein fluorescence developed in parallel with lipid hydroperoxides, yet the reaction was attributed to aldehydes. Thus, although mechanisms have not been elucidated, it is likely that LOOH always contribute at least in part to the development of fluorescent protein products, and this needs to be considered when interpreting damage mechanisms and causality.

The potential to form multiple fluorescent products as lipid oxidation progresses argues strongly for not using fluorescence to attribute damage sources without detailed identification of fluorescent products and their structures. When interpreting mechanisms from fluorescence, it is important to remember that fluorescence characteristics vary with the amino acid and lipid involved in the complex (Table 8.2), the degree of lipid oxidation, and reaction conditions (particularly pH, solvent, temperature, presence of metals, and sensitivity to reduction) (Kikugawa and Beppu, 1987; Kagan, 1988; Li et al., 2006; Veberg et al., 2006). As shown in Maillard reaction chemistry, the lipid carbonyl-protein amine condensation cannot occur at acid pH where the amine is protonated; it accelerates as pH increases and is very rapid in alkaline solutions (Hidalgo et al., 1999). In model systems of BSA reacted with LOOH and its secondary products, fluorescence maxima occurred at pH 10; maximum arginine loss was also at pH 10, but highest lysine loss and color formation was observed at pH 7 (Hidalgo and Zamora, 1993; Zamora and Hidalgo, 1995). Fluorescence production maxed at 37°C and decreased at higher temperatures, even though pyrroles usually assumed to be fluorescent continued to increase. These observations emphasize the complexity of oxidizing lipid reactions with proteins, showing clearly that multiple reactions occur simultaneously and that attribution of macroscopic behaviors to products without isolation and structural identification can lead to serious errors.

One additional cautionary comment must be offered. Only recently have products begun to be quantitated, and it is not uncommon to find that fluorescent products that seem to be major because of their intense emissions actually account for a very small percentage of total products. For example, in one study, 2-heptenal was reacted with RNAse and strong fluorescence was observed. However, heptenal complexed only 1.5% of the lysine after being incubated with RNAse at 25°C for 10 days, and the reactions required very high aldehyde concentrations. Even so, the aldehyde-amine reaction was proposed as the major damaging reaction in that model system. With lipid oxidation products, attention focused too intently and exclusively on single products or processes loses all the action in other reactions that may be as much or even more important in overall damage.

## Inhibition of Enzyme Activity

The chemical and physical modifications of proteins described previously—amino acid radical formation, oxidation, or adduct formation; covalent and non-covalent lipid binding; and changes in protein conformation, denaturation, crosslinking, and scission—all impair protein function. Indeed, one of the earliest observations of lipid oxidation effects on proteins was marked loss of activity in enzymes in both model systems and in tissues. Sulfhydryl enzymes are the most sensitive (Tsen and Tappel, 1958; Little and O'Brien 1967, 1968); thiols react rapidly with lipid radicals and hydroperoxides, and they also undergo nucleophilic additions to double bonds (Gardner, 1979). Metallo-enzymes, enzymes of the mitochondria, microsomes, endoplasmic reticulum, and respiratory proteins such as cytochromes, are also extensively damaged by oxidizing lipids (Table 8.3).

Extensive studies of damage to RNAse and lysozyme as model enzymes have revealed some factors that critically influence the type and extent of damage by oxidizing lipids (Menzel, 1967; Little and O'Brien, 1968; Chio and Tappel, 1969a;

Matsushita et al., 1970). Both pH, which affects the protein charge and conformation, and lipid hydroperoxide concentration (which influences contact, potential for hydrogen bonding, and oxidizing potential) are important. RNAse is inactivated at acidic but not alkaline pH, presumably due to oxidation of the histidine residue and potentially lysine and methionine in the active site (Matsushita et al., 1970). LOOH binding to proteins, which caused changes in conformation and increased amino acid accessibility, was considered critical. Binding effects are concentration dependent (Little and O'Brien, 1968). At low concentration LOOH impedes –SH oxidation by hydrophobic bonding to proteins, but at high concentrations LOOH binding leads to denaturation and exposure of new reactive sites.

The speed with which enzymes are inactivated by oxidizing lipids in both model systems and tissue studies suggests that reactions of LOO• and LO• radical attack on cysteine, tryptophan, lysine, and histidine side chains is the major damaging process for actively oxidizing lipids. Obviously, reactions with carbonyl products can also inhibit enzymes at later stages of lipid oxidation. However, even though model system studies have suggested a number of feasible explanations to inactivate a few enzymes, the exact oxidant and mechanisms responsible in complex systems where lipids are actively oxidizing and multiple oxidation products are present at different times remain to be identified.

## Alteration or Loss of Biological Function

It seems obvious that any functions, both immediate and downstream, that are dependent on protein composition, structure, conformation, or binding properties will be affected by lipid-induced modifications. Just a few examples are described here to illustrate how the impact of modified proteins in most cases reaches far beyond the immediate interaction chemistry.

Covalent attachment of HNE to myoglobin or oxymyoglobin increases oxidation to metMyb and inhibits enzymatic reduction and recycling back to myoglobin (Bolgar and Gaskell, 1996; Faustman et al., 1999; Lynch and Faustman, 2000; Alderton et al., 2003). The tertiary structure of the apoprotein apparently is not affected (Alderton et al., 2003), but HNE binding to critical histidine residues via Michael addition (Bolgar and Gaskell, 1996; Alderton et al., 2003) alters heme binding or orientation in a way that increases heme catalysis of lipid oxidation (Faustman et al., 1999; Lynch and Faustman, 2000). In vivo, rapid conversion of myoglobin to metmyoglobin and inability to repair and recycle oxidized metMb disrupts oxygen transport to tissues and forces muscles into anoxic respiration, markedly impairing muscle function and increasing exercise fatigue rate.

Enzyme inactivation by oxidizing lipids can be devastating in cells far beyond the damage to the individual protein molecules and reactions. To cite just a few longer-range implications, lipid reactions with mitochondrial enzymes inhibit or interrupt metabolic pathways at multiple points (McKnight and Hunter, 1966; Benedetti et al., 1979; Thomas and Poznansky, 1990; Tsuchiya et al., 2005), while reaction with cytochromes in mitochondria leads to loss of respiratory control and decouples phos-

Table 8.3. Examples of Enzymes Damaged by Various Forms of Oxidizing Lipids.

| Enzyme | Damage Agent[1] | Reference |
|---|---|---|
| **SH enzymes** | | |
| Succinoxidase, amino oxidase | Light-oxidized ML and MLn | (Ottolenghi et al., 1955) |
| Succinoxidase, amino oxidase, urease, papain, glyoxylase, choline oxidase, β-amylase, cytochrome c oxidase | oxidizing L and Ln | (Wills, 1961a, 1961b) |
| Catalase | oxidizing L | (Roubal and Tappel, 1966a) |
| Papain | oxidizing ML | (Lewis and Wills, 1962) |
| Papain | MDA | (Shin et al., 1972) |
| Isocitrate DHG | LOOH | (Green et al., 1971) |
| Alcohol DHG, papain, chymotrypsin, carboxypeptidase A, urease | oxidizing EtLn or EtAn | (Chio and Tappel, 1969a) |
| Protein disulfide isomerase | HNE | (Carbone et al., 2005) |
| **Mitochondrial enzymes** | | |
| NADP$^+$-isocitrate DHG | HNE, isolated rat heart mitochondria | (Benderdour et al., 2003) |
| Isocitrate, succinate, and malate DHG | oxidizing mitochondrial membranes | (McKnight and Hunter, 1966) |
| Aldehyde and glyceraldehyde phosphate DHG | trans-HNE | (Mitchell and Petersen, 1991) |
| Aldehyde DHG, glucokinase, glyceraldehyde PO$_4$ DHG | LOOH and SP (secondary products) | (Kanazawa and Ashida, 1991) |
| **Microsomal Enzymes** | | |
| Endoplasmic reticulum enzymes | LOOH | (Hochstein and Ernster, 1964) |
| Glucose-6-phosphatase | oxidizing microsomal lipids | (Hruszkewycz et al., 1978; Benedetti et al., 1979) |
| Glycerol-3 phosphate acyl transferase | oxidizing microsomal lipids in situ | (Thomas and Poznansky, 1990) |
| Gulonolactone oxidase | oxidizing microsomal lipids | (McCay, 1966) |
| Glyceraldehyde-6-phosphate DHG | HNE, HHE | (Tsuchiya et al., 2005) |

Table 8.3., cont. Examples of Enzymes Damaged by Various Forms of Oxidizing Lipids.

| Enzyme | Damage Agent[1] | Reference |
|---|---|---|
| Glucose-6-phosphate DHG | HNE | (Friguet et al., 1994a) |
| Glucose-6-phosphate DHG | LOOH and SP (secondary products) | (Kanazawa and Ashida, 1991) |
| **Respiratory enzymes** | | |
| Cytochrome c | LOOH | (O'Brien and Frazer, 1966; Little and O'Brien, 1968) |
| Cytochrome c | peroxidizing Ln | (Desai and Tappel, 1963) |
| Cytocrome P450 | oxidized microsomal lipids | (Hruszkewycz et al., 1978) |
| Cytochromes | oxidized microsomal lipids | (McKnight and Hunter, 1966) |
| **General** | | |
| Lysozyme | oxidizing ML | (Funes et al., 1982) |
| Lysozyme | MDA | (Chander et al., 1981) |
| RNAse, lysozyme, creatine kinase, lactate dehydrogenase | peroxidizing MLn; high conc. MDA | (Chio and Tappel, 1969a) |
| RNAse, trypsin, pepsin, chymotrypsin | LOOH and SP (secondary products) | (Matsushita et al., 1970; Matsushita, 1975) |
| Cathepsin B | HNE | (Crabb et al., 2002) |
| Lecithin:cholesterol acyl transferase | PL-OOH | (Mickel et al., 1972; Bielicki and Forte, 1999) |

[1] Abbreviations: DHG, dehydrogenase; EtLn, ethyl linolenate; EtAn, ethyl arachidonate; HNE, hydroxynonenal; HHE, hydroxyhexenal; LOOH, linolenic acid or ester hydroperoxide; ML, methyl linoleate; MLn, methyl linolenate; MDA, malonaldehyde; PL-OOH, phospholipid hydroperoxides.

phorylation (Hruszkewycz et al., 1978). Reaction of hydroxynonenal with membrane proteins decreases neuronal plasticity, disrupts mitochondria, and impairs glutamate transport (Keller et al., 1997). Inhibition of protein disulfide isomerase via cysteine modification results in incorrect disulfide formation in newly synthesized proteins (Carbone et al., 2005), an effect that is especially important when glutathione is depleted under stress conditions. Oxidation of methionines in the binding site of HDL apo A1 and A2 impairs lipid binding and pick up, thus depressing HDL ability to promote efflux of cholesterol from cells; alterations in secondary structure of the apoprotein further affect the ability to interact with lipids critical for cholesterol removal and activation of LCAT (Garner et al., 1998a). Similarly, complexation of lysine in LDL impedes LDL binding to the epithelial receptor and subsequent LDL removal from plasma (Steinbrecher, 1987).

The type of protein damage generated by lipids markedly affects hydrolysis and recycling of proteins. Proteases are specifically induced to remove damaged proteins in vivo. Disruption of secondary and tertiary structure, increased hydrophobicity, or moderate oxidation of side chains all act as signals to activate proteases and increase proteolytic susceptibility (Davies, 1987b), particularly by multicatalytic proteinase or proteasomes (Friguet and Szweda, 1997). The ubiquitin/proteasome system, for example, selectively degrades oxidized proteins (Grune et al., 1997; Grune and Davies, 2003). Reaction of hydroxynonenal and hydroxyhexenal with glyceraldehyde-6-phosphate dehydrogenase inhibits enzyme activity and renders the enzyme susceptible to proteolysis by a giant chymotrypsin-like serine protease TPP II and lysosomal enzymes (Tsuchiya et al., 2005).

In contrast, aggregation, crosslinking and high lipid binding reduce degradability of the proteins, as has been demonstrated with G6PDH (Friguet and Szweda, 1997), muscle myofibrillar proteins (Morzel et al., 2006), BSA (Zamora and Hidalgo, 2001), and other proteins. This should be expected since proteases with specific cleavage sites will be inhibited if the target amino acids are occluded, blocked, or are in or near scission or crosslink sites. For example, papain (a cysteine protease) does not degrade muscle proteins in which the cysteine has been oxidized into disulfide crosslinks (Morzel et al., 2006). Analysis of crosslink or binding sites thus will require use of multiple or non-specific proteases to cleave peptides at points other than the damage sites.

## Abnormal Function, Immunochemistry, and Contributions to Disease

The converse of losing biological function is a shift to non-productive or toxic activity. Detailed information about the roles of oxidized proteins in a wide range of pathologies has been accumulating for several decades, but the participation of oxidizing lipids as one of the strongest biological oxidants is only slowly gaining attention. Although a thorough review of the mechanisms by which lipid oxidation of proteins is involved in disease processes is beyond the scope of this chapter, at least some mention is warranted to show why lipid co-oxidation pathways are becoming so important in oxidative stress. For the most part, current information demonstrates that lipid-protein co-oxidation products are present in target tissues or associated with pathological changes, and there is growing evidence for signaling activity of both the oxidized lipids and proteins (Page et al., 1999; Prescott, 1999; Uchida et al., 1999; Leonarduzzi et al., 2000; Petersen and Doorn, 2004; Laurora et al., 2005; Zhang et al., 2005; Valko et al., 2007). Despite these encouraging suggestions, considerable research is still needed to elucidate specific causal roles of lipid-protein oxidation in individual diseases.

*Atherosclerosis: Oxidation of LDL and HDL.*
Atherosclerosis is truly a disease of lipid and protein oxidation. Development of atherosclerosis is most often associated with LDL, but HDL and two enzymes—lecithin-cholesterol acyl transferase (LCAT) and cholesterol ester transfer protein (CETP)—also play surprisingly critical roles. In each case, lipid oxidation of a protein is intimately involved in the transformations leading to full-blown atherosclerosis.

Consider first the notorious LDL. LDL and monocytes enter endothelial cells to collect and recycle cholesterol and phospholipids and to remove cellular debris, respectively. Normally, they perform their jobs and leave the cell. However, oxidation of the LDL lipid and protein components stimulates a series of dramatic changes that accelerate the process of atherosclerosis (Uchida, 2000). Oxidized LDL signal endothelial cells to express monocyte chemotactic protein 1 (MCP-1), which sits on the cell surface and attracts monocytes from the vessel lumen into the subendothelial space where they are converted into macrophages by accumulation of cholesterol (Barter, 2001). At the same time, oxidized phospholipids form protein adducts that anchor the complex to the epithelial surface, promoting oxidative stress and receptor-independent endocytosis that leads to accumulation of oxidized intracellular lipids (Weisgraber et al., 1978; Januszewski et al., 2005). Inside the endothelial cell, monocyte/macrophages normally do not ingest cellular lipids or lipoproteins. However, they very strongly and selectively bind and engulf oxidized lipoproteins and cholesterol by separate receptors 7- to 10-fold faster than native LDL (Fogelman et al., 1980; Haberland, 1982), particularly when oxidized phospholipids are covalently attached to apolipoprotein B (apoB) of LDL (Boullier et al., 2000 345; Podrez et al., 2002b; Januszewski et al., 2005). Mild oxidation of LDL by lipid hydroperoxides results in denaturation of apo B (Nishida and Kummerow, 1960) followed by decomposition of LOOH to reactive aldehydes that covalently link to apoB, shut down the normal LDL receptor, and increase recognition and uptake of LDL by acetyl-LDL macrophage receptors (Mahley et al., 1977; Steinbrecher, 1987; Haberland et al., 1988; Hoff et al., 1992; Kim et al., 1999; Friedman et al., 2002). The switch of receptors occurs when 15% of the lysine has been oxidized ("lost"), causing a change in specific lysine residues (Haberland, 1982, 1984; Steinbrecher et al., 1987) that alters conformation of the recognition site or alignment of reactive residues or H-bond through ε-amino to receptor (Weisgraber et al., 1978). Oxidation of arginine (Mahley et al., 1977; Weisgraber et al., 1978) and tryptophan (Shoukry et al., 1994) residues also appears to be involved. Recent evidence suggests that receptors recognize both the modified lipid moieties and the modified protein moieties (Bird et al., 1999) and that both oxidized lipids and individual lipid and protein oxidation products may activate multiple receptors with different specificities so that uptake of oxidized LDL and cholesterol becomes even more efficient as oxidation proceeds (Gillotte et al., 2000).

More extensive oxidation by lipid hydroperoxides or products such as MDA and HNE produces protein aggregates that are viewed by macrophages as particles; they too are absorbed, but by endocytosis (Steinbrecher, 1987; Viita et al., 1999). While these multiple pathways were probably designed to be protective, clearing oxidized materials out of arteries and arterial walls, the paradoxical result with continued uncontrolled absorption of oxidized lipoproteins is to overload macrophages, causing a conversion to foam cells that play a major role in plaque formation and acceleration of pathogenic processes in atherosclerosis (Podrez et al., 2002a). Malondialdehyde-altered protein has been found in plaque deposits in experimental atherosclerosis in rabbits (Haberland et al., 1988).

Normally, oxidation in LDL can be counteracted by HDL which remove accumulating lipid and cholesterol from macrophages and peripheral cells and transport them to the liver for detoxification and disposal. Most LOOH in the bloodstream are carried by HDL rather than LDL. Paradoxically, LDL is loaded with high levels of antioxidants, particularly $CoQH_2$, that protects it and prevents lipid oxidation, while HDL have no antioxidants, except perhaps component apo and other proteins (Bowry et al., 1992). Lipid and cholesterol hydroperoxides are transferred from LDL to HDL, where they are reduced and then cleared rapidly by liver HepG2 cells that selectively remove oxidized HDL from the bloodstream. As lipid oxidation levels increase, however, lipid hydroperoxides undergo concerted two-electron reactions with met[112] and met[148] in Apo A1 and met[26] of A2, oxidizing them to methionine sulfoxide without covalent lipid binding (early stages) (Garner et al., 1998b). These residues are all in hydrophobic regions of active binding sites, not on the surface (Garner et al., 1998a), so the reaction is highly oriented and specific. In addition, oxidized phospholipid bound by HDL damage cys 31 and cys 184 in the HDL binding site. LOOH were active at very low concentrations, breaking down and generating free radicals in situ, while aldehydes were effective only at high concentrations (0.16 mM) (McCall et al., 1995). Destruction of either met or cys residues reduces LOOH binding to HDL as well as liver receptor recognition and clearance rate; lipid hydroperoxides then accumulate in plasma and in endothelial cells, increasing the potential for extensive protein co-oxidation and providing fuel for the macrophages.

Transport of oxidized fatty acids and cholesterol from LDL to HDL and from the endothelial cells to the plasma and liver also is impaired as lipids oxidize the transfer enzymes LCAT (lecithin:cholesterol acyl transferase) and CETP (cholesterol ester transfer protein). Phospholipid hydroperoxides (PL-OOH) in LDL modify free cysteine (or other catalytic residues) on LCAT transferase (Mickel et al., 1972; Bielicki and Forte, 1999), impairing HDL-cholesterol transport so cholesterol is not removed from arterial walls and accumulation accelerates transformation of macrophages to foam cells. As little as 0.2 and 1.0 mole% PL-OOH in plasma reduced LCAT activity by 20 and 50% in 2 hours, respectively (Bielicki and Forte, 1999). At low levels of oxidation, phospholipid hydroperoxides attacked cysteine residues in enzyme active sites, while at higher oxidation levels increasing levels of aldehydes inhibited the enzyme by reactions outside the active site. Similarly, oxidation of cholesterol ester transport protein prevents transport of oxidized lipids from LDL to HDL, so hydroperoxides accumulate in the cell and help accelerate macrophage transformation.

Paradoxically, there is also some evidence for cell signaling and induced protection via modulation of gene expression and biochemical pathways by lipid-mediated protein oxidations (Uchida et al., 1999; Uchida 2000; Leonarduzzi et al., 2000; Laurora et al., 2005; Zhang et al., 2005), particularly at the low physiological concentrations of lipid aldehydes. HNE inhibits SK-N-BE cell proliferation by up-regulating p53 family gene expression (Laurora et al., 2005) and prevents NF-κB activation and tumor necrosis factor expression by inhibiting I-κB phosphorylation and proteolysis (Page et al., 1999). The cytokine-induced expression of adhesion molecules in endothelial cells has been shown in vitro and more recently in vivo to be inhibited by HDL, in a process that potentially blocks a very early inflammatory stage in the development of atherosclerosis. Increased titers of antibodies to oxidation-specific epitopes of oxidized LDL in patients with advanced atherosclerosis is a protective response aimed at eliminating modified LDL (Friedman et al., 2002).

## Alzheimer's Disease

The etiology and mechanisms of progression for Alzheimer's disease are still poorly understood, but it is clearly recognized as a process involving extensive oxidative degradation of proteins, and oxidizing lipids are involved in some way. Early lipid radicals and hydroperoxides induce conformation changes in Alzheimer-specific epitopes of Tau (Liu et al., 2005), and oxidation of met to met sulfoxide in β-amyloid peptide converts it to the toxic form, in which it is no longer able to penetrate membranes or make pleated sheets but still binds $Cu^{2+}$ and makes $H_2O_2$ (Barnham et al., 2003).

As lipid oxidation progresses in oxidative stress, 4-HNE is significantly increased and is thought to play a role in the formation of β-amyloid (Sayre et al., 1997). Over one-half of all-paired helical filament (PHF)-1-labeled neurofibrilary tangles and dystrophic neuritis surrounding the β-amyloid core were found to contain protein-bound acrolein (Calingasan et al., 1991 272); the β-amyloid core itself also had adducts. Other proteins in the brain cortex are also modified by lipid oxidation (Pamplona et al., 2005), but this may be selective rather than generalized. About 100 oxidized proteins, mostly involved in signaling processes, have been identified in aged mice. Alterations in fatty acid synthetic enzymes cause shifts in brain fatty acids that affect brain function and susceptibility to further oxidative stress (Soreghan et al., 2003).

## Age-Related Macular Degeneration

DHA accounts for ~80 mol% of the lipids in the photoreceptor outer segments of the retina. The high unsaturation of DHA provides critical fluidity in the retina, but its oxidation also contributes to development of age-related macular degeneration (Gu et al., 2003; Ebrahem et al., 2006), at least in part through indirect effects in which protein co-oxidation products induce release of angiogenic factors. Carboxyethylpyridinium (CEP) protein adducts from DHA stimulate overproduction of new blood vessels that are tortuous and leaky, and they exacerbate choroidal neovascularization, abnormal vessel growth from capillaries through Bruch's membrane. 2-(ω-carboxyheptyl)pyrroles from linolenic acid and 2-(ω-carboxypropyl)pyrroles from arachidonic acid have also been isolated from the retina.

ω-6 polyunsaturated fatty acids also play an important role in damaging retinal proteins by providing a ready source of HNE. Three classes of proteins appear to be particularly sensitive to HNE modification: chaperone/cell protection (heat shock cognate; αA, αB, and βB2 crystallins); energy metabolism (triose phosphate isomerase, α enolase, aldolase C); and fatty acid transport (α enolase and βB2 crystallin) (Kapphahn et al., 2006). Most of these proteins have reactive cysteine residues that provide reducing equivalents to the cells as well as nucleophilic targets for HNE binding. Since the affected enzymes are part of the glycolytic pathway cascade, their inhibition has a critical affect on retinal function, where >50% of the ATP is produced via glycolysis. However, whether this is a negative or protective effect is being debated. Protein binding of toxic HNE and other aldehydes may be viewed as a "molecular sponge" for removing oxidant species, and protein modifications leading to reprogramming of metabolism to the pentose phosphate pathway for production of NADPH aids in recovery from oxidative damage (Kapphahn et al., 2006).

*Other Diseases*
Involvement of lipid oxidation in a wide range of oxidative pathologies is suspected, but isolating causative lipid oxidation products and determining their specific roles is extremely difficult in actively metabolizing tissues where abnormal compounds are rapidly cleared. With advances in instrumental analysis and immunological techniques, stable lipid-protein adducts are now being tracked to establish lipid oxidation associations with diseases (although presence of uncleared adducts is not proof that they actually caused the damage). Certainly, it is easy to envision how the protein changes described previously can play key roles in many forms of tissue degradation. Lysine-pyridinium adducts have been identified in proteins from patients with amyotrophic lateral sclerosis (Ichihashi et al., 2001), a disease that involves progressive deterioration of the myelin sheath and loss of nerve function. Alexander's disease is a progressive neurological disorder in which formation of fibrous, eosinophilic deposits called Rosenthal fibers leads to destruction of white matter in the brain (Castellani et al., 1998).

Systemic lupus erythematosus (SLE or lupus) is an auto-immune disease in which normal cells are mistakenly identified as foreign and labeled with antibodies that set up a cascade of radical-generating reactions that lead to inflammation. Modification of lupus-associated 60-kDa Ro protein with 4-hydroxy-2-nonenal increases recognition of cells as abnormal and facilitates spreading of the marker epitope and associated inflammation to other sites (Scofield et al., 2005). Crotonaldehyde is a strong tissue irritant in humans and carcinogen in male rats. Schiff-base pyridinium adducts to proteins have been identified in tissues exposed to this agent (Ichihashi et al., 2001).

# Conclusions

The past 20 years have seen great advances in the understanding of reactions between lipid oxidation products and proteins. While it is now clear that lipid oxyl radicals, hydroperoxides, epoxides, and carbonyl products all damage proteins, there is still no clear elucidation of the relative importance of various lipid oxidation products in damage to proteins. Indeed, perhaps even more questions have been raised about which lipid species is the dominant oxidant and which causes the most significant modifications contributing to pathologies in vivo or degrading protein function in foods.

This may seem to be a strange statement considering the detailed chemistry of individual reactions that has been reviewed in this chapter. What must be remembered, however, is that lipid oxidation is a dynamic process, constantly changing and contributing new oxidants; the mechanisms of interaction, amino acid target preferences, and reaction rates with proteins change with the extent of oxidation, and so do the reactivity and stability of products. The critical question is not which oxidant causes "the" damage to proteins, but how the various oxidants each contribute over time in actively oxidizing foods or biological systems, and perhaps more importantly, how damage from each of the oxidants interacts to change the overall functional, physiological, and pathological outcomes.

Damage to proteins (and other biomolecules) occurs at all stages of lipid oxidation. Protein degradation begins with the first lipid radicals long before aldehyde concentrations rise, and it continues after development of carbonyl secondary products. In general, lipid radicals and peroxides tend to induce protein oxidation and crosslinking without incorporation of lipid, and intermediate products on proteins are not stable. There are no definitive, easily detectable molecular "flags" in the products, so consequences of radical reactions are often difficult to recognize. All changes (e.g., loss of amino acids; production of oxidation products including new amino acids, crosslinking, and scission; lipid binding; etcetera) must be followed to assess reaction effects. In contrast, carbonyls in lipid secondary oxidation products react exclusively by addition reactions. Their relatively stable protein adducts are easily distinguishable by physical properties, such as fluorescence, and by reactions with chemical, optical, and immunological labels. Even so, definitive analysis is complicated by the reversibility of reactions in water or acid and by transformation of initial simple adducts to cyclic products over time, especially as aldehyde concentrations increase.

Protein reaction with oxidizing lipids has traditionally been monitored by global effects on chemical properties (particularly production of protein carbonyls) and macroscopic properties, such as enzyme inhibition, crosslinking or scission, or development of fluorescence in proteins. Traditionally, such measurements in model systems have been used to infer the active lipid oxidant and protein damage mechanisms. However, as this chapter has shown, lipid radicals, hydroperoxides, and carbonyls all produce similar global changes, particularly the protein carbonyls, fluorescence, and crosslinking that are analyzed universally (Fig. 8.19). A major message of this chapter, therefore, is that in systems with actively oxidizing lipids (not isolated products), oxidant sources of protein damage and reaction mechanisms cannot be concluded from macroscopic, global behaviors alone.

Such a flat statement is not meant to trivialize the issue. Indeed, the problem of identifying the lipid oxidants damaging proteins in real systems is as complex as the lipid oxidation reaction itself. Reactions occurring in isolated model systems with high concentrations of single oxidants are probably not the reactions occurring in situ in complex media with multiple oxidants forming at different times, competing for reactive sites, and altering availability (both chemical and physical) of reactive sites. Biological tissues pose an additional challenge because enzymes mediate secondary oxidations, reductions, and conjugations, and damaged molecules are rapidly cleared or repaired.

Monitoring protein products that have accumulated at a given time detects primarily products that are stable and have not been cleared by reaction in foods or by enzymes. But are these the products that have caused the most changes in system properties? Does identification of trace amounts (e.g., nanomolar) of stable aldehyde-protein adducts, for example, prove that these are the active damage agents or just the ones that are most long-lived or not removed by reaction or cell recycling and protection mechanisms? Of what impact are the protein products from radicals or hydroperoxides that have already degraded or been transformed further and cannot be detected or easily identified? Cleavage of histidine, arginine, lysine, and proline side chains to other amino acids, for example, would never be detected unless the individual protein was isolated and hydrolyzed, but it could have huge effects on protein functionality in foods and biological activity and recognition in vivo.

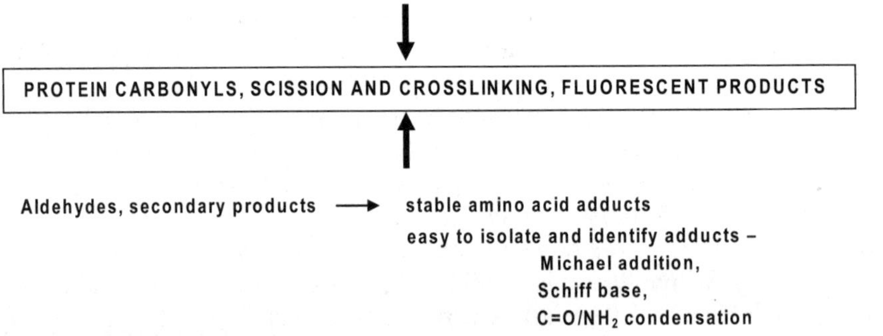

**LIPID RADICAL VERSUS LIPID CARBONYL REACTION PRODUCTS WITH PROTEINS**

LO(O)• /LOOH  →  protein free radicals,
not easy to trace source,
attack site migrates
some adducts form by radical addition

↓

| PROTEIN CARBONYLS, SCISSION AND CROSSLINKING, FLUORESCENT PRODUCTS |

↑

Aldehydes, secondary products  →  stable amino acid adducts
easy to isolate and identify adducts –
Michael addition,
Schiff base,
$C=O/NH_2$ condensation

**Fig. 8.19.** Despite differences in initial reactions, lipid radical, hydroperoxide, and carbonyl reactions with proteins yield products that may be indistinguishable by global analyses of macroscopic behaviors. Hence, measurement of protein carbonyls, crosslinking and scission, or fluorescence can be used to assess extent of protein degradation but not to deduce lipid oxidant sources.

A second message, therefore, is it is useless to argue that any one lipid oxidation product is the factor mediating oxidative changes. Research on oxidative damage to proteins induced by oxidizing lipids needs to move away from battles between the oxidants and claims of exclusive supremacy to focus instead on how timing of analysis, rates of reactions, effects of concentrations on direction of reactions, and fates of intermediates may affect detectability of changes and interpretation of reaction mechanisms, and how reactions of the various lipid oxidants are balanced and interact in different environments or with different proteins.

Three critical areas of information are still missing: detailed structural analysis of intermediates and products; quantitative analysis of individual lipid oxidation species, their protein interaction products, and comparisons between classes; and effects of reaction conditions on dominant pathways and individual products. The tremendous advantages in product separation and analysis offered by LC-MS are reflected in the increasing numbers of studies identifying intermediates, products, and mechanisms with precision. What must be added to this data base is information about individual reaction conditions, rates, and yields to put current studies in perspective. A critical mass of fundamental data is available, so now details are more important.

Model systems may reveal individual reaction products that are possible, but what are the actual yields of individual products under various conditions? Few papers actually report yields. How will a particular reaction product compete in a complex system if the rates of production are slow and the yields are low? Are there products whose low yields are counterbalanced by extraordinarily high reactivity so that traces overwhelm other products in damaging proteins? Reactions and products of aldehydes with unmodified proteins are reasonably well documented in defined model systems; how are these changed if the protein has previously been modified by lipid radicals? How does the dominant mechanism change when the reaction moves from neutral aqueous phases (intracellular, extracellular, or emulsion) to acid compartments to lipid phases of membranes, local environments, or emulsions?

In foods and model systems it is clear that low levels of lipid radicals and hydroperoxides rapidly damage proteins, and that aldehyde reactions are slow and require levels 100-1000x higher for reaction. Even so, aldehydes, by virtue of their longer lifetimes, presumed greater diffusibility, stability of protein adducts, and thus ease of detection are considered the most toxic lipids in vivo. Considering the chemistry covered in this chapter, the issue of aldehyde reactions is very intriguing and puzzling, and it raises many questions. How can model system studies demonstrating that aldehyde reactions with proteins are very slow on a physiological time scale and orders of magnitude slower than LOOH and radicals, and require 1:1 molar ratios and concentrations orders of magnitude higher than what are found in vivo be reconciled with claims of cytotoxicity from nanomolar (or lower) levels of aldehydes and their protein products in vivo? How can aldehyde concentrations in vivo accumulate to levels high enough to match the concentrations required for reaction in model systems? Are there catalysts that facilitate or speed up aldehyde reactions in vivo? Are aldehydes cleared less slowly so they accumulate over time, or are they detected just because they are stable, while earlier lipid oxidation species have already mediated damage? Do aldehyde products derive from direct dominant reaction with native proteins, or are there reactions of secondary products that are facilitated by preliminary damage from radicals but are slow or absent without conformation or other changes induced by hydroperoxides? Kanazawa and Ashida (1998) claim that lipid hydroperoxides in foods are decomposed in the stomach to aldehydes that are then absorbed. If the absorbed aldehydes are not metabolized or detoxified, deposition in tissues and incorporation into membranes could lead to build-up of toxic concentrations. Is this an accurate explanation?

These questions have serious implications for how experiments are designed, data is interpreted, and protective measures are structured, both for food preservation methods and for lipid-protein interactions in vivo. The issue of data interpretation thus deserves much debate and discussion.

Analyzing products in situ provides footprints of reactions and clues about causal agents, but more definitive studies in model systems closely coordinated with in situ oxidations in foods, cells, or tissues are needed to determine causality conclusively. This chapter has reviewed the extensive efforts focused on determining products of individual lipid oxidation products with intact proteins and component amino acids, and much has been learned about breakdown pathways. The time has now come to apply this knowledge to detailed analysis of integrated oxidation sequences in complex model systems, to replace global characteristic analysis with detailed determination of protein properties and amino acid changes step by step as lipids oxidize.

Lipid oxidation damage to proteins also must be juxtaposed with protein damage from other oxidant sources. Data cited in this chapter has shown repeatedly that protein changes induced by oxidizing lipids are the same as or comparable to reactions of hydroxyl radicals, which are blamed for most of the oxidant damage in vivo. Lipid-protein products, lipid reaction kinetics, and lipid peroxide and aldehyde concentrations that can accumulate provide overwhelming evidence that oxidizing lipids are competitive with other biological oxidants and should be included as biological oxidants along with the other agents normally cited as reactive oxidant species (Stadtman, 2004). For proteins in the endoplasmic reticulum or other membranes or closely associated with other lipid structures (e.g., blood lipoproteins), lipids are probably the dominant or most important oxidant. Biomedical research is just now beginning to recognize this, and hopefully the future will see definitive research focused on distinguishing specific roles of oxygen radicals versus lipid oxidation species in both physiological and pathological processes.

# References

Adams, J.Q. Electron Paramagnetic Resonance of tert-Butoxy Radical Reactions with Sulfides and Disulfides. *J. Am. Chem. Soc.* **1970,** *92,* 4535-4537.

Akagawa, M.; D. Sasaki; Y. Ishii; Y. Kurota; M. Yotsu-Yamashita; K. Uchida; and K. Suyama. New Method for the Quantitative Determination of Major Carbonyls, α-Aminoadipic and γ-Glutamic Semialdehydes: Investigation of the Formation Mechanism and Chemical Nature in vitro and in vivo. *Chem. Res. Toxicol.* **2006**, *19*, 1059-1065.

Alaiz, M. and J. Girón. Modification of Histidine Residues in Bovine Serum Albumin by Reaction with (E)-2-Octenal. *J. Agric. Food Chem.* **1994**, *42*, 2094-2098.

Alderton, A.L.; C. Faustman; D.C. Liebler; and D.W. Hill. Induction of Redox Instability of Bovine Myoglobin by Adduction with 4-Hydroxy-2-Nonenal. *Biochemistry* **2003**, *42*, 4398-4405.

Amarnath, V.; K. Amarnath; W.M. Valentine; M.A. Eng; and D.G. Graham. Intermediates in the Paal-Knorr Synthesis of Pyrroles. 2-Oxoaldehydes. *Chem. Res. Toxicol.* **1995**, *8*, 234-238.

Amarnath, V.; W.M. Valentine; K. Amarnath; M.A. Eng; and D.G. Graham. The Mechanism of Nucleophilic Substitution of Alkylpyrroles in the Presence of Oxygen. *Chem. Res. Toxicol.* **1994**, *7*, 56-61.

Amarnath, V.; W.M. Valentine; T.J. Montine; W.H. Patterson; K. Amarnath; C.N. Bassett; and D.G. Graham. Reactions of 4-Hydroxy-2(E)-Nonenal and Related Aldehydes with Proteins Studied by Carbon-13 Nuclear Magnetic Resonance Spectroscopy. *Chem. Res. Toxicol.* **1998**, *11*, 317-328.

ATSDR. Agency for Toxic Substances and Disease Registry ToxFAQs for crotonaldehyde, 2002, http://www.atsdr.cdc.gov/tfacts180.html#bookmark04.

Aubourg, S.P.; C.G. Sotelo; and R. Perez-Martin. Assessment of Quality Changes in Frozen Sardine (*Sardina pilchardus*) by Fluorescence Detection. *J. Am. Oil Chem. Soc.* **1998**, *75*, 575-580.

Augusto, A.; M.G. Bonini; and D.F. Trindade. Spin Trapping of Glutathiyl and Protein Radicals Produced from Nitric Oxide-Derived Oxidants. *Free Radic. Biol. Med.* **2004**, *36*, 1224-1232.

Avdulov, N.A.; S.V. Chochina; U. Igbavboa; E.O. O'Hare; F. Schroeder; J.P. Cleary; and W.G. Wood. Amyloid Beta-Peptides Increase Annular and Bulk Fluidity and Induce Lipid Peroxidation in Brain Synaptic Plasma Membranes. *J. Neurochem.* **1997**, *68*, 2086–2091.

Badghisi, H.; and D.C. Liebler. Sequence Mapping of Epoxide Adducts in Human Hemoglobin with LC-Tandem MS and the Salsa Algorithm. *Chem. Res. Toxicol.* **2002**, *15*, 799-805.

Barnham, K.J.; G.D. Ciccotosto; A.K. Tickler; F.E. Ali; D.G. Smith; N.A. Williamson; Y.-H. Lam; D. Carrington; D. Tew; G. Kocak, et al. Neurotoxic, Redox-Competent Alzheimer's β-Amyloid Is Released from Lipid Membrane by Methionine Oxidation. *J. Biol. Chem.* **2003**, *278*, 42959-42965.

Barter, P. Role of lipoproteins in inflammation, 2001, http://www.lipidsonline.org/slides.

Benderdour, M.; G. Charron; D. deBlois; B. Comte; and C. Des Rosiers. Cardiac Mitochondrial NADP+-Isocitrate Dehydrogenase Is Inactivated Through 4-Hydroxynonenal Adduct Formation. *J. Biol. Chem.* **2003**, *278*, 45154-45159.

Benedetti, A.; A.F. Casini; and M. Ferrali. Extraction and Partial Characterization of Dialyzable Products Originating from the Peroxidation of Liver Microsomal Lipids and Inhibiting Microsomal Glucose-6-Phosphatase Activity. *Biochem. Pharmacol.* **1979**, *28*, 2909-2918.

Beppu, B. Role of Heme Compounds in the Erythrocyte Membrane Damage Induced by Lipid Hydroperoxide. *Chem. Pharm. Bull.* **1986**, *34*, 781-788.

Berger, P.; N.K. Vel Leitner; M. Doré; and B. Legube. Ozone and Hydroxyl Radicals Induced Oxidation of Glycine. *Water Res.* **1999**, *33*, 433-441.

Bielicki, J.K.; and T.M. Forte. Evidence that Lipid Hydroperoxides Inhibit Plasma Lecithin:

Cholesterol Acytransferase Activity. *J. Lipid Res.* **1999**, *40*, 948-954.
Bird, D.A.; K.L. Gillotte; S. Hörkkö; P. Friedman; E.A. Dennis; J.L. Witztum; and D. Steinberg. Receptors for Oxidized Low-Density Lipoprotein on Elicited Mouse Peritoneal Macrophages Can Recognize Both the Modified Lipid Moieties and the Modified Protein Moieties: Implications with Respect to Macrophage Recognition of Apoptotic Cells. *Proc. Nat. Acad. Sci. USA* **1999**, *96,* 6347-6352.
Blair, I.A. Lipid Hydroperoxide-Mediated DNA Damage. *Exp. Gerontol.* **2001**, *36,* 1473-1481.
Boatright, W.L.; and N.S. Hettiarachchy. Effect of Lipids on Soy Protein Isolate Solubility. *J. Am. Oil Chem. Soc.* **1995**, *72,* 1439-1444.
Bobrowski, K.; and C. Schöneich. Decarboxylation Mechanism of the N-Terminal Glutamyl Moiety in γ-Glutamic Acid and Methionine Containing Peptides. *Radiat. Phys. Chem.* **1996**, *47,* 507-510.
Bolgar, M.S.; and S.J. Gaskell. Determination of the Sites of 4-Hydroxy-2-Nonenal Adduction to Protein by Electrospray Tandem Mass Spectrometry. *Anal. Chem.* **1996,** *68,* 2325-2330.
Borg, D.C.; and K.M. Schaich. Cytotoxicity from Coupled Redox Cycling of Autoxidizing Xenobiotics and Metals. *Isr. J. Chem.* **1984,** *24,* 38-53.
Borovyagin, V.L.; A.F. Muronov; V.D. Rumyantseva; Y.S. Tarakhovskii; and I.A. Vasilenko. Model Membrane Morphology and Crosslinking of Oxidized Lipids with Proteins. *J. Ultrastruct. Res.* **1984,** *89,* 261-273.
Bossman, S.H.; E. Oliveros; S. Göb; S. Siegwart; E.P. Dahlen; L.J. Payawan; M. Straub; M. Wörner; and A.M. Braun. New Evidence Against Hydroxyl Radicals as Reactive Intermediates in the Thermal and Photochemically Enhanced Fenton Reactions. *J. Phys. Chem.* **1998,** *102,* 5542-5550.
Boullier, A.; K.L. Gillotte; S. Hörkkö; S.R. Green; P. Friedman; E.A. Dennis; J.L. Witztum; D. Steinberg; and O. Quehenberger. The Binding of Oxidized Low Density Lipoprotein to Mouse CD36 Is Mediated in Part by Oxidized Phospholipids that Are Associated with Both the Lipid and Protein Moieties of the Lipoprotein. *J. Biol. Chem.* **2000,** *275,* 9163-9169.
Bowry, V.W.; K.K. Stanley; and R. Stocker. High Density Lipoprotein Is the Major Carrier of Lipid Hydroperoxides in Human Blood Plasma from Fasting Donors. *Proc. Nat. Acad. Sci. USA* **1992,** *89,* 10316-10320.
Brame, C.J.; O. Boutaud; S.S. Davies; T. Yang; J.A. Oates; D. Roden; and L.J.I. Roberts. Modification of Proteins by Isoketal-Containing Oxidized Phospholipids. *J. Biol. Chem.* **2004,** *279,* 13447-13451.
Brame, C.J.; R.G. Salomon; J.D. Morrow; and L.J.I. Roberts. Identification of Extremely Reactive γ–Ketoaldehydes (Isolevuglandins) as Products of the Isoprostane Pathway and Characterization of Their Lysyl Protein Adducts. *J. Biol. Chem.* **1999,** *274,* 13139-13146.
Brock, J.W.C.; J.M. Ames; S.R. Thorpe; and J.W. Baynes. Formation of Methionine Sulfoxide During Glycoxidation and Lipoxidation of Ribonuclease A. *Arch. Biochem. Biophys.* **2007,** *457,* 170-176.
Bruenner, B.A.; A.D. Jones; and J.B. German. Direct Characterization of Protein Adducts of the Lipid Peroxidation Product 4-Hydroxy-2-Nonenal Using Electrospray Mass Spectrometry. *Chem. Res. Toxicol.* **1995,** *8,* 552-559.
Brunner, B.; N. Stogatis; and M. Lautens. Synthesis of 1,2-Dihyropyridine Using Vinyloxiranes as Masked Dienolates in Imino-Aldol Reactions. *Org. Lett.* **2006,** *8,* 3473-3476.

Burcham, P.C.; and Y.T. Kuhan. Introduction of Carbonyl Groups into Proteins by the Lipid Peroxidation Product, Malondialdehyde. *Biochem. Biophys. Res. Commun.* **1996**, *220,* 996-1001.

Buttery, R.G.; L.C. Ling; R. Teranishi; and T.R. Mon. Roasted Fat: Basic Volatile Components. *J. Agric. Food Chem.* **1977**, *25,* 1227-1229.

Buttkus, H. The Reaction of Myosin with Malonaldehyde. *J. Food Sci.* **1967**, *32,* 432-434.

Buttkus, H. Reaction of Cysteine and Methionine with Malonaldehyde. *J. Am. Oil Chem. Soc.* **1968**, *46,* 88-93.

Buttkus, H. Accelerated Denaturation of Myosin in Frozen Solution. *J. Food Sci.* **1970**, *35,* 558-562.

Buttkus, H. The Reaction of Malonaldehyde or Oxidized Linolenic Acid with Sulfhydryl Compounds. *J. Am. Oil Chem. Soc.* **1972**, *49,* 613-614.

Buttkus, H.A. Fluorescent Lipid Autoxidation Products. *J. Agric. Food Chem.* **1975**, *23,* 823-825.

Cagen, L.M.; H.M. Fales; and J.J. Pisano. Formation of Glutathione Conjugates of Prostaglandin A in Human Red Blood Cells. *J. Biol. Chem.* **1976**, *251,* 6550-6554.

Calingasan, N.Y.; K. Uchida; and G.E. Gibson. Protein-Bound Acrolein: A Novel Marker of Oxidative Stress in Alzheimer's Disease. *J. Neurochem.* **1991**, *72,* 751-756.

Carbone, D.L.; J.A. Doorn; Z. Kiebler; and D.R. Petersen. Cysteine Modification by Lipid Peroxidation Products Inhibits Protein Disulfide Isomerase. *Chem. Res. Toxicol.* **2005**, *18,* 1324-1331.

Castellani, R.J.; G. Perry; P.L.R. Harris; M.L. Cohen; L.M. Sayre; R.G. Salomon; and M.A. Smith. Advanced Lipid Peroxidation End-Products in Alexander's Disease. *Brain Res.* **1998**, *787,* 15-18.

Chander, R.; S.V. Sherekar; and M.S. Gore. Studies on the Inactivation of Lysozyme by Malonaldehyde. *J. Food Biochem.* **1981**, *5,* 313-324.

Chelh, I.; P. Gatellier; and V. Sante-Lhoutellier. Characterisation of Fluorescent Schiff Bases Formed During Oxidation of Pig Myofibrils. *Meat Sci.* **2007**, *76,* 210-215.

Chen, H.-J.C.; and F.-L. Chung. Epoxidation of *trans*-4-Hydroxy-2-Nonenal by Fatty Acid Hydroperoxides and Hydrogen Peroxide. *Chem. Res. Toxicol.* **1996**, *9,* 306-312.

Chen, P.; D. Wiesler; J. Chmelik; and M. Novotny. Substituted 2-Hydroxy-1,2-Dihydropyrrol-3-Ones: Fluorescent Markers Pertaining to Oxidative Stress and Aging. *Chem. Res. Toxicol.* **1996**, *9,* 970-979.

Chio, K.S.; and A.L. Tappel. Inactivation of Ribonuclease and Other Enzymes by Peroxidizing Lipids and by Malonaldehyde. *Biochemistry* **1969a**, *8,* 2827-2832.

Chio, K.S.; and A.L. Tappel. Synthesis and Characterization of the Fluorescent Products Derived from Malondialdehyde and Amino Acids. *Biochemistry* **1969b**, *8,* 2821-2827.

Chung, F.-L.; H.-J.C. Chen; J.B. Guttenplan; A. Nishikawa; and G.C. Hard. 2,3-Epoxy-4-Hydroxynonanal as a Potential Tumor-Initiating Agent of Lipid Peroxidation. *Carcinogenesis* **1993**, *14,* 2073-2077.

Cohn, J.A.; L. Tsai; B. Friguet; and L.I. Szweda. Chemical Characterization of a Protein-4-Hydroxy2-Nonenal Crosslink: Immunochemical Detection in Mitochondria Exposed to Oxidative Stress. *Arch. Biochem. Biophys.* **1996**, *328,* 158-164.

Connell, J.J. Studies on the Proteins of Fish Skeletal Muscle. 7. Denaturation and Aggregation of Cod Myosin. *Biochem. J.* **1960**, *75,* 530-538.

Connell, J.J.; and P.F. Howgate. Studies on the Proteins of Fish Skeletal Muscle. 6. Amino-Acid Composition of Cod Fibrillar Proteins. *Biochem. J.* **1959**, *71,* 83-86.

Crabb, J.W.; J. O'Neill; M. Miyagi; K. West; and H.F. Hoff. Hydroxynonenal Inactivates Cathepsin B by Forming Michael Adducts with Active Site Residues. *Protein Sci.* **2002,** *11,* 831-840.

Crawford, D.L.; T.C. Yu; and R.O. Sinnhuber. Reaction of Malonaldehyde with Glycine. *J. Agric. Food Chem.* **1966,** *14,* 182-184.

Culbertson, S.M.; G.D. Enright; and K.U. Ingold. Synthesis of a Novel Radical Trapping and Carbonyl Group Trapping Anti-Age Agent: A Pyridoxamine Analogue for Inhibiting Advanced Glycation (Age) and Lipoxidation (ALE) End Products. *Org. Lett.* **2003,** *5,* 2659-2662.

Dalsgaard, T.K.; J.H. Nielsen; and L.B. Larsen. Characterization of Reaction Products Formed in a Model Reaction Between Pentanal and Lysine-Containing Oligopeptides. *J. Agric. Food Chem.* **2006,** *54,* 6367-6373.

Davies, K.A.J. Protein Damage and Degradation by Oxygen Radicals. I. General Aspects. *J. Biol. Chem.* **1987a,** *262,* 9895-9901.

Davies, K.A.J. Protein Damage and Degradation by Oxygen Radicals. IV. Degradation of Denatured Proteins. *J. Biol. Chem.* **1987b,** *262,* 9914-9920.

Davies, K.A.J.; and M.E. Delsignore. Protein Damage and Degradation by Oxygen Radicals. III. Modification of Secondary and Tertiary Structure. *J. Biol. Chem.* **1987,** *262,* 9908-9913.

Davies, M.J. The Oxidative Environment and Protein Damage. *Biochim. Biophys. Acta* **2005,** *1703,* 93-109.

Davies, M.J.; S. Fu; and R.T. Dean. Protein Hydroperoxides Can Give Rise to Reactive Free Radicals. *Biochem. J.* **1995,** *305,* 643-649.

Davies, M.J.; S. Fu; H. Wang; and R.T. Dean. Stable Markers of Oxidant Damage to Proteins and Their Application in the Study of Human Disease. *Free Radic. Biol. Med.* **1999** *27,* 1151-1163.

Davies, M.J.; and C.L. Hawkins. EPR Spin Trapping of Protein Radicals. *Free Radic. Biol. Med.* **2004,** *36,* 1072-1086.

Davies, S.S.; V. Amarnath; and L.J.I. Roberts. Isoketals: Highly Reactive γ-Ketoaldehydes Formed from the $H_2$-Isoprostane Pathway. *Chem. Phys. Lipids* **2004,** *128,* 85-99.

Desai, I.D.; and A.L. Tappel. Damage to Proteins by Peroxidized Lipids. *J. Lipid Res.* **1963,** *4,* 204-207.

Dillard, C.J.; and A.L. Tappel. Fluorescent Products of Lipid Peroxidation of Mitochondria and Microsomes. *Lipids* **1971,** *6,* 715-721.

Dillard, C.J.; and A.L. Tappel. Fluorescent Products from Reaction of Peroxidizing Polyunsaturated Fatty Acids with Phosphatidyl Ethanolamine and Phenylalanine. *Lipids* **1973,** *8,* 183-189.

Dillard, C.J.; and A.L. Tappel. Fluorescent Damage Products of Lipid Peroxidation. *Methods Enzymol.* **1984,** *105,* 337-341.

Doorn, J.A.; and D.R. Petersen. Covalent Modification of Amino Acid Nucleophiles by the Lipid Peroxidation Products 4-Hydroxy-2-Nonenal and 4-Oxo-2-Nonenal. *Chem. Res. Toxicol.* **2002,** *15,* 1445-1450.

Doorn, J.A.; and D.R. Petersen. Covalent Adduction of Nucleophilic Amino Acids by 4-Hydroxynonenal and 4-Oxononenal. *Chem. Biol. Interact.* **2003,** *143-144,* 93-100.

Dvorak, Z. Availability of Essential Amino Acids from Proteins. II. Food Proteins. *J. Sci. Food Agric.* **1968,** *19,* 77-82.

Ebrahem, Q.; K. Ranganathan; J. Sears; A. Vasanji; X. Gu; L. Lu; R.G. Salomon; J.W. Crabb;

and B. Anand-Apte. Carboxyethylpyrrole Oxidative Protein Modifications Stimulate Neovascularization: Implications for Age-Related Macular Degeneration. *Proc. Nat. Acad. Sci. USA* **2006**, *103,* 13480-13484.

Ege, S.N. Organic Chemistry: Structure and Reactivity, Houghton: Boston, pp. 517-518, 631-654, 692-693, 725-733, 1999.

El-Gharbawi, M.I.; and L.R. Dugan, Jr. Stability of Nitrogenous Compounds and Lipids during Storage of Freeze-Dried Raw Beef. *J. Food Sci.* **1965,** *30,* 817-822.

Elliot, A.J.; R.J. McEachern; and D.A. Armstrong. Oxidation of Amino-Containing Disulfides by $Br_2^-$ and OH. A Pulse-Radiolysis Study. *J. Phys. Chem.* **1981,** *85,* 68-75.

Esterbauer, H.; A. Ertl; and N. Scholz. The Reaction of Cysteine with α,β–Unsaturated Aldehydes. *Tetrahedron* **1976,** *32,* 285-289.

Esterbauer, H.; R.J. Schaur; and H. Zollner. Chemistry and Biochemistry of 4-Hydroxynonenal, Malonaldehyde, and Related Aldehydes. *Free Radic. Biol. Med.* **1991,** *11,* 81-128.

Esterbauer, H.; H. Zollner; and N. Scholz. Reaction of Glutathione with Conjugated Carbonyls. *Z. Naturforsch., C: Biosci.* **1975,** *30,* 466-473.

Faustman, C.; D.C. Liebler; T.D. McClure; and Q. Sun. α,β-Unsaturated Aldehydes Accelerate Oxymyoglobin Oxidation. *J. Agric. Food Chem.* **1999,** *47,* 3140-3144.

Fenaille, F.; P.A. Guy; and J.-C. Tabet. Study of Protein Modification by 4-Hydroxy-2-Nonenal and Other Short Chain Aldehydes Analyzed by Electrospray Ionization Tandem Mass Spectrometry. *J. Am. Soc. Mass Spectrom.* **2002**, *14,* 215-226.

Fenaille, F.; J.-C. Tabet; and P.A. Guy. Study of Peptides Containing Modified Lysine Residues by Tandem Mass Spectrometry: Precursor Ion Scanning of Hexanal-Modified Peptides. *Rapid Communications in Mass Spectrometry* **2004a,** *18,* 67-76

Fenaille, F.; J.-C. Tabet; and P.A. Guy. Identification of 4-Hydroxy-2-Nonenal-Modified Peptides Within Unfractionated Digests Using Matrix-Assisted Laser Desorption/Ionization Time-Of-Flight Mass Spectrometry. *Anal. Chem.* **2004b,** *76* (4), 867-873.

Finley, J.W.; and R.E. Lundin. Lipid Hydroperoxide Induced Oxidation of Cysteine in Peptides. In *Autoxidation in Food and Biological Systems,* M.G. Simic, and M. Karel, Eds. Plenum: New York, 1980, pp. 223-236.

Fletcher, B.L.; C.J. Dillard; and A.L. Tappel. Measurement of Fluorescent Lipid Peroxidation Products in Biological Systems and Tissues. *Anal. Biochem.* **1973,** *52,* 1-9.

Fletcher, B.L.; and A.L. Tappel. Fluorescent Modification to Serum Albumin by Lipid Peroxidation. *Lipids* **1970,** *6,* 172-175.

Fogelman, A.M.; I. Schecter; J. Seager; M. Hokom; J.S. Child; and P.A. Edwards. Malondialdehyde Alteration of Low Density Lipoproteins Leads to Cholesteryl Ester Accumulation in Human Monocyte-Macrophages. *Proc. Nat. Acad. Sci. USA* **1980,** *77,* 2214-2218.

Fong, L.G.; S. Parthasarathy; J.L. Witztum; and D. Steinberg. Nonenzymatic Oxidative Cleavage of Peptide Bonds in Apoprotein B-100. *J. Lipid Res.* **1987,** *28,* 1466-1477.

Freeman, T.L.; A. Haver; M.J. Duryee; D.J. Tuma; L.W. Klassen; F.G. Hamel; R.L. White; S.I. Rennard; and G.M. Thiele. Aldehydes in Cigarette Smoke React with the Lipid Peroxidation Product Malonaldehyde to Form Fluorescent Protein Adducts on Lysines. *Chem. Res. Toxicol.* **2005,** *18,* 817-824.

Friedman, P.; S. Hörkkö; D. Steinberg; J.L. Witztum; and E.A. Dennis. Correlation of Antiphospholipid Antibody Recognition with the Structure of Synthetic Oxidized Phospholipids. Importance of Schiff Base Formation and Aldol Condensation. *J. Biol. Chem.* **2002,** *277,* 7010-7020.

Friguet, B.; E.R. Stadtman; and L.I. Szweda. Modification of Glucose-6-Phosphate Dehydrogenase by 4-Hydroxy-2-Nonenal. Formation of Cross-Linked Protein that Inhibits the Multicatalytic Protease. *J. Biol. Chem.* **1994a**, *269*, 21639-21643.

Friguet, B.; and L.I. Szweda. Inhibition of the Multicatalytic Proteinase (Proteasome) by 4-Hydroxy-2-Nonenal Cross-Linked Protein. *FEBS Lett.* **1997**, *405*, 21-25.

Friguet, B.; L.I. Szweda; and E.R. Stadtman. Susceptibility of Glucose-6-Phosphate Dehydrogenase Modified by 4-Hydroxy-2-Nonenal and Metal-Catalyzed Oxidation to Proteolysis by the Multicatalytic Protease. *Arch. Biochem. Biophys.* **1994b**, *311*, 168-173.

Fruebis, J.; S. Parsasarathy; and D. Steinberg. Evidence for a Concerted Reaction Between Lipid Hydroperoxides and Polypeptides. *Proc. Nat. Acad. Sci. USA* **1992**, *89*, 10588-10592.

Fu, S.; L.A. Hick; M.M. Sheil; and R.T. Dean. Structural Identification of Valine Hydroperoxides and Hydroxides on Radical-Damaged Amino Acid, Peptide, and Protein Molecules. *Free Radic. Biol. Med.* **1995**, *19*, 281-292.

Funes, J.; and M. Karel. Free Radical Polymerization and Lipid Binding of Lysozyme Reacted with Peroxidizing Linoleic Acid. *Lipids* **1981**, *16*, 347-350.

Funes, J.A.; U. Weiss; and M. Karel. Effects of Reaction Conditions and Reactant Concentrations on Polymerization of Lysozyme Reacted with Peroxidizing Lipids. *J. Agric. Food Chem.* **1982**, *29*, 404-407.

Gamage, P.T.; T. Mori; and A. Matsushita. Mechanism of Polymerization of Proteins by Autoxidized Products of Linoleic Acid. *J. Nutr. Sci. Vitaminol. (Tokyo).* **1973**, *19*, 173-182.

Gardner, H.W. Lipid Hydroperoxide Reactivity with Proteins and Amino Acids: A Review. *J. Agric. Food Chem.* **1979**, *27*, 220-229.

Gardner, H.W. Oxygen Radical Chemistry of Polyunsaturated Fatty Acids. *Free Radic. Biol. Med.* **1989**, *7*, 65-86.

Gardner, H.W.; R. Kleiman; D. Weisleder; and G.E. Inglett. Cysteine Adds to Lipid Hydroperoxide. *Lipids* **1977**, *12*, 655-660.

Gardner, H.W.; and R. Kleiman. Degradation of Linoleic Acid Hydroperoxides by a Cysteine-$FeCl_3$ Catalyst as a Model for Similar Biochemical Reactions. II. Specificity in Formation of Fatty Epoxides. *Biochim. Biophys. Acta* **1981**, *665* (1), 113-125.

Gardner, H.W.; and D. Weisleder. Addition of N-Acetylcysteine to Linoleic Acid Hydroperoxide. *Lipids* **1976**, *11*, 127-134.

Garner, B.; A.R. Waldeck; P.K. Witting; K.-A. Rye; and R. Stocker. Oxidation of High Density Lipoproteins. II. Evidence for Direct Reduction of Lipid Hydroperoxides by Methionine Residues of Apolipoproteins A1 and A2. *J. Biol. Chem.* **1998a**, *273*, 6088-6095.

Garner, B.; P.K. Witting; A.R. Waldeck; J.K. Christison; M. Raftery; and R. Stocker. Oxidation of High Density Lipoproteins. I. Formation of Methionine Sulfoxide in Apolipoproteins A1 and A2 Is an Early Event that Accompanies Lipid Peroxidation and Can Be Enhanced by α-Tocopherol. *J. Biol. Chem.* **1998b**, *273*, 6080-6087.

Garrison, W.M. Reaction Mechanisms in the Radiolysis of Peptides, Polypeptides, and Proteins. *Chem. Rev.* **1987**, *87*, 381-398.

Gebicki, S.; and J.M. Gebicki. Crosslinking of DNA and Proteins Induced by Protein Hydroperoxides. *Biochemical J.* **1999**, *338*, 629-636.

Gerrard, J.A.; P.K. Brown; and S.E. Fayle. Maillard Crosslinking of Food Proteins I: The Reaction of Glutaraldehyde, Formaldehyde, and Glyceraldehyde with Ribonuclease. *Food Chem.* **2002**, *79*, 343-349.

Gillotte, K.L.; S. Hörkkö; J.L. Witztum; and D. Steinberg. Oxidized Phospholipids, Linked to Apolipoprotein B of Oxidized LDL, Are Ligands for Macrophage Scavenger Receptors. *J.*

*Lipid Res.* **2000**, *41,* 824-833.
Giron-Calle, J.; M. Alaiz; F. Millan; V. Ruiz-Guiterrez; and E. Vioque. Bound Malondialdehyde in Foods: Bioavailability of N,N'-Di-(4-Methyl-1,4-Dihydropyridine-3,5-Dicarbaldehyde)Lysine. *J. Agric. Food Chem.* **2003**, *51,* 4799-4803.
Giulivi, C.; and K.A.J. Davies. Dityrosine and Tyrosine Oxidation Products are Endogenous Markers for the Selective Proteolysis of Oxidatively Modified Red Blood Cell Hemoglobin by (the 19S) Proteosome. *J. Biol. Chem.* **1993**, *268,* 8752-8759.
Giulivi, C.; N.J. Traaseth; and K.A.J. Davies. Tyrosine Oxidation Products: Analysis and Biological Relevance. *Amino Acids* **2003**, *25,* 227-232.
Green, R.C.; C. Little; and P.J. O'Brien. The Inactivation of Isocitrate Dehydrogenase by a Lipid Peroxide. *Arch. Biochem. Biophys.* **1971**, *142,* 598-605.
Grune, T.; and K.A.J. Davies. The Proteasomal System and HNE-Modified Proteins. *Mol. Aspects Med.* **2003**, *24,* 195-204.
Grune, T.; T. Reinheckel; and K.A.J. Davies. Degradation of Oxidized Proteins in Mammalian Cells. *FASEB J.* **1997**, *11,* 526-534.
Gu, X.; S.G. Meer; M. Miyagi; M.E. Rayborn; J.G. Hollyfield; J.W. Crabb; and R.G. Salomon. Carboxyethylpyrrole Protein Adducts and Autoantibodies, Biomarkers for Age-Related Macular Degeneration. *J. Biol. Chem.* **2003**, *278,* 42027-42035.
Guitton, J.; F. Tinardon; R. Lamrini; P. Lacan; M. Desage; and A. Francina. Decarboxylation of [1-$^{13}$C]Leucine by Hydroxyl Radicals. *Free Radic. Biol. Med.* **1998**, *25,* 340-345.
Haberland, M.E. Specificity of Receptor-Mediated Recognition of Malondialdehyde-Modified Low Density Lipoproteins. *Proc. Nat. Acad. Sci. USA* **1982**, *79,* 1712-1716.
Haberland, M.E. Role of Lysines in Mediating Interaction of Modified Low Density Lipoproteins with the Scavenger Receptor of Human Monocyte Macrophages. *J. Biol. Chem.* **1984**, *259,* 11305-11311.
Haberland, M.E.; D. Fong; and I. Cheng. Malondialdehyde-Altered Protein Occurs in Atheromas of Watanabe Heritable Hyperlipidemic Rabbits. *Science* **1988**, *241,* 215-217.
Halligudi, N.N.; S.M. Desai; S.K. Mavalangi; and S.T. Nandibewoor. Kinetics of the Oxidative Degradation of *rac*-Serine by Aqueous Alkaline Permanganate. *Monatsh. Chem.* **2000**, *131,* 321-332.
Ham, E.A.; H.G. Oien; E.H. Ulm; and F.A. Kuehl, Jr. The Reaction of PGA1 with Sulfhydryl Groups: A Component in the Binding of A-Type Prostaglandins to Proteins. *Prostaglandins* **1975**, *10,* 217-229.
Hamberg, M.; and B. Gotthammar. A New Reaction of Unsaturated Fatty Acid Hydroperoxides: Formation of 11-Hydroxy-12,13-Epoxy-9-Octadecenoic Acid from 13-Hydroperoxy-9,11-Octadecadienoic Acid. *Lipids* **1973**, *8,* 737-744.
Harris, L.; and H.S. Olcott. Reaction of Aliphatic Tertiary Amines with Hydroperoxides. *J. Am. Oil Chem. Soc.* **1966**, *43,* 11-15.
Hasegawa, K.; and L.K. Patterson. Pulse Radiolysis Studies in Model Lipid Systems: Formation and Behavior of Peroxy Radicals in Fatty Acids. *Photochem. Photobiol.* **1978**, *28,* 817-823.
Headlam, H.A.; and M.J. Davies. Cell-Mediated Reduction of Protein and Peptide Hydroperoxides to Reactive Free Radicals. *Free Radic. Biol. Med.* **2003**, *34,* 44-55.
Headlam, H.A.; A. Mortimer; C.J. Easton; and M.J. Davies. β-Scission and C-3(β-Carbon) Alkoxyl Radicals on Peptides and Proteins: A Novel Pathway Which Results in the Formation of α-Carbon Radicals and the Loss of Amino Acid Side Chains. *Chem. Res. Toxicol.* **2000**, *13,* 1087-1095.

Hendley, D.D.; A.S. Mildvan; M.D. Reporter; and B.L. Strehler. The Properties of Isolated Human Cardiac Age Pigment. I. Preparation and Physical Properties. *J. Gerontol.* **1963a**, *18*, 144-450.

Hendley, D.D.; A.S. Mildvan; M.D. Reporter; and B.L. Strehler. The Properties of Isolated Human Cardiac Age Pigment. II. Chemical and Enzymatic Properties. *J. Gerontol.* **1963b**, *18*, 250-258.

Hidalgo, F.J.; M. Alaiz; and R. Zamora. Effect of pH and Temperature on Comparative Nonenzymatic Browning of Proteins Produced by Oxidized Lipids and Carbohydrates. *J. Agric. Food Chem.* **1999**, *47*, 742-747.

Hidalgo, F.J.; E. Gallardo; and R. Zamora. Strecker Type Degradation of Phenylalanine by 4-Hydroxy-2-Nonenal in Model Systems. *J. Agric. Food Chem.* **2005**, *53*, 10254-10259.

Hidalgo, F.J.; and J.E. Kinsella. Changes in β-Lactoglobulin B Following Interactions with Linoleic Acid 13-Hydroperoxide. *J. Agric. Food Chem.* **1989**, *37*, 860-866.

Hidalgo, F.J.; and R. Zamora. Fluorescent Pyrrole Products From Carbonyl-Amine Reactions. *J. Biol. Chem.* **1993**, *268*, 16190-16197.

Hidalgo, F.J.; and R. Zamora. Modification of Bovine Serum Albumin Structure Following Reaction with 4,5(E)-Epoxy-2-(E)-Heptenal. *Chem. Res. Toxicol.* **2000**, *13*, 501-508.

Hochstein, P.; and L. Ernster. Microsomal Peroxidation of Lipids and Its Possible Role in Cellular Injury. In *Ciba Foundation Symposium on Cellular Injury,* A.V.S. DeReuck, and J. Knight, Eds.; Little and Brown: Boston, MA, 1964, pp. 123-135.

Hochstein, P.; and S.K. Jain. Association of Lipid Peroxidation and Polymerization of Membrane Proteins with Erythrocyte Aging. *Fed. Proc.* **1981**, *40*, 183-188.

Hoff, H.F.; T.E. Whitaker; and J.O. O'Neill. Oxidation of Low Density Lipoprotein Leads to Particle Aggregation and Altered Macrophage Recognition. *J. Biol. Chem.* **1992**, *267*, 602-609.

Horigome, T.; and M. Miura. Interaction of Protein and Oxidized Ethyl Linolenate in the Dry State and Nutritive Value of Reacted Protein. *J. Agric. Chem. Soc. Jpn.* **1974**, *48*, 437-444.

Horigome, T.; T. Yanagida; and M. Miura. Nutritive Value of Proteins Prepared by the Reaction with Oxidized Ethyl Linoleate in Aqueous Medium. *J. Agric. Chem. Soc. Jpn.* **1974**, *48*, 195-199.

Hruszkewycz, A.M.; E.A. Glende, Jr.; and R.O. Recknagel. Destruction of Microsomal Cytochrome P-450 and Glucose-6-Phosphatase by Lipids Extracted from Peroxidized Microsomes. *Toxicol. Appl. Pharmacol.* **1978**, *46*, 695-702.

Huang, Y.; Y. Hua; and A. Qui. Soybean Protein Aggregation Induced by Lipoxygenase Catalyzed Linoleic Acid Oxidation. *Food Res. Int.* **2006**, *39*, 240-249.

Huerta, F.; E. Morallóna; F. Casesb; A. Rodesa; J.L. Vázquez; and A. Aldaza. Electrochemical Behaviour of Amino Acids on Pt(III). A Voltammetric and in situ FTIR Study. Part II. Serine and Alanine on Pt(III). *J. Electroanal. Chem.* **1997**, *431*, 269-275.

Hunt, J.V.; J.A. Simpson; and R.T. Dean. Hydroperoxide-Mediated Fragmentation of Proteins. *Biochem. J.* **1988**, *250*, 87-93.

Hunter, E.P.L.; M.F. Desrosiers; and M.G. Simic. The Effect Of Oxygen, Antioxidants, and Superoxide Radical on Tyrosine Phenoxyl Radical Dimerization. *Free Radic. Biol. Med.* **1989**, *6*, 581-585.

Huss, H.H. Quality and Quality Changes in Fresh Fish--4. Chemical Composition, FAO Fisheries Technical Papers, 1995, www.fao.org/docrep/v7180e/v7180e05.htm.

Ichihashi, K.; T. Osawa; S. Toyokuni; and K. Uchida. Endogenous Formation of Protein Adducts with Carcinogenic Aldehydes. *J. Biol. Chem.* **2001,** *276,* 23903-23913.

Iio, T.; and K. Yoden. Formation of Fluorescent Substances from Degradation Products of Methyl Linoleate Hydroperoxides with Amino Compounds. *Lipids* **1988,** *23,* 1069-1072.

Ingold, K.U. *Structure and Mechanism in Organic Chemistry, 2nd edition,* Cornell University Press: Ithaca, NY, 1969,pp. 457-463.

Itakura, K.; and K. Uchida. Evidence that Malondialdehyde-Derived Aminoenimine Is Not a Fluorescent Age Pigment. *Chem. Res. Toxicol.* **2001,** *14,* 473-475.

Iyer, R.S.; S. Ghosh; and R.G. Salomon. Levuglandin E2 Crosslinks Proteins. *Prostaglandins* **1989,** *37,* 471-480.

Januszewski, A.S.; N.L. Alderson; A.J. Jenkins; S.R. Thorpe; and J.W. Baynes. Chemical Modification of Proteins during Peroxidation of Phospholipids. *J. Lipid Res.* **2005,** *46,* 1440-1449.

Jirousek, M.R.; K.K. Murthi; and R.G. Salomon. Electrophilic Levuglandin $E_2$-Protein Adducts Bind Glycine: A Model for Protein Cross-Linking. *Prostaglandins* **1990,** *40,* 187-203.

Kagan, V.E. Lipid Peroxidation. In *Biomembranes,* Ed.; CRC Press: Boca Raton, FL, 1988, pp. 13-54.

Kanazawa, K.; and H. Ashida. Target Enzymes on Hepatic Dysfunction Caused by Dietary Products of Lipid Peroxidation. *Arch. Biochem. Biophys.* **1991,** *288,* 71-78.

Kanazawa, K.; and H. Ashida. Dietary Hydroperoxides of Linoleic Acid Decompose to Aldehydes in Stomach Before Being Absorbed into the Body. *Biochim. Biophys. Acta* **1998,** *1393,* 349-361.

Kanazawa, K.; H. Ashida; and M. Natake. Autoxidizing Process Interaction of Linoleic Acid with Casein. *J. Food Sci.* **1987,** *52,* 475-479.

Kanazawa, K.; G. Danno; and M. Natake. Lysozyme Damage Caused by Secondary Degradation Products during the Autoxidation Process of Linoleic Acid. *J. Nutr. Sci. Vitaminol. (Tokyo)* **1975,** *21,* 373-382.

Kapphahn, R.J.; B.M. Giwa; K.M. Berg; H. Roehrich; X. Feng; T.W. Olsen; and D.A. Ferrington. Retinal Proteins Modified by 4-Hydroxynonenal: Identification of Molecular Targets. *Exp. Eye Res.* **2006,** *83,* 165-175.

Karel, M.; K.M. Schaich; and R.B. Roy. Interaction of Peroxidizing Methyl Linoleate with Some Proteins and Amino Acids. *J. Agric. Food Chem.* **1975,** *23,* 159-163.

Kato, Y.; Y. Mori; Y. Makino; S. Morimitsu; S. Hiroi; T. Ishikawa; T. Osawa. Formation of $N^{\varepsilon}$-(hexanonyl)lysine in protein exposed to lipid hydroperoxide. A plausible marker for lipid hydroperoxide-derived protein modification. *J. Biol. Chem.* **1999,** *274,* 20406-20414.

Kato, Y.; K. Uchida; and S. Kawakishi. Oxidative Fragmentation of Collagen and Prolyl Peptide by Cu(II)/$H_2O_2$: Conversion of Proline Residue to 2-Pyrrolidone. *J. Biol. Chem.* **1992,** *267,* 23646-23651.

Kawai, Y.; Y. Kato; H. Fujii; Y. Makino; Y, Mori, Y.; M. Naito, and T. Osawa. Immunochemical detection of a novel lysine adduct using an antibody to linoleic acid hydroperoxide-modified protein. *J. Lipid Res.* **2003,** *44,* 1124-1131.

Kehrer, J.P.; and S.S. Biswal. The Molecular Effects of Acrolein. *Toxicol. Sci.* **2000,** *57,* 6-15.

Keller, J.N.; R.J. Mark; A.J. Bruce; E. Blanc; J.D. Rothstein; K. Uchida; G. Waeg; and M.P. Mattson. 4-Hydroxynonenal, an Aldehydic Product of Membrane Lipid Peroxidation, Impairs Glutamate Transport and Mitochondrial Function in Synaptosomes. *Neuroscience* **1997,** *80,* 685-696.

Kikugawa, K. Examination of the Extraction Methods and Re-Evaluation of Blue Fluorescence Generated in Rat Tissues in situ. *Biol. Pharm. Bull.* **1994**, *17*, 9-15.

Kikugawa, K.; and M. Beppu. Involvement of Lipid Oxidation Products in the Formation of Fluorescent and Cross-Linked Proteins. *Chem. Phys. Lipids* **1987**, *44*, 277-296.

Kikugawa, K.; M. Beppu; T. Kato; S. Yamaki; and H. Kasai. Accumulation of Autofluorescent Yellow Lipofuschin in Rat Tissues Estimated by Sodium Dodecylsulfate Extraction. *Mech. Ageing Dev.* **1994**, *74*, 135-148.

Kikugawa, K.; and Y. Ido. Studies on Peroxidized Lipids. V. Formation and Characterization of 1,4-Dihydropyridine-3,5-Dicarbaldehydes as Model of Fluorescent Components in Lipofuschin. *Lipids* **1984**, *19*, 600-608.

Kikugawa, K.; Y. Ido; and J. Mikami. Studies on Peroxidized Lipids. VI. Fluorescent Products Derived from the Reaction of Primary Amines, Malondialdehyde, and Monofunctional Aldehydes. *J. Am. Oil Chem. Soc.* **1984**, *61*, 1574-1581.

Kikugawa, K.; K. Takayanagi; and S. Watanabe. Polylysines Modified with Malonaldehyde, Hydroxylinoleic Acid and Monofunctional Aldehydes. *Chem. Pharm. Bull.* **1985**, *33*, 5437-5444.

Kikugawa, K.; T. Kato; A. Hayasaka. Formation of Dityrosine and Other Fluorescent Amino Acids by Reaction of Amino Acids with Lipid Hydroperoxides. *Lipids* **1991**, *26* (11), 922-92.

Kim, J.G.; W.R. Taylor; and S. Parthasarathy. Demonstration of the Presence of Lipid Peroxide-Modified Proteins in Human Atherosclerotic Lesions Using a Novel Lipid Peroxide-Modified Anti-Peptide Antibody. *Atherosclerosis* **1999**, *143*, 335-340.

King, A.J.; and S.J. Li. Association of Malonaldehyde with Rabbit Myosin Subfragment 1. In *Quality Attributes of Muscle Foods*, Y.L. Xiong, F. Shahidi, and C.T. Ho Eds.; Kluwer Academic/Plenum Publishing: New York, 1999, pp. 277-286.

Kowalik-Jankowska, T.; M. Ruta; K. Winiewska; L. Lankiewicz; and M. Dyba. Products of Cu(II)-Catalyzed Oxidation in the Presence of Hydrogen Peroxide of the 1–10, 1–16 Fragments of Human and Mouse ß-Amyloid Peptide. *J. Inorg. Biochem.* **2004**, *98*, 940-950.

Krogull, M.K.; and O. Fennema. Oxidation of Tryptophan in the Presence of Oxidizing Methyl Linoleate. *J. Agric. Food Chem.* **1987**, *35*, 66-70.

Laurora, S.; E. Tamagno; F. Briatore; P. Bardini; S. Pizzimenti; C. Toaldo; P. Reffo; P. Costelli; M.U. Dianzani; O. Danni; et al. 4-Hydroxynonenal Modulation of p53 Family Gene Expression in the SK-N-BE Neuroblastoma Cell Line. *Free Radic. Biol. Med.* **2005**, *38*, 215-225.

Lea, C.H. Deteriorative Reactions Involving Phospholipids and Lipoproteins. *J. Sci. Food Agric.* **1957**, *8*, 1-14.

Leake, L.; and M. Karel. Nature of Fluorescent Compounds Generated by Exposure of Protein to Oxidizing Lipids. *J. Food Biochem.* **1985**, *9*, 117-136.

Leaver, J.; A.J.R. Law; and E.Y. Brechany. Covalent Modification of Emulsified β-Casein Resulting from Lipid Peroxidation. *J. Colloid Interface Sci.* **1999a**, *210*, 207-214.

Leaver, J.; A.J.R. Law; E.Y. Brechany; and C.H. McCrae. Chemical Changes in β-Lactoglobulin Structure during Ageing of Protein-Stabilized Emulsions. *Int. J. Food Sci. Nutr.* **1999b**, *34*, 503-508.

Lederer, M.O. Reactivity of Lysine Moieties Toward γ-Hydroxy-α,β-Unsaturated Epoxides: A Model Study on Protein-Lipid Oxidation Product Interaction. *J. Agric. Food Chem.* **1996**, *44*, 2531-2537.

Lederer, M.O.; A. Schuler; and M. Ohmenhäuser. Reactivity of Lysine Moieties Toward an Epoxyhydroxylinoleic Acid Derivative: Aminolysis vs Hydrolysis. *J. Agric. Food Chem.* **1998,** *47,* 4611-4620.

Lee, S.H.; T. Oe; and I.A. Blair. Vitamin C-Induced Decomposition of Lipid Hydroperoxides to Endogeneous Genotoxins. *Science* **2001,** *292,* 2083-2086.

Lee, S.H.; T. Oe; and I.A. Blair. 4,5-Epoxy-2(*E*)-Decenal-Induced Formation of 1,*N*(6)-Etheno-2'-Deoxyadenosine and 1,*N*(2)-Etheno-2'-Deoxyguanosine Adducts. *Chem. Res. Toxicol.* **2002,** *15,* 300-304.

Leonarduzzi, G.; M.C. Arkan; H. Basaga; E. Chiarpotto; A. Sevanian; and G. Poli. Lipid Oxidation Products in Cell Signaling. *Free Radic. Biol. Med.* **2000,** *28,* 1370-1378.

Lewis, S.E.; and E.D. Wills. Destruction of Sulfhydryl Groups of Proteins and Amino Acids by Peroxides of Unsaturated Fatty Acids. *Biochem. Pharmacol.* **1962,** *11,* 901-912.

Li, G.; Y. Liao; X. Wang; S. Sheng; and D. Yin. In situ Estimation of the Entire Color and Spectra of Age Pigment-Like Materials: Application of a Front-Surface 3D-Fluorescence Technique. *Exp. Gerontol.* **2006,** *41,* 328-336.

Li, S.J.; and A.J. King. Structural Changes of Rabbit Myosin Subfragment 1 Altered by Malonaldehyde, a Byproduct of Lipid Oxidation. *J. Agric. Food Chem.* **1999,** *47,* 3124-3129.

Lin, D.; H. Lee; Q. Liu; G. Perry; M.A. Smith; and L.M. Sayre. 4-Oxo-2-Nonenal Is Both More Neurotoxic and More Protein Reactive Than 4-Hydroxy-2-Nonenal. *Chem. Res. Toxicol.* **2005,** *18,* 1219-1231.

Lion, Y.; M. Kuwabara; and P. Riesz. Spin-Trapping and ESR Studies of the Direct Photolysis of Aromatic Amino Acids, Dipeptides, Tripeptides and Polypeptides in Aqueous Solutions. I. Phenylalanine and Related Compounds. *Photochem. Photobiol.* **1981,** *34,* 297-307.

Little, C.; and P.J. O'Brien. Products of Oxidation of a Protein Thiol Group After Reaction with Various Oxidizing Agents. *Arch. Biochem. Biophys.* **1967,** *122,* 406-410.

Little, C.; and P.J. O'Brien. The Effectiveness of a Lipid Peroxide in Oxidizing Protein and Non-Protein Thiols. *Biochem. J.* **1968,** *106,* 419-423.

Liu, Q.; M.A. Smith; J. Avilá; J. DeBernardis; M. Kansal; A. Takeda; X. Zhu; A. Nunomura; K. Honda; P.I. Moreira; et al. Alzheimer-Specific Epitopes of Tau Represent Lipid Peroxidation-Induced Conformations. *Free Radic. Biol. Med.* **2005,** *38,* 746-754.

Liu, W.; and J.-Y. Wang. Modifications of Protein by Polyunsaturated Fatty Acid Ester Peroxidation Products. *Biochim. Biophys. Acta* **2005,** *1752,* 93-98.

Liu, Y.; G. Sun; A. David; and L.M. Sayre. Model Studies on the Metal-Catalyzed Protein Oxidation: Structure of a Possible His-Lys Cross-Link. *Chem. Res. Toxicol.* **2004,** *17,* 110-118.

Liu, Z.; P.E. Minkler; and L.M. Sayre. Mass Spectroscopic Characterization of Protein Modification by 4-Hydroxy-2-(*E*)-Nonenal and 4-Oxo-2-(*E*)-Nonenal. *Chem. Res. Toxicol.* **2003,** *16,* 901-911.

Liu, Z.; and L.M. Sayre. Model Studies on the Modification of Proteins by Lipoxidation-Derived 2-Hydroxyaldehydes. *Chem. Res. Toxicol.* **2003,** *16,* 232-241.

Lqari, H.; J. Pedroche; J. Giron-Calle; J. Vioque; and F. Millan. Interaction of Lupinus angustifolius L. α and γ Conglutins with 13-Hydroperoxide-11,9-Octadienoic Acid. *Food Chem.* **2003,** *80,* 517-523.

Luxford, C.; R.T. Dean; and M.J. Davies. Radicals Derived from Histone Hydroperoxides Damage Nucleobases in RNA and DNA. *Chem. Res. Toxicol.* **2000,** *13,* 665-672.

Lynch, M.P.; and C. Faustman. Effect of Aldehyde Lipid Oxidation Products on Myoglobin.

*J. Agric. Food Chem.* **2000**, *48*, 600-604.
Maga, J.A. Pyridines in Foods. *J. Agric. Food Chem.* **1981**, *29*, 895-898.
Mahley, R.W.; T.L. Innerarity; R.E. Pitas; K.H. Weisgraber; J.H. Brown; and E. Gross. Inhibition of Lipoprotein Binding to Cell Surface Receptors of Fibroblasts Following Selective Modification of Arginyl Residues in Arginine-Rich and B Apoproteins. *J. Biol. Chem.* **1977**, *252*, 7279-7287.
Malshet, M.G.; and A.L. Tappel. Fluorescent Products of Lipid Peroxidation. I. Structural Requirement for Fluorescence in Conjugated Schiff Bases. *Lipids* **1973**, *8*, 194-198.
Mason, R.P. Using Anti-5,5'-Dimethyl-1-[Pyrroline *N*-Oxide] (Anti-DMPO) to Detect Protein Radicals in Time and Space with Immuno-Spin Trapping. *Free Radic. Biol. Med.* **2004**, *36*, 1214-1223.
Matoba, T.; O. Kurita; and D. Yonezawa. Changes in Molecular Size and Chemical Properties of Gelatin Caused by the Reaction with Oxidizing Methyl Linoleate. *Agric. Biol. Chem.* **1984a**, *48*, 2633-2638.
Matoba, T.; D. Yonezawa; B.M. Nair; and M. Kito. Damage of Amino Acid Residues of Proteins after Reaction with Oxidizing Lipids: Estimation by Proteolytic Enzymes. *J. Food Sci.* **1984b**, *49*, 1082-1084.
Matsushita, S. Specific Interactions of Linoleic Acid Hydroperoxides and Their Secondary Oxidation Products with Enzyme Proteins. *J. Agric. Food Chem.* **1975**, *23*, 150-154.
Matsushita, S.; M. Kobayashi; and Y. Nitta. Inactivation of Enzymes by Linoleic Acid Hydroperoxides and Linoleic Acid. *Agric. Biol. Chem.* **1970**, *34*, 817-824.
Matysik, J.; P.S. Alia; B. Bhalu; and P. Mohanty. Molecular Mechanisms of Quenching of Reactive Oxygen Species by Proline under Stress in Plants. *Current Sci.* **2002**, *82* (5), 525-532.
McCall, M.R.; J.Y. Tang; J.K. Bielicki; and T.M. Forte. Inhibition of Lecithin:Cholesterol Acyl Transferase and Modification of HDL Apolipoproteins by Aldehydes. *Arterioscler. Thromb. Vasc. Biol.* **1995**, *15*, 1599-1605.
McCay, P.B. Studies on Microsomal Phospholipids That Inhibit Gulonolactone Oxidase. *J. Biol. Chem.* **1966**, *241*, 2333-2339.
McKnight, R.C.; and F.E. Hunter, Jr. Mitochondrial Membrane Ghosts Produced by Lipid Peroxidation Induced by Ferrous Ion. II. Composition and Enzymatic Activity. *J. Biol. Chem.* **1966**, *241*, 2757-2765.
McMurray, J. *Organic Chemistry*, Brooks/Cole: Pacific Grove, CA, 2000, pp.718-724, 770-774, 955-960.
Menzel, D.B. Reaction of Oxidizing Lipids with Ribonuclease. *Lipids* **1967**, *2*, 83-84.
Mickel, H.S.; E.L. Foulds, Jr.; and D.A. Clark. Inhibition of the Plasma Lecithin-Cholesterol Acytransferase Reaction by Hydrogen Peroxide and Peroxidized Lecithin. *Lipids* **1972**, *7*, 121-124.
Mitchell, D.Y.; and D.R. Petersen. Inhibition of Rat Hepatic Mitochondrial Aldehyde Dehydrogenase Mediated Acetaldehyde Oxidation by trans-4-Hydroxy-2-Nonenal. *Hepatology* **1991**, *13*, 728-734.
Moll, T.S. Characterization of the Reactivity, Regioselectivity, and Stereoselectivity of the Reactions of Butadiene Monoxide with Valinamide and the N-Terminal Valine of Mouse and Rat Hemoglobin. *Chem. Res. Toxicol.* **1999**, *12*, 679-689.
Moll, T.S.; A.C. Harms; and A.A. Elfarra. A Comprehensive Structural Analysis of Hemoglobin Adducts Formed after in vitro Exposure of Erythrocytes to Butadiene Monoxide. *Chem. Res. Toxicol.* **2000**, *13*, 1103-1113.

Montine, T.J.; V. Amarnath; M.E. Martin; W.J. Strittmatter; and D.E. Graham. (*E*)-4-Hydroxy-2-Nonenal Is Cytotoxic and Crosslinks Cytoskeletal Proteins in P19 Neuroglial Cultures. *Am. J. Pathol.* **1996,** *148,* 89-93.

Morzel, M.; P. Gatellier; T. Sayd; M. Renerre; and E. Laville. Chemical Oxidation Decreases Proteolytic Susceptibility of Skeletal Muscle Myofibrillar Proteins. *Meat Science* **2006,** *73,* 536-543.

Musatov, A.; C.A. Carroll; Y.C. Liu; G.I. Henderson; S.T. Weintraub; and N.C. Robinson. Identification of Bovine Heart Cytochrome c Oxidase Subunit Modified by the Lipid Peroxidation Product 4-Hydroxy-2-Nonenal. *Biochemistry* **2002,** *41,* 8212-8220.

Nadkarni, D.V.; and L.M. Sayre. Structural Definition of Early Lysine and Histidine Adduction Chemistry of 4-Hydroxynonenal. *Chem. Res. Toxicol.* **1995,** *8,* 284-291.

Nair, V.; R.J. Offerman; G.A. Turner; A.N. Pryor; and N.C. Baenziger. Fluorescent 1,4-Dihydropyridines--The Malondialdehyde Connection. *Tetrahedron* **1988,** *44,* 2793-2803.

Nair, V.; D.E. Vietti; and C.S. Cooper. Degenerative Chemistry of Malonaldehyde: Structure, Stereochemistry, and Kinetics of Formation of Enaminals from Reaction with Amino Acids. *J. Am. Chem. Soc.* **1981,** *103,* 3030-3036.

Narayan, K.A.; and F.A. Kummerow. Oxidized Fatty Acid-Protein Complexes. *J. Am. Oil Chem. Soc.* **1958,** *35,* 52-56.

Nazir, H.; M. Yildiz; H. Yilmaz; M.N. Tahir; and D. Ülkü. Intramolecular Hydrogen Bonding and Tautomerism in Schiff bases. Structure of N-(2-Pyridil)-2-Oxo-1-Naphthylidenemethylamine. *J. Mol. Struct.* **2000,** *524,* 241-250.

Nielsen, H.K. Covalent Binding of Cardiolipin to γ-Globulin and Albumin Depends on Peroxidation. *Chem. Phys. Lipids* **1978,** *22,* 339-340.

Nielsen, H.K.; P.A. Finot; and R.F. Hurrell. Reactions of Proteins with Oxidizing Lipids. 2. Influence on Protein Quality and on the Bioavailability of Lysine, Methionine, Cyst(e)ine and Tryptophan as Measured in Rat Assays., *Br. J. Nutr.* **1985a,** *53,* 75-86.

Nielsen, H.K.; J. Löliger; and R.F. Hurrell. Reactions of Proteins with Oxidizing Lipids. 1. Analytical Measurements of Lipid Oxidation and of Amino Acid Losses in a Whey Protein-Methyl Linolenate Model System. *Br. J. Nutr.* **1985b,** *53,* 61-73.

Nishida, T.; and F.A. Kummerow. Interaction of Serum Lipoproteins with the Hydroperoxide of Methyl Linoleate. *J. Lipid Res.* **1960,** *1,* 450-458.

Obanu, Z.A.; D.A. Ledward; and R.A. Lawrie. Lipid Protein Interactions as Agents of Quality Deterioration in Intermediate Moisture Meats: An Appraisal. *Meat Science* **1980,** *4,* 79-88.

O'Brien, P.J.; and A.C. Frazer. The Effect of Lipid Peroxides on the Biochemical Constituents of the Cell. *Proc. Nutr. Soc.* **1966,** *25,* 9-18.

Ooizumi, T. Effect of Peroxidized Lipid as an Oxidant on Biochemical Properties of Fish Myofibrillar Proteins. *Erisorubinsan no Kenkyu* **1999,** *5,* 1-7.

Ottolenghi, A.; F. Bernheim; and K.M. Wilbur. The Inhibition of Certain Mitochondrial Enzymes by Fatty Acids Oxidized by Ultraviolet Light or Ascorbic Acid. *Arch. Biochem. Biophys.* **1955,** *56,* 157-164.

Page, S.; C. Fischer; B. Baumgartner; M. Haas; U. Kreusel; G. Loidl; M. Hayni; H.W.L. Ziegler-Heitbrock; D. Neumeier; and K. Brand. 4-Hydroxynonenal Prevents NF-κB Activation and Tumor Necrosis Factor Expression by Inhibiting I-κB Phosphorylation and Subsequent Proteolysis. *J. Biol. Chem.* **1999,** *274,* 11611-11618.

Pamplona, R.; E. Dalfó; V. Ayala; M.J. Bellmunt; J. Prat; I. Ferrer; and M. Portero-Otín. Proteins in Human Brain Cortex Are Modified by Oxidation, Glycoxidation, and Lipoxidation. *J. Biol. Chem.* **2005**, *280*, 21522-21530.

Patterson, L.K.; and K. Hasegawa. Pulse Radiolysis Studies in Model Lipid Systems. The Influence of Aggregation on Kinetic Behavior of OH Induced Radicals in Aqueous Sodium Linoleate. *Ber. Bunsenges. Phys. Chem.* **1978**, *82*, 951-956.

Petersen, D.R.; and J.A. Doorn. Reactions of 4-Hydroxynonenal with Proteins and Cellular Targets. *Free Radic. Biol. Med.* **2004**, *37*, 937-945.

Pietzsch, J. Measurement of 5-Hydroxy-2-Aminovaleric Acid as a Specific Marker of Iron-Mediated Oxidation of Proline and Arginine Side-Chain Residues of Low-Density Lipoprotein Apolipoprotein B-100. *Biochem. Biophys. Res. Commun.* **2000**, *270*, 852-857.

Player, T.J.; and H.O. Hultin. The Effect of Lipid Peroxidation on the Calcium-Accumulating Ability of the Microsomal Fraction Isolated from Chicken Breast Muscle. *Biochem. J.* **1978**, *174*, 17-22.

Podrez, E.A.; E. Poliakov; Z. Shen; R. Zhang; Y. Deng; M. Sun; P.J. Finton; L. Shan; M. Febbraio; D.P. Hajjar; et al. A Novel Family of Atherogenic Oxidized Phospholipids Promotes Macrophage Foam Cell Formation Via the Scavenger Receptor CD36 and Is Enriched in Atherosclerotic Lesions. *J. Biol. Chem.* **2002a**, *277*, 38517-38523.

Podrez, E.A.; E. Poliakov; Z. Shen; R. Zhang; Y. Deng; M. Sun; P.J. Finton; L. Shan; B. Gugiu; P.L.H. Fox; et al. Identification of a Novel Family of Oxidized Phospholipids That Serve as Ligands for the Macrophage Scavenger Receptor CD36. *J. Biol. Chem.* **2002b**, *277*, 38503-38516.

Pokorny, J.; S. Klein; and J. Koren. Reaction of Oxidized Lipids with Protein. II. Reactions of Albumin with Epoxy Derivatives. *Nahrung* **1966**, *10*, 321-325.

Pokorny, J.; M. Kminek; W. Janitz; E. Novotna; and J. Davidek. Reactions of Oxidized Lipids with Protein. Part 13. Autoxidation of Hexanal in Presence of Nonlipid Substances. *Die Nahrung* **1985**, *29*, 459-465.

Prescott, S.M. A Thematic Series on Oxidation of Lipids as a Source of Messengers. *J. Biol. Chem.* **1999**, *274*, 22901.

Pryor, W.A. *Free Radicals*, McGraw-Hill: New York, 1966. pp. 118-126.

Ran, C. Synthesis and Protein Reactivity of 2*E*,4*E*,6*E*,-Dodecatrienal. *Tetrahedron Lett.* **2004**, *45*, 7851-7853.

Refsgaard, H.H.F.; L. Tsai; and E.R. Stadtman. Modifications of Proteins by Polyunsaturated Fatty Acid Peroxidation Products. *Proc. Nat. Acad. Sci. USA* **2000**, *97*, 611-616.

Requena, J.R.; C.-C. Chao; R.L. Levine; E.R. Stadtman. Glutamic and Aminoadipic Semialdehydes are the Main Carbonyl Products of Metal-Catalyzed Oxidation of Proteins. *Proc. Nat. Acad. Sci. USA* **2001** *98*, 69-74.

Reubsaet, J.L.E.; J.H. Beijnen; A. Bult; R.J. van Maanen; J.A.D. Marchal; and W.J.M. Underberg. Analytical Techniques Used to Study the Degradation of Proteins and Peptides: Chemical Instability. *J. Pharm. Biomed. Anal.* **1998**, *17*, 955-978.

Riley, M.; and J.J. Harding. The Reaction of Malonaldehyde with Lens Proteins and the Protective Effect of Aspirin. *Biochim. Biophys. Acta* **1993**, *1158*, 107-112.

Roubal, W.T. Trapped Radicals in Dry Lipid-Protein Systems Undergoing Oxidation. *J. Am. Oil Chem. Soc.* **1970**, *47*, 141-144.

Roubal, W.T.; and A.L. Tappel. Damage to Proteins, Enzymes, and Amino Acids by Peroxidizing Lipids. *Arch. Biochem. Biophys.* **1966a**, *113*, 5-8.

Roubal, W.T.; and A.L. Tappel. Polymerization of Proteins Induced by Free-Radical Lipid Peroxidation. *Arch. Biochem. Biophys.* **1966b**, *113*, 150-155.

Roy, R.B. and M. Karel. Reaction Products of Histidine with Autoxidized Methyl Linoleate. *J.*

*Food Sci.* **1973**, *38* (5), 896-897.

Saeed, S.; S.A. Fawthrop; and N. Howell. Electron Spin Resonance (ESR) Study on Free Radical Transfer in Fish Lipid-Protein Interaction. *J. Sci. Food Agric.* **1999**, *79*, 1809-1816.

Saeed, S.; D. Gillies; G. Wagner; and N.K. Howell. ESR and NMR Spectroscopy Studies on Protein Oxidation and Formation of Dityrosine in Emulsions Containing Oxidised Methyl Linoleate. *Food Chem. Toxicol.* **2006**, *44*, 1385-1392.

Salomon, R.G.; K. Kaur; E. Podrez; H.F. Hoff; A.V. Krushinsky; and L.M. Sayre. HNE-Derived 2-Pentylpyrroles Are Generated during Oxidation of LDL, Are More Prevalent in Blood Plasma from Patients with Renal Disease or Atherosclerosis, and Are Present in Atherosclerotic Plaques. *Chem. Res. Toxicol.* **2000**, *13*, 557-564.

Sánchez-Vioque, R.; J. Vioque; A. Clemente; J. Pedroche; J. Bautista; and F. Millán. Interaction of Chickpea (*Cicer arietinum* L.) Legumin with Oxidized Linoleic Acid. *J. Agric. Food Chem.* **1999**, *47*, 813-818.

Sayre, L.M.; P.K. Arora; R.S. Iyer; and R.G. Salomon. Pyrrole Formation from 4-Hydroxynonenal and Primary Amines. *Chem. Res. Toxicol.* **1993**, *6*, 19-22.

Sayre, L.M.; D.A. Zelasko; P.L. Harris; G. Perry; R.G. Salomon; and M.A. Smith. 4-Hydroxynonenal-Derived Advanced Lipid Peroxidation End Products Are Increased in Alzheimer's Disease. *J. Neurochem.* **1997**, *68*, 2092-2097.

Schaich, K.M. Free Radical Initiation in Proteins and Amino Acids by Ionizing and Ultraviolet Radiations and Lipid Oxidation--Part I: Ionizing Radiation. *CRC Crit. Rev. Food Sci. Nutr.* **1980a**, *13*, 89-129.

Schaich, K.M. Free Radical Initiation in Proteins and Amino Acids by Ionizing and Ultraviolet Radiations and Lipid Oxidation--Part III: Free Radical Transfer from Oxidizing Lipids. *CRC Crit. Rev. Food Sci. Nutr.* **1980b**, *13*, 189-244.

Schaich, K.M. EPR Methods for Detecting and Identifying Free Radicals in Foods. In *Free Radicals in Foods: Chemistry, Nutrition, and Health*, C.T. Ho, and M. Moreno, M. Eds.; American Chemical Society: Washington, D.C., 2002a, pp. 12-34.

Schaich, K.M. Free Radical Generation during Extrusion: A Critical Contributor to Texturization. In *Free Radicals in Foods: Chemistry, Nutrition, and Health*, C.T. Ho, and M. Moreno, Eds.; American Chemical Society: Washington, D.C., 2002b, pp. 35-48.

Schaich, K.M. NO Production during Thermal Processing of Beef: Evidence for Protein Oxidation. In *Free Radicals in Foods: Chemistry, Nutrition, and Health*, C.T. Ho, and M. Morello, Eds.; American Chemical Society: Washington, D.C., 2002c, pp. 151-161.

Schaich, K.M. A Spin Label Study of Water Binding and Protein Mobility in a Lysozyme Model System. In *Free Radicals in Foods: Chemistry, Nutrition, and Health*, C.T. Ho, and M. Moreno, Eds.; American Chemical Society: Washington, D.C., 2002d, pp. 98-113.

Schaich, K.M. Lipid Oxidation in Fats and Oils: An Integrated View. In *Bailey's Industrial Fats and Oils, 6th Edn.*, F. Shahidi, Ed.; John Wiley: New York, 2005, pp. 2681-2767.

Schaich, K.M.; and M. Karel. Free Radicals in Lysozyme Reacted with Peroxidizing Methyl Linoleate. *J. Food Sci.* **1975**, *40*, 456-459.

Schaich, K.M.; and M. Karel. Free Radical Reactions of Peroxidizing Lipids with Amino Acids and Proteins: An ESR Study. *Lipids* **1976**, *11*, 392-400.

Schauenstein, E.; M. Taufer; H. Esterbauer; A. Kylianek; and T. Seelich. Reaction of Protein -SH Groups with 4-Hydroxy-2-Nonenal. *Monatsch. Chem.* **1971**, *102*, 517-529.

Schaur, R.J. Basic Aspects of the Biochemical Reactivity of 4-Hydroxynonenal. *Mol. Aspects Med.* **2003**, *24*, 149-159.

Schmolka, I.R.; and P.E. Spoerri. Thiazolidine Chemistry. II. The Preparation of 2-Substituted

Thiazolidine-4-Carboxylic Acids. *J. Org. Chem.* **1957**, *22*, 943-946.

Schöneich, C.; and V.S. Sharov. Mass Spectrometry of Protein Modifications by Reactive Oxygen and Nitrogen Species. *Free Radic. Biol. Med.* **2006**, *41*, 1507-1520.

Schuessler, H.; K. Schilling. Oxygen Effect in the Radiolysis of Proteins. *Int. J. Radiat. Biol.* **1984**, *45*, 267-281.

Scofield, R.H.; B.T. Kurien; S. Ganick; M.T. McClain; Q. Pye; J.A. Kames; R.I. Schneider; R.H. Broyles; M. Bachmann; and K. Hensley. Modification of Lupus-Associated 60-kDa Ro Protein with the Lipid Oxidation Product 4-Hydroxy-2-Nonenal Increases Antigenicity and Facilitates Epitope Spreading. *Free Radic. Biol. Med.* **2005**, *38*, 719-728.

Shi, H.; S. Shen; H. Suna; Z. Liua; and L. Lia. Oxidation of L-Serine and L-Threonine by *bis*(Hydrogen Periodato)Argentate(III) Complex Anion: A Mechanistic Study. *J. Inorg. Biochem.* **2007**, *101*, 165-172.

Shimasaki, H.; N. Ueta; and O.S. Privett. Covalent Binding of Peroxidized Linoleic Acid to Protein and Amino Acids as Models for Lipofuschin Formation. *Lipids* **1982**, *17*, 878-883.

Shin, B.C.; J.W. Huggins; and K.L. Carraway. Effects of pH, Concentration and Aging on the Malonaldehyde Reaction with Protein. *Lipids* **1972**, *7*, 229-233.

Shoukry, M.I.; E.L. Gong; and A.V. Nichols. Apolipoprotein-Lipid Association in Oxidatively Modified HDL and LDL. *Biochim. Biophys. Acta* **1994**, *1210*, 355-360.

Siems, W.; and T. Grune. Intracellular Metabolism of 4-Hydroxynonenal. *Mol. Aspects Med.* **2003**, *24*, 167-175.

Simpson, J.A.; S. Narita; S. Gieseg; S. Gebicki; and J.M. Gebicki. Long-lived Reactive Species on Free-Radical-Damaged Proteins. *Biochem. J.* **1992**, *282*, 621-624.

Slatter, D.A.; M. Murray; and A.J. Bailey. Formation of a Dihydropyridine Derivative as a Potential Cross-Link Derived from Malondialdehyde in Physiological Systems. *FEBS Lett.* **1998**, *421*, 180-184.

Smith, D.M.; S.H. Noormarji; J.F. Price; M.R. Bennink; and T.J. Herald. Effect of Lipid Oxidation on the Functional and Nutritional Properties of Washed Chicken Myofibrils Stored at Different Water Activities. *J. Agric. Food Chem.* **1990**, *38*, 1307-1312.

Soreghan, B.A.; F. Yang; S.N. Thomas; J. Hsu; and A.J. Yang. High-Throughput Proteomic-Based Identification of Oxidatively Induced Protein Carbonylation in Mouse Brain. *Pharm. Res.* **2003**, *20*, 1713-1720.

Soszylqski, M.; A. Filipiak; G. Bartosz; and J.M. Gebicki. Effect of Amino Acid Peroxides on the Erythrocyte. *Free Radic. Biol. Med.* **1996**, *20*, 45-51.

Soyer, A.; and H.O. Hultin. Kinetics of Oxidation of the Lipids and Proteins of Cod Sarcoplasmic Reticulum. *J. Agric. Food Chem.* **2000**, *48*, 2127-2134.

Spiteller, G. Peroxyl radicals: Inductors of neurodegenerative and other inflammatory diseases. Their origin and how they transform cholesterol, phospholipids, plamalogens, polyunsaturated fatty acids, sugars, and proteins into deleterious products. *Free Radic. Biol. Med.* **2006**, *41*, 362-387.

Stadtman, E.R. Role of Oxidant Species in Aging. *Curr. Med. Chem.* **2004**, *11*, 1105-1112.

Stadtman, E.R.; and B.S. Berlett. Reactive Oxygen-Mediated Protein Oxidation in Aging and Disease. *Chem. Res. Toxicol.* **1997**, *10*, 485-494.

Stadtman, E.R.; and R.L. Levine. Free Radical-Mediated Oxidation of Free Amino Acids and Amino Acid Residues in Proteins. *Amino Acids* **2003**, *25*, 207-218.

Stapelfeldt, H.; and L.H. Skibsted. Kinetics of Formation of Fluorescent Products from Hexanal and L-Lysine in a Two-Phase System. *Lipids* **1996**, *31*, 1125-1132.

Steinbrecher, U.P. Oxidation of Human Low Density Lipoprotein Results in Derivatization of Lysine Residues of Apolipoprotein B by Lipid Peroxide Decomposition Products. *J. Biol. Chem.* **1987,** *262,* 3603-3608.

Steinbrecher, U.P.; J.L. Witztum; S. Parthasarathy; and D. Steinberg. Decrease in Reactive Amino Groups during Oxidation or Endothelial Cell Modification of LDL. Correlation with Changes in Receptor-Mediated Catabolism. *Arterioscler. Thromb. Vasc. Biol.* **1987,** *7,* 135-143.

Suyama, K.; and A. Adachi. Origin of Alkyl-Substituted Pyridines in Food Flavor: Formation of the Pyridines from the Reaction of Alkanals with Amino Acids. *J. Agric. Food Chem.* **1980,** *28,* 546-549.

Suyama, K.; and S. Adachi. Reaction of Alkanals and Amino Acids or Primary Amines. Synthesis of 1,2,3,5- and 1,3,4,5-Substituted Quarternary Pyridinium Salts. *J. Org. Chem.* **1979,** *44,* 1417-1420.

Suyama, K.; T. Arakawa; and S. Adachi. Free Fatty Aldehydes and Their Aldol Condensation Products in Heated Meat. *J. Agric. Food Chem.* **1981,** *29,* 875-878.

Szweda, L.I.; K. Uchida; L. Tsai; and E.R. Stadtman. Inactivation of Glucose-6-phosphate Dehydrogenase by 4-hydroxy-2-nonenal. Selective Modification of an Active-site Lysine. *J. Biol. Chem.* **1993**, *268* (5), 3342-3347.

Tanczos, A.; C. Mendez; and K.M. Schaich. An Investigation of Free Radicals in Foods, Introduction to Scientific Research project report, Douglass Project for Women in Science, Technology, Engineering, and Math, Douglass College, Rutgers University, New Brunswick, NJ, 2002.

Tappel, A.L. Studies of the Mechanism of Vitamin E Action. III. In vitro Copolymerization of Oxidized Lipids with Proteins. *Arch. Biochem. Biophys.* **1955,** *54,* 266-280.

Tappel, A.L. Free-Radical Lipid Peroxidation Damage and Its Inhibition by Vitamin E and Selenium. *Fed. Proc.* **1965,** *24,* 73-78.

Tappel, A.L. Lipid Peroxidation Damage to Cell Components. *Fed. Proc.* **1973,** *32,* 1870-1874.

Thomas, P.D.; and M.J. Poznansky. Lipid Peroxidation Inactivates Liver Microsomal Glycerol-3-Phosphate Acyl Transferase. *J. Biol. Chem.* **1990,** *265,* 2684-2691.

Tiku, M.L.; R. Shah; and G.T. Allison. Evidence Linking Chondrocyte Lipid Peroxidation to Cartilage Matrix Protein Degradation. *J. Biol. Chem.* **2000,** *275,* 20069-20076.

Tsai, L.; P.A. Szweda; O. Vinogradova; and L.I. Szweda. Structural Characterization and Immunochemical Detection of a Fluorophore Derived from 4-Hydroxy-2-Nonenal and Lysine. *Proc. Nat. Acad. Sci. USA* **1998,** *95,* 7975-7980.

Tsen, C.C.; and A.L. Tappel. Oxygen Lability of Cysteine in Hemoglobin. *Arch. Biochem. Biophys.* **1958,** *75,* 243-250.

Tsuchiya, Y.; M. Yamaguchi; T. Chikuma; and H. Hojo. Degradation of Glyceraldehyde-3-Phosphate Dehydrogenase Triggered by 4-Hydroxy-2-Nonenal and 4-Hydroxy-2-Hexenal. *Arch. Biochem. Biophys.* **2005,** *438,* 217-222.

Uchida, K. Current Status of Acrolein as a Lipid Peroxidation Product. *Trends Cardiovasc. Med.* **1999,** *9,* 109-113.

Uchida, K. Role of Reactive Aldehyde in Cardiovascular Diseases. *Free Radic. Biol. Med.* **2000,** *28,* 1685-1696.

Uchida, K.; K. Itakura; H. Kawakishi; H. Hiai; S. Toyokuni; and E.R. Stadtman. Characterization of Epitopes Recognized by 4-Hydroxy-2-Nonenal Specific Antibodies. *Arch. Biochem. Biophys.* **1995,** *324,* 241-248.

Uchida, K.; M. Kanematsu; Y. Morimutsu; T. Osawa; N. Noguchi; and E. Niki. Acrolein Is a Product of Lipid Peroxidation Reaction. Formation of Free Acrolein and Its Conjugate with Lysine Residues in Oxidized Low Density Lipoproteins. *J. Biological Chemistry* **1998a**, *273*, 16058-16066.

Uchida, K.; M. Kanematsu; K. Sakai; T. Matsuda; N. Hattori; Y. Mizuno; D. Suzuki; T. Miyata; N. Noguchi; E. Niki; et al. Protein-Bound Acrolein: Potential Markers for Oxidative Stress. *Proc. Nat. Acad. Sci. USA* **1998b**, *95*, 4882-4887.

Uchida, K.; M. Shiraishi; Y. Naito; Y. Torii; Y. Nakamura; and T. Osawa. Activation of Stress Signaling Pathways by the End Product of Lipid Peroxidation. 4-Hydroxy-2-Nonenal Is a Potential Inducer of Intracellular Peroxide Production. *J. Biol. Chem.* **1999**, *274*, 2234–2242.

Uchida, K.; and E.R. Stadtman. Modification of Histidine Residues in Proteins by Reaction with 4-Hydroxynonenal. *Proc. Nat. Acad. Sci. USA* **1992**, *89*, 4544-4548.

Uchida, K.; and E.R. Stadtman. Covalent Attachment of 4-Hydroxynonenal to Glyceraldehyde-3-Phosphate Dehydrogenase: A Possible Involvement of Intramolecular and Intermolecular Crosslinking Reactions. *J. Biol. Chem.* **1993**, *268*, 6388-6393.

Uchida, K.; L.I. Szweda; H.Z. Chae; and E.R. Stadtman. Immunochemical Detection of 4-Hydroxynonenal Protein Adducts in Oxidized Hepatocytes. *Proc. Nat. Acad. Sci. USA* **1993**, *90*, 8742-8746.

Valko, M.; D. Leibfritz; J. Moncol; M.T.D. Cronin; M. Mazur; and J. Telser. Free Radicals and Antioxidants in Normal Physiological Functions and Human Disease. *Int. J. Biochem. Cell Biol.* **2007**, *39*, 44-84.

Veberg, A.; G. Vogt; and J.P. Wold. Fluorescence in Aldehyde Model Systems Related to Lipid Oxidation. *LWT* **2006**, *39*, 562-570.

Viita, H.; O. Narvanen; and S. Yla-Herttuala. Different Apolipoprotein B Breakdown Patterns in Models of Oxidized Low Density Lipoprotein. *Life Sci.* **1999**, *65*, 783-793.

Wainwright, T.; J.F McMahon; and J. McDowell. Formation of methional and methanethiol from methionine. *J. Sci. Food Agric.* **1972**, *23*, 911-914.

Weisgraber, K.H.; T.L. Innerarity; and R.W. Mahley. Role of Lysine Residues of Plasma Lipoproteins in High Affinity Binding to Cell Surface Receptors on Human Fibroblasts. *J. Biol. Chem.* **1978**, *253*, 9053-9062.

Wills, E.D. Effect of Unsaturated Fatty Acids and Their Peroxides on Enzymes. *Biochem. Pharmacol.* **1961a** *7*, 7-16.

Wills, E.D. Enzyme Inhibition by Oxidized Unsaturated Fatty Acids. In The Enzymes of Lipid Metabolism, P. Desnuelle, Ed.; Pergamon: New York, **1961b**, pp. 74-77.

Xu, G.; Y. Liu; M.M. Kansal; and L.M. Sayre. Rapid Crosslinking of Proteins by 4-Keto Aldehydes and 4-Hydroxy-Alkenals Does Not Arise from the Lysine-Derived Monoalkylpyrroles. *Chem. Res. Toxicol.* **1999a**, *12*, 855-861.

Xu, G.; Y. Liu; and L.M. Sayre. Mechanism of Protein Lysine Crosslinking by the Lipid Peroxidation Product 4-Hydroxy-2-Nonenal (HNE). *Book of Abstracts, 216th ACS National Meeting*, Boston, August 23-27, 1998. ORGN-281, American Chemical Society, Washington, D. C.

Xu, G.; Y. Liu; and L.M. Sayre. Independent Synthesis, Solution Behavior, and Studies on the Mechanism of Formation of the Primary Amine-Derived Fluorophore Representing Cross-Linking of Proteins by (E)-4-Hydroxy-2-Nonenal. *J. Org. Chem.* **1999b**, *64*, 5732-5745.

Xu, G.; Y. Liu; and L.M. Sayre. Polyclonal Antibodies to a Fluorescent 4-Hydroxy-2-Nonenal

(HNE)-Derived Lysine-Lysine Cross-Link: Characterization and Application to HNE-Treated Protein and in vitro Oxidized Low-Density Lipoprotein. *Chem. Res. Toxicol.* **2000**, *13,* 406-413.

Xu, G.; and L.M. Sayre. Structural Characterization of a 4-Hydroxy-2 Alkenal-Derived Fluorophore that Contributes to Lipoperoxidation-Dependent Protein Crosslinking in Aging and Degenerative Disease. *Chem. Res. Toxicol.* **1998**, *11,* 247-251.

Xu, G.; and L.M. Sayre. Structural Elucidation of a 2:2 4-Ketoaldehyde-Amine Adduct as a Model for Lysine-Directed Crosslinking of Proteins by 4-Ketoaldehydes. *Chem. Res. Toxicol.* **1999**, *12,* 862-868.

Yamada, S.; S. Kumazawa; T. Ishii; T. Nakayama; K. Itakura; N. Shibata; M. Kobayashi; K. Sakai; T. Osawa; and K. Uchida. Immunochemical Detection of a Lipofuschin-Like Fluorophore Derived from Malondialdehyde and Lysine. *J. Lipid Res.* **2001**, *42,* 1187-1196.

Yamaki, S.; T. Kato; and K. Kikugawa. Characteristics of Fluorescence Formed by the Reaction of Proteins with Unsaturated Aldehydes, Possible Degradation Products of Lipid Radicals. *Chem. Pharm. Bull.* **1992**, *40,* 2138-2142.

Yanagita, T.; H. Orioka; and M. Sugano. Influence of Amino Acid Supplementation to Egg Albumin Reacted with Oxidized Lipids on Liver Lipids of Rats. *Agric. Biol. Chem.* **1976**, *40,* 1751-1756.

Yanagita, T.; and M. Sugano. Effect of Egg Albumin Reacted with Oxidized Lipid on Growth and Liver Components of Rat. *J. Jpn. Soc. Food Nutr.* **1974**, *27,* 281-287.

Yanagita, T.; and M. Sugano. Liver and Plasma Lipids in Rats Fed Casein Reacted with Oxidized Ethyl Linoleate. *Agric. Biol. Chem.* **1975**, *39,* 63-69.

Yang, M.-H. Damage of DNA and Its Constituents by Oxidizing Lipids. Ph.D. thesis, Rutgers University, New Brunswick, NJ, 1993, pp. 1-91.

Yang, M.-H. and K.M. Schaich. Factors Affecting DNA Damage by Lipid Hydroperoxides and Aldehydes. *J. Free Radical Biol. Med.* **1996** *20,* 225-236.

Yildiz, M.; Z. Kiliç, and T. Hökelek. Intramolecular Hydrogen Bonding and Tautomerism in Schiff Bases. Part I. Structure of 1,8-Di[N-2-Oxyphenyl-Salicylidene]-3,6-Dioxaoctane. *J. Mol. Struct.* **1998**, *441,* 1-10.

Yong, S.H.; and M. Karel. Reaction of Histidine with Methyl Linoleate: Characterization of the Histidine Degradation Products. *J. Am. Oil Chem. Soc.* **1978**, *55* (3), 352-7.

Yong, S.H.; and M. Karel. Cleavage of the Imidazole Ring in Histidyl Residue Analogs Reacted with Peroxidizing Lipids. *J. Food Sci.* **1979**, *22* (2), 568-74.

Yong, S.H.; S. Lau; Y. Hsieh; and M. Karel. Degradation Products of L-Tryptophan Reacted with Peroxidizing Methyl Linoleate. In *Autoxidation in Foods and Biological Systems*; Simic, M.G.,Karel, M., Ed.; Plenum Press: New York, 1980, pp. 237-247.

Yuan, Q.; X. Zhu; and L.M. Sayre. Chemical Nature of Stochastic Generation of Protein-Based Carbonyls: Metal-Catalyzed Oxidation Versus Modification by Products of Lipid Oxidation. *Chem. Res. Toxicol.* **2007**, *20,* 129-139.

Zamora, R.; M. Alaiz; and F.J. Hidalgo. Modification of Histidine Residues by 4,5-Epoxy-2-Alkenals. *Chem. Res. Toxicol.* **1999**, *12,* 654-660.

Zamora, R.; M. Alaiz; and F.J. Hidalgo. Contribution of Pyrrole Formation and Polymerization to the Nonenzymatic Browning Produced by Amino-Carbonyl Reactions. *J. Agric. Food Chem.* **2000**, *48,* 3152-3158.

Zamora, R.; E. Gallardo; and F.J. Hidalgo. Amine Degradation by 4,5-Epoxy-2-Decenal in Model Systems. *J. Agric. Food Chem.* **2006**, *54,* 2398-2404.

Zamora, R.; and F.J. Hidalgo. Linoleic Acid Oxidation in the Presence of Amino Compounds

Produces Pyrroles by Carbonylamine Reactions. *Biochim. Biophys. Acta* **1995,** *1258,* 319-327.

Zamora, R.; and F.J. Hidalgo. Inhibition of Proteolysis in Oxidized Lipid-Damaged Proteins. *J. Agric. Food Chem.* **2001,** *49,* 6006-6011.

Zamora, R.; and F.J. Hidalgo. Comparative Methyl Linoleate and Methyl Linolenate Oxidation in the Presence of Bovine Serum Albumin at Several Lipid/Protein Ratios. *J. Agric. Food Chem.* **2003a,** *51,* 4661-4667.

Zamora, R.; and F.J. Hidalgo. Phosphatidylethanolamine Modification by Oxidative Stress Product 4,5(*E*)-Epoxy-2(*E*)-Heptenal. *Chem. Res. Toxicol.* **2003b,** *16,* 1632-1641.

Zamora, R.; and F.J. Hidalgo. 2-Alkylpyrrole Formation from 4,5-Epoxy-2-Alkenals. *Chem. Res. Toxicol.* **2005,** *18,* 342-348.

Zamora, R.; R.F. Millán; F.J. Hidalgo; M. Alaiz; M.P. Maza; J.M. Olías; and E. Vioque. Interaction Between the Peptide Glutathione and Linoleic Acid Hydroperoxide. *Food/Nahrung* **1989,** *33,* 283-288.

Zamora, R.; J.L. Navarro; and F.J. Hidalgo. Determination of Lysine Modification Product, ε-N-Pyrrolylnorleucine in Hydrolyzed Proteins and Trout Muscle Microsomes by Micellar Electrokinetic Capillary Chromatography. *Lipids* **1995,** *30,* 477-483.

Zhang, H.; D.A. Dickinson; R.-M. Liu; and H.J. Forman. 4-Hydroxynonenal Increases γ-Glutamyl Transpeptidase Gene Expression Through Mitogen-Activated Protein Kinase Pathways. *Free Radic. Biol. Med.* **2005,** *38,* 463-471.

Zhang, W.-H.; J. Liu; G. Xu; Q. Yuan; and L.M. Sayre. Model Studies on Protein Side Chain Modification by 4-Oxo-2-Nonenal. *Chem. Res. Toxicol.* **2003,** *16,* 512-523.

Zhu, M.; D.C. Spink; B. Yan; S. Bank; and A.P. DeCaprio. Formation and Structure of Cross-Linking and Monomeric Pyrrole Autoxidation Products in 2,5-Hexanedione-Treated Amino Acids, Peptides, and Protein. *Chem. Res. Toxicol.* **1994,** *7,* 551-558.

Zirlin, A.; and M. Karel. Oxidation Effects in a Freeze-Dried Gelatin-Methyl Linoleate System. *J. Food Sci.* **1969,** *34,* 160-164.

# 9

# Lipid Oxidation in Food Dispersions

Eric A. Decker, Wilailuk Chaiyasit, Min Hu, Habibollah Faraji, and D. Julian McClements
Department of Food Science, University of Massachusetts, Amherst, Massachusetts, 01003

## Introduction

The concentrations of polyunsaturated fatty acids in food products are increasing. This is due to the desire to improve the nutritional profile of foods by replacing atherogenic saturated fatty acids with unsaturated fatty acids and incorporating bioactive fatty acids, such as the ω-3 fatty acids, into functional foods. In addition, recent trans fatty acid labeling requirements throughout the world have resulted in a reduction of the use of hydrogenated fats leading to an increase in utilization of oils with higher levels of polyunsaturated fatty acids. Increasing the concentrations of polyunsaturated fatty acids in foods leads to increased lipid oxidation susceptibility which can result in the formation of undesirable rancid odors and flavors as well as changes in texture, color, and nutritional quality. Therefore, effective methods are needed to control lipid oxidation. These methods include oxygen removal, light exclusion, temperature reduction, and utilization of antioxidants. Unfortunately, these methods are not suitable for all food products and antioxidant additives are often label unfriendly. Therefore, new technologies are needed to control lipid oxidation in food products.

Most lipids in foods exist as food dispersions or emulsions. Food emulsions consist of oil dispersed in water (an oil-in-water emulsion) or water dispersed in oil (a water-in-oil emulsion). The dispersed phase exists as small spherical droplets ranging in size from 0.1 to 100 μm (Dickinson and Stainsby, 1982; Dickinson, 1992). The positive free energy needed to increase the surface area between the oil and water phases and the density difference between oil and water make emulsions thermodynamically unstable systems (Dickinson, 1992; McClements, 1999). To form emulsions that are kinetically stable over the shelf life of the food products (a few weeks, months, or years), emulsifiers must be utilized. Emulsifiers are molecules that have the ability to absorb to the lipid-water interface allowing them to decrease surface tension, form a physical barrier between oil and water, and impart a charge onto the emulsion droplet; all are factors that inhibit emulsion destabilization. Proteins, phospholipids, and small molecule surfactants are the most common emulsifiers used in foods.

An emulsion can be described as having three distinct regions: the droplet interior, the continuous phase, and the emulsion-droplet interface that separates the oil and water. The interface region consists mainly of the emulsifier but can also contain other surface-active molecules, such as antioxidants and lipid hydroperoxides as well as molecules that may be attracted towards a charged interface (e.g., transition metals). The properties of the emulsion-droplet interface (e.g., thickness and charge) are a function of the type and concentration of the surface-active molecules present. Lipid oxidation chemistry in emulsions is highly dependent on the properties of the emulsion-droplet interface since factors such as interfacial thickness and charge will impact the physical location of prooxidative and antioxidative factors and thus the ability of compounds to either promote or inhibit lipid oxidation. By understanding how the physical properties of emulsions impact lipid oxidation chemistry, it is possible to develop emulsion technologies to inhibit lipid oxidation.

## Lipid Oxidation in Emulsions

Lipid oxidation reactions are dependent on the chemical reactivity of numerous components including reactive oxygen species, prooxidants, and antioxidants. However, research over the past few decades has shown that the physical properties of food systems are extremely important to the chemistry of lipid oxidation (Abdalla and Roozen, 1999; Frankel et al., 1994; Fritsch, 1994; Halliwell et al., 1995; McClements and Decker, 2000; Naz et al., 2005). The most extensive research on the impact of physical properties on lipid oxidation has been conducted in oil-in-water emulsions. This is because numerous methods are available to characterize the physical properties and location of compounds in oil-in-water emulsions.

Both lipid hydroperoxides and transition metals exist in foods. Most fats and oils undergo oxidation during oil extraction and refining. Bleaching, deodorization, and physical refining can remove many oxidation products, but in reality most commercial fats and oils contain lipid oxidation products. For example lipid hydroperoxide concentrations range from <1.0 to over 15.0 meq/kg in commercially available oils (Chaiyasit et al., 2007). These lipid hydroperoxide concentrations have been estimated to be over 10,000 times higher than the lipid hydroperoxide concentrations found in living tissues (Decker and McClements, 2001). Metals, such as iron and copper, are present naturally in oilseeds and fruits and thus end up in crude oil. Typical iron and copper concentrations in refined oils are <0.1 and 0.02 ppm, respectively (Chaiyasit et al., 2007).

Transition metals react with lipid hydroperoxides to produce high-energy free radicals (e.g., alkoxyl radicals) that can promote the oxidation of unsaturated fatty acids and can cause the scission of fatty acids to produce the volatile compounds responsible for rancidity. Thus the oxidative stability of many food lipids is dependent on both hydroperoxide and metal concentrations. The interactions between metals and lipid hydroperoxides are dependent on the physical properties of foods, since surface active lipid hydroperoxides tend to migrate to the water-oil interface (Decker and McClements, 2001). For example, the ability of iron to promote lipid oxidation in oil-in-water emulsions is influenced by the net charge of the emulsion-droplet interface. In corn oil-in-water emulsion stabilized with anionic (SDS), cationic (DTAB), or nonionic (Brij 35) surfactants, oxidation rates were highest for negatively charged droplets, intermediate for uncharged droplets, and lowest for positively charged droplets (Mancuso et al., 1999). The observed alterations in oxidation rates are likely due to increased iron-lipid hydroperoxide interactions when positively charged iron ions were electrostatically attracted to the surface of the negatively charged emulsion droplets thus increasing metal-lipid interactions. Conversely, lipid oxidation was retarded when the iron ions were electrostatically repelled from the surface of the positively charged droplets. Another potential physical property that influences iron-lipid hydroperoxide interactions in oil-in-water is the presence of a thick barrier at the lipid-droplet interface. The ability of iron to promote hydroperoxide decomposition as well as oxidation of salmon oil was lower in emulsion droplets stabilized by nonionic surfactants that have a large polar head group that forms a thick interfacial membrane (Silvestre et al., 2000).

## Ability of Proteins to Impact Lipid Oxidation in Oil-in-Water Emulsions

The previously mentioned research with synthetic surfactants has shown that emulsion droplets with thick, cationic interfacial membranes have improved oxidative stability due to decreased interactions between aqueous phase prooxidants and lipid phase oxidation substrates. Unfortunately, many of the synthetic surfactants used in these studies are not approved for food applications. Proteins represent an emulsifier that could be used to produce cationic emulsion droplets with thick interfacial membranes. When salmon or corn oil emulsions are stabilized with proteins, oxidation rates are dramatically slower when the pH is below the pI of the protein; thus the emulsion droplet is cationic (Hu et al., 2003a, 2003b). While the existence of a cationic charge is critical to decrease lipid oxidation rates, the charge density does not seem to be directly related to oxidative stability. For instance, the cationic charge density of whey protein-stabilized salmon oil emulsions at pH 3.0 was in order: β-lactoglobulin > α-lactalbumin > whey protein isolate > sweet whey, while inhibition of lipid oxidation was in order: β-lactoglobulin ≥ sweet whey > whey protein isolate ≥ α-lactalbumin. Similarly, the fact that the cationic charge (as determined by zeta potential) of corn oil emulsion droplets stabilized by whey protein isolate (+55.9 mV) was almost twice as high as casein- and soy-protein-isolate-stabilized emulsions droplets ( +29.9 and +29.4 mV, respectively), while the oxidative stability of the whey-protein-isolate-stabilized emulsions was intermediate among the 3 proteins. This suggests that the magnitude of the positive charge of the emulsion droplet charge did not have a major impact on lipid oxidation rates.

The lack of correlation between emulsion droplet charge density and oxidative stability suggests that additional factors impact lipid oxidation rates in protein-stabilized emulsions. As described above, increasing the thickness of the interfacial membrane of emulsion droplets decreases oxidation rates. This factor may help explain why casein, which can form a thick interfacial layer around dispersed oil droplets of up to 10 nm compared to 1-2 nm for whey proteins (Dickinson and McClements, 1995), was more effective at decreasing lipid oxidation rates than whey proteins when it was used to stabilize corn oil-in-water emulsions (Hu et al., 2003b). An additional factor that could account for the observed differences in the oxidative stability of the emulsions is differences in amino acid composition between the proteins. The free sulfhydryl group of cysteine can inhibit lipid oxidation. When whey protein isolate was treated with N-ethylmaleimide to block free sulfhydryls prior to the formation of emulsions, no alteration in oxidation rates was observed suggesting that free sulfhydryls at the emulsion interface do not inhibit lipid oxidation rates (Hu et al., 2003a). It is possible that other antioxidative amino acids, such as tyrosine, phenylalanine, tryptophan, proline, methionine, lysine, and histidine, could be responsible for differences in the oxidative stability of emulsions stabilized by various proteins.

In addition to the impact of proteins at the interface of oil-in-water emulsion droplets, aqueous phase proteins can also influence lipid oxidation rates. Addition of whey proteins to the continuous phase of Tween 20-stabilized salmon oil-in-water emulsions results in inhibition of lipid oxidation (Tong et al., 2000a). The free sulfhydryls of the continuous phase whey proteins are involved in this antioxidant activity since blocking sulfhydryls with N-ethylmaleimide decreased antioxidant activity. Proteins can also change the physical location of iron in emulsions suggesting that chelation could also be involved in the antioxidant activity of continuous phase proteins (Tong et al., 2000b).

## Role of Other Emulsifiers in Lipid Oxidation Chemistry

The impact of emulsifier type on lipid oxidation can be seen further in experiments where oil-in-water emulsions were made with whey protein isolate (WPI), gum arabic (GA), and modified starch (MS) (Purity Gum BE, National Starch Co.). Figure 9.1 shows that menhaden oil-in-water emulsions stabilized by these emulsifiers had large variations in oxidative stability. At both pH 7.0 and 3.0, the emulsion stabilized with modified starch was the least oxidatively stable while gum-arabic- and whey-protein-stabilized emulsions had similar oxidative stability when lipid oxidation was measured with headspace propanal. When emulsions are made, the emulsifier absorbs to the emulsion-droplet surface until the surface becomes saturated. Remaining emulsifiers partition into the continuous phase where they impact lipid oxidation by chelating metals and scavenging free radicals. At pH 7.0, the continuous phase WPI could be inhibiting lipid oxidation through its ability to more strongly chelate metals (Fig. 9.2) and scavenge free radicals (as determined by the oxygen radical absorbance capacity (ORAC) assay, Fig. 9.3) than GA or MA. The low oxidation rates in the WPI-stabilized emulsions at pH 3.0 could be due to the ability of the proteins to produce a cationic emulsion droplet (Fig. 9.4) that can repel prooxidant metals. While GA was effective at decreasing lipid oxidation rates (Fig. 9.1), it did not have strong metal chelating (Fig. 9.2) or free radical scavenging (Fig. 9.3) properties. This suggest that

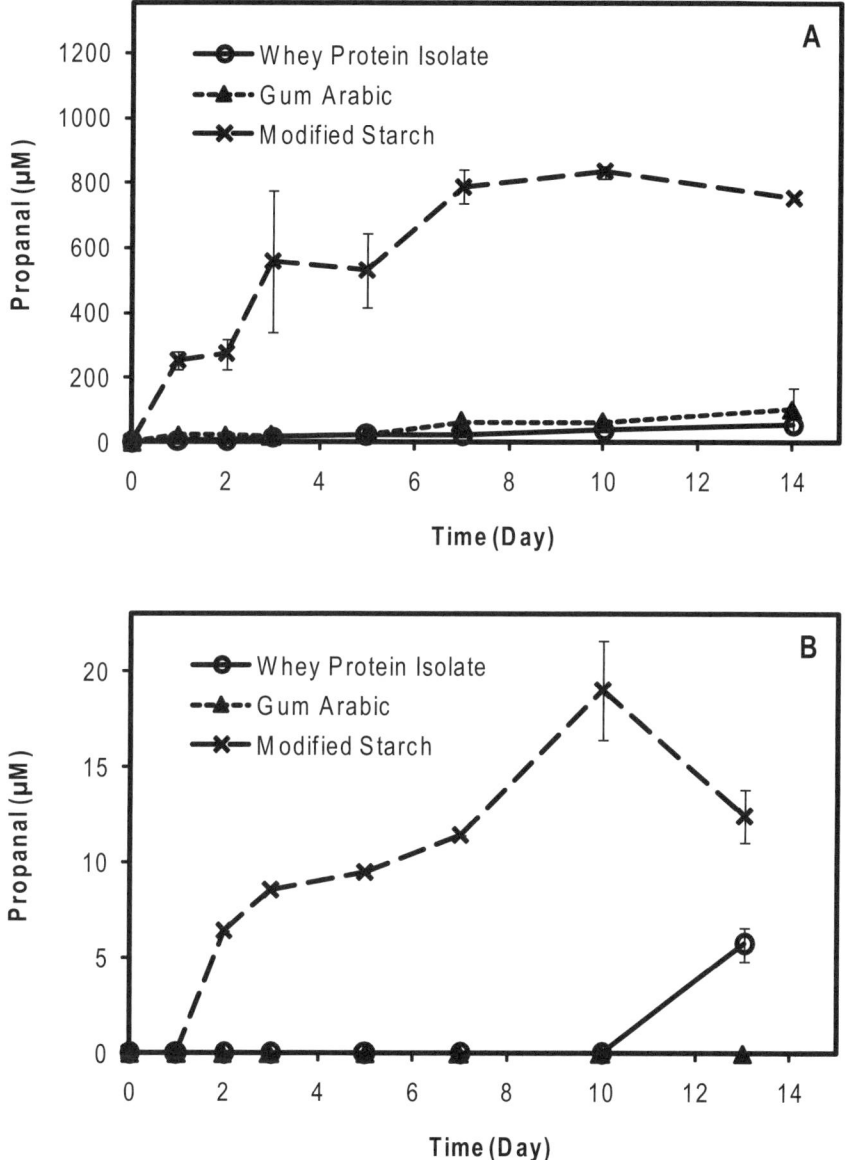

**Fig. 9.1.** Influence of emulsifier type on the formation of headspace propanal in 7.0 % Menhaden oil-in-water emulsions stabilized by whey protein isolate, gum arabic, or modified starch (stored at 5.0°C in the dark) (A) at pH 7.0 and (B) at pH 3.0. Emulsions were washed prior to storage to remove continuous phase components.

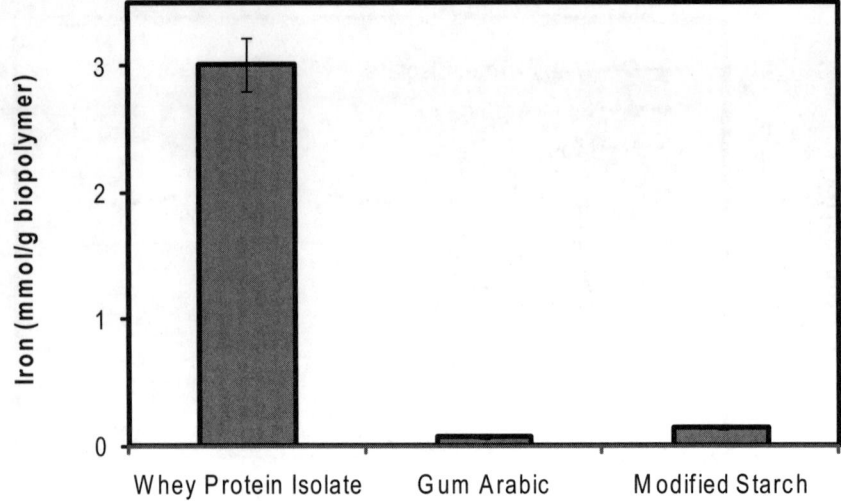

**Fig. 9.2.** Ability of whey protein isolate, gum arabic, and modified starch to bind iron as determined by the method of Diaz et al., 2003.

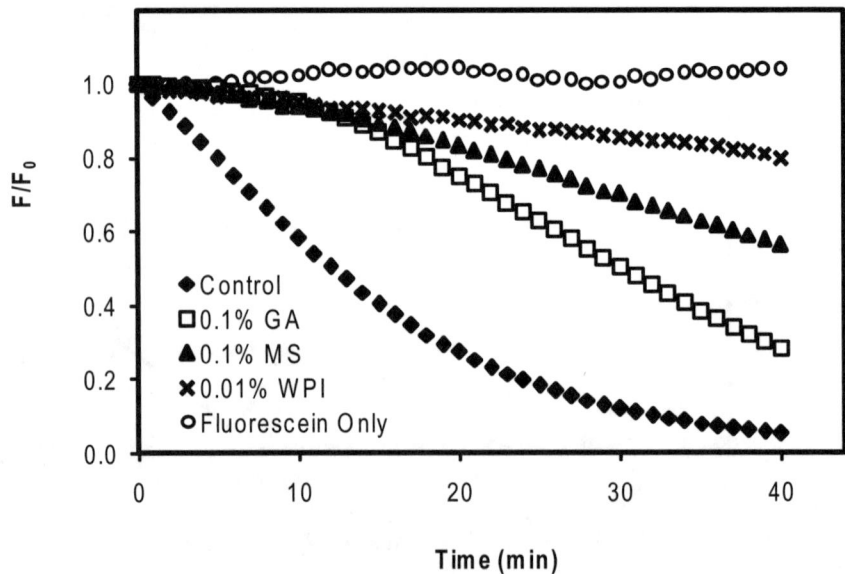

**Fig. 9.3.** Effect of 0.01% Whey protein isolate, gum arabic, or modified starch on the relative fluorescence intensity of 45 nM Fluorescein ($\lambda$EX 493 nm; $\lambda$EM 515 nm) in the presence of 20 mM AAPH at 37°C. Fluorescence values (F) are given relative to the initial time values (F0). The blank (fluorescein only) was prepared without AAPH and the control was prepared without antioxidant.

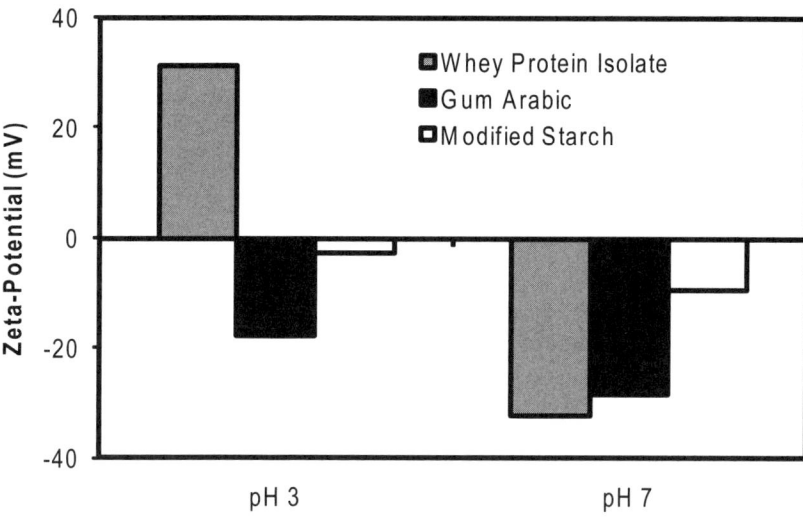

**Fig. 9.4.** Influence of emulsifier type on electrical charge of emulsion droplets as determined by ξ-potential (Hu et al., 2003b).

its ability to decrease lipid oxidation rates could be through the formation of an emulsion-droplet interfacial region that inhibits lipid-prooxidant interactions since GA forms a thick birefringent gel-like mechanical barrier at oil-water interfaces (Garti and Leser, 2001).

## Impact of Free Radical Scavenging Antioxidants on Lipid Oxidation in Oil-in-Water Emulsions

The impact of physical properties on lipid oxidation in oil-in-water emulsions can also be seen with the effectiveness of free radical scavenging antioxidants. For example, hydrophilic antioxidants are less effective in oil-in-water emulsions than lipophilic antioxidants (Frankel, 1998; Porter, 1980). The differences in the effectiveness of antioxidants of varying polarity in oil-in-water emulsions is thought to be due to the physical location of the antioxidants, since lipophilic antioxidants are retained in the oil droplet where they can scavenge free radicals whereas hydrophilic antioxidants partition into the aqueous phase where they are ineffective (Frankel, 1998; Porter, 1980).

The effectiveness of a lipid-soluble antioxidant is also dependent on the concentration since antioxidant activity does not increase linearly with concentration. In some cases high antioxidant concentrations can decrease effectiveness since they are able to reduce metals and increase prooxidant activity. Therefore, addition of a lipid-soluble antioxidant to an emulsion system will not always result in increased oxidative stability if the naturally occurring concentration of antioxidant in the oil is already high. For example, in emulsions made with menhaden oil that did not contain α-tocopherol, addition of α-tocopherol (100-1000 ppm) resulted in a decrease in propanal formation (Fig. 9.5). Conversely, addition of α-tocopherol to algae oil-in-water emulsions did not decreased propanal formation (Fig. 9.6) since the algae oil contained 860 ppm α-tocopherol. Even addition of other free radical scavengers, such as ascorbyl palmitate and rosemary extract, to the algae oil-in-water emulsion containing 860 ppm α-tocopherol also did not increase oxidative stability (Fig. 9.7). These data suggest that the preexisting α-tocopherol in the algae oil was sufficient to obtain maximum free radical scavenging at the site where lipid oxidation was occurring and that additional antioxidants were not able to gain access to free radicals and thus could not further inhibit lipid oxidation.

## Do Physical Structures Impact Lipid Oxidation in Bulk Oils?

During the past 50 years, most studies on lipid oxidation in bulk oils considered the oil matrix to be a homogeneous liquid phase. However, commercial oils contain numerous surface-active compounds such as mono- and diacylglycerols, phospho-

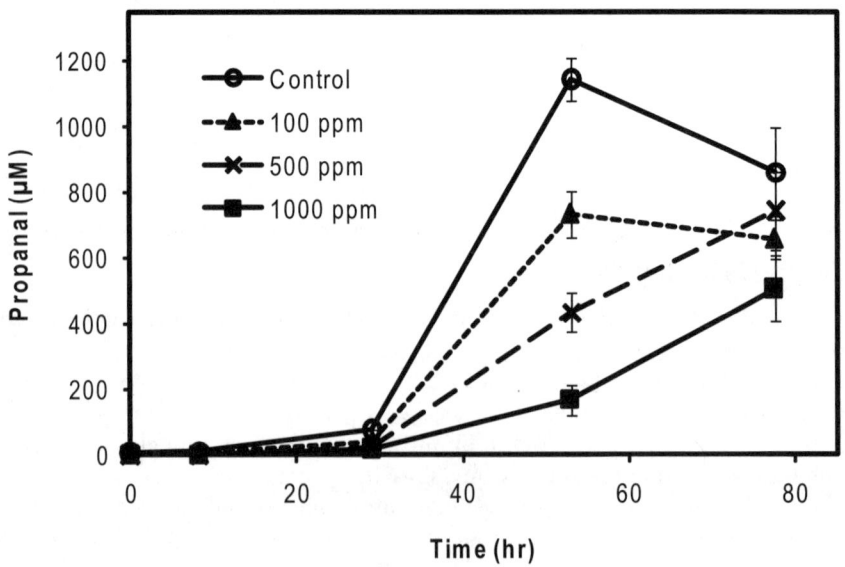

**Fig. 9.5.** Influence of added α-tocopherol (0-1000 ppm) on the formation of headspace propanal in 5.0% Menhaden oil-in-water emulsions stabilized by 0.5% Whey protein isolate at ph 3.0 During storage in the dark at 37.0°C.

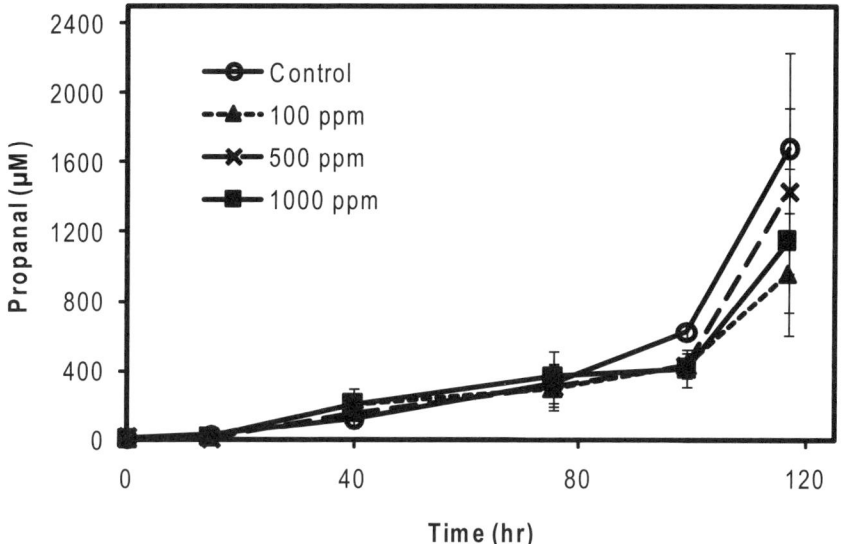

**Fig. 9.6.** Influence of added α-tocopherol (0-1000 ppm) on the formation of headspace propanal in 5.0% Menhaden oil-in-water emulsions stabilized by 0.5% Whey protein isolate at ph 3.0 During storage in the dark at 37.0°C. The algae oil used in this experiment contained 860 ppm α-tocopherol.

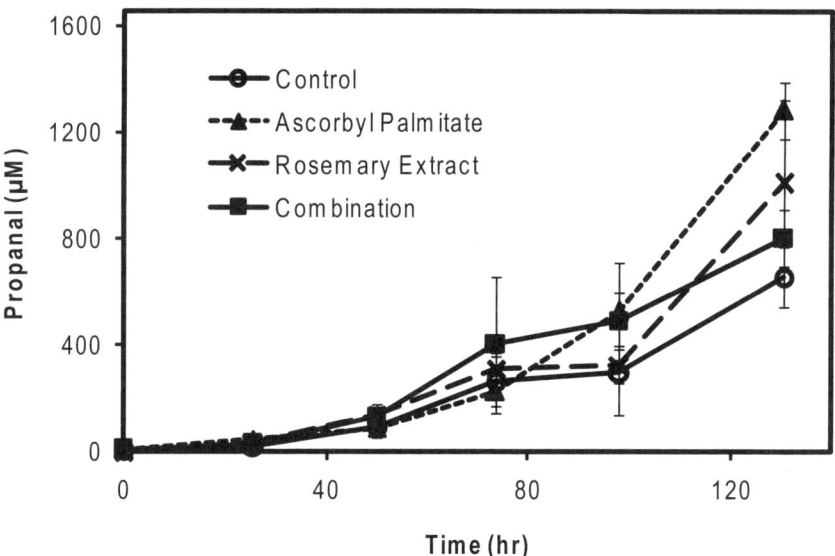

**Fig. 9.7.** Influence of ascorbyl palmitate (1000 ppm), rosemary extract (1000 ppm), and the combination of ascorbyl palmitate and rosemary extract (1000 ppm each) on the formation of headspace propanal in 5.0% Menhaden oil-in-water emulsions stabilized by 0.5% Whey protein isolate at ph 3.0 During storage in the dark at 37.0°C. The algae oil used in this experiment contained 860 ppm α-tocopherol.

lipids, and free fatty acids. When commercial oils are stripped of their minor components, the resulting stripped oil has a higher interfacial tension (less surface-active compounds) than the original refined oil (Table 9.1) indicating that commercial oils contain surface-active compounds. In addition, Dana and Saguy (2006) also found that during frying, a process well know to produce surface-active lipids, such as free fatty acids, interfacial tension decreased from 24.4 to 13.0 mN/m but surface tension of that oil was stable at 32.6 mN/m (Dana and Saguy, 2006).

Table 9.1. Surface and Interfacial Tension of Commercial Corn Oils as Determined by Du Noüy Ring Method at 30°C After 24 hr of Equilibration

| Oil | Surface Tension (mN/m) | Interfacial Tension (mN/m) |
| --- | --- | --- |
| Corn Oil | 31.8 ± 0.1 | 16.4 ± 0.0 |
| Stripped Corn Oil | 31.8 ± 0.1 | 29.6 ± 0.0 |

The combination of these surface-active compounds and the water naturally found in commercial oils can form physical structures known as association colloids. Evidence for the existence of association colloids in commercial algae oil can be seen in the X-ray diffraction pattern of algal oil (Fig. 9.8). This X-ray diffraction pattern of commercial oil is very complicated suggesting that the physical structures are a mixture of reverse micelles and lamellar structures. This diffraction pattern strongly suggests that the surface-active components and water in commercial oil can form association colloids. Addition of water to a commercial algae oil increases X-ray scattering intensity indicating that the oil contained excess surface-active compounds that form additional physical structures upon the addition of water.

Physical structures, such as reverse micelles and lamellar structures, can increase chemical reactions by creating surfaces that increase interactions between lipid- and water-soluble compounds. Surface-active compounds with low hydrophilic-lipophilic balances (HLB) typically form reverse micelles and lamellar structures in bulk oils. Examples of surface-active compounds in bulk oils with low HLB numbers that could form association colloids include free fatty acids (HLB ≈ 1.0), diacylglycerols (≈ 1.8), and monoacylglycerols (≈3.4-3.8) (McClements, 2004). Phospholipids which have intermediate HLB values (≈ 8, McClements, 2004) can form lamellar structures as well as reverse micelles (Gupta et al., 2001). The complex X-ray diffraction patterns seen in commercial algal oil suggest that the wide variety of surface-active components found in a commercial oil form a mixture of reverse micelles and lamellar structures.

The hypothesis that association colloids are the site of lipid oxidation in bulk oils is supported by several lines of experimental evidence. In bulk oil, polar antioxidants are more effective than lipophilic antioxidants as described by the "antioxidant polar paradox" (Frankel, 1998). The increased effectiveness of polar antioxidants in bulk oils has been suggested to be due to their ability to concentrate at the oil-air interface

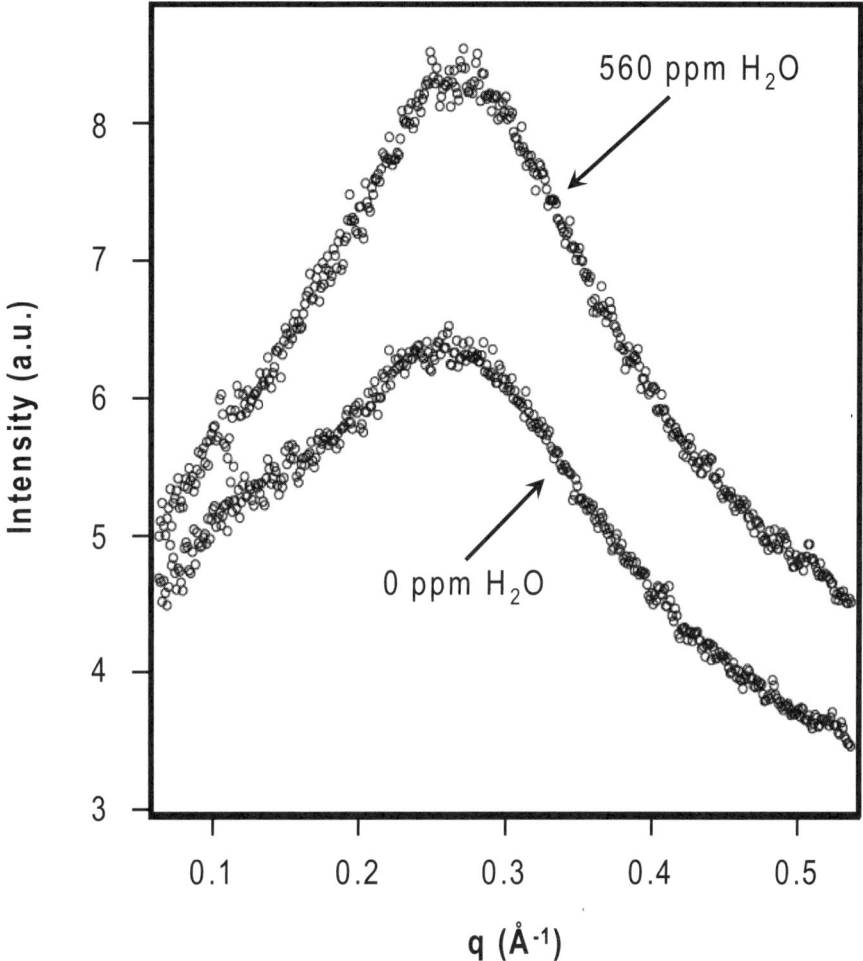

**Fig. 9.8.** Experimental X-ray scattering curves of algae oil with and without added water (560 ppm).

where oxidation is prevalent. However, it is unclear why polar antioxidants would migrate to the air-oil interface since air is less polar than oil (dielectric constant of air is 1.0 compared to approximately 3.0 for food oils, CRC Press, 1982). If antioxidants do migrate and concentrate at the air-oil interface they would be able to decrease surface tension (the air-oil interface). However, this is not the case as can be seen in Table 9.2. Polar antioxidants are more likely to concentrate at oil-water interfaces as can be seen by their ability to decrease interfacial tension (Table 9.2). The ability of polar antioxidants to concentrate at oil-water interfaces suggests that in bulk oils they are more likely to partition in association colloids. Thus, the increased antioxidant activity of polar antioxidants in bulk oil seems to be due to their concentrations in association colloids where oxidative reactions are more prevalent.

**Table 9.2.** Influence of Antioxidants (1 mmol/kg lipid) on Surface and Interfacial Tension of Hexadecane as Determined by Du Noüy Ring Method at 30°C after 24 hr of Equilibration

| Antioxidant | Surface Tension (mN/m) | Interfacial Tension (mN/m) |
| --- | --- | --- |
| Control (no Antioxidant) | 26.0 + 0.1 | 43.3 + 0.4 |
| α-Tocopherol | 26.0 + 0.1 | 41.5 + 0.0 |
| δ-Tocopherol | 26.0 + 0.1 | 37.7 + 0.3 |
| BHT | 26.0 + 0.1 | 43.7 + 0.2 |
| TBHQ | 26.0 + 0.1 | 40.0 + 0.0 |
| Trolox | 26.0 + 0.1 | 28.0 + 0.1 |
| Propyl Gallate | 26.0 + 0.1 | 34.6 + 0.1 |

The metal chelator, citric acid, is an effective antioxidant in bulk oils suggesting that metal-catalyzed decomposition of lipid hydroperoxides is an important pathway for lipid oxidation in bulk oils. Interfacial tension data shows that lipid hydroperoxides, such as cumene hydroperoxide, are surface active in stripped corn oil (Fig. 9.9) suggesting that they would likely migrate to the water-lipid interface of the association colloids where they would interact with metals in the water or at the water-oil interface. Free fatty acids also decrease the interfacial tension of stripped corn oil (Fig. 9.9) suggesting that they could also migrate to the water-lipid interface of the association colloids. Concentration of free fatty acids at the water-oil interface of association colloids would impart a negative charge to the interface that could attract prooxidant metals, such as iron, thus increasing iron-promoted decomposition of lipid hydroperoxides leading to increased development of oxidative rancidity. This could help explain why free fatty acids are prooxidative in bulk oils (Chaiyasit et al., 2007).

Phospholipids are more surface active than free fatty acids and cumene hydroperoxides (Fig. 9.10) suggesting that they would also concentrate in association colloids. In bulk oils, a unique property of phospholipids is their ability to increase the antioxidant activity of tocopherols (Koga and Terao, 1995). Phospholipids increase interactions between α-tocopherol and water-soluble free radicals (generated from 2,2′-azobis(2-amidinopropyl) dihydrochloride, AAPH). In addition, free radical and α-tocopherol interactions increased as the number of association colloids produced by phospholipids increased. Koga and Terao (1995) suggested that the increased activity of α-tocopherol in the presence of phospholipids was due to the ability of phospholipids to concentrate α-tocopherol at the water-lipid interface of association colloids where oxidative stress is greatest.

Differences in the ability of surface-active compounds to decrease surface and interfacial tension are related to their surface activity, as well as their physical structure that allow them to pack at the interface (Cercaci et al., 2007). Increasing concentrations of surface-active compounds will decrease interfacial tension until the water-oil surfaces are completely saturated. This was seen in the stripped corn oil where concentrations of phosphatidylcholine up to 0.75 mmol/kg oil decreased interfacial tension while concentrations from 0.75 to 1.0 mmol/kg oil did not further change ($p > 0.05$)

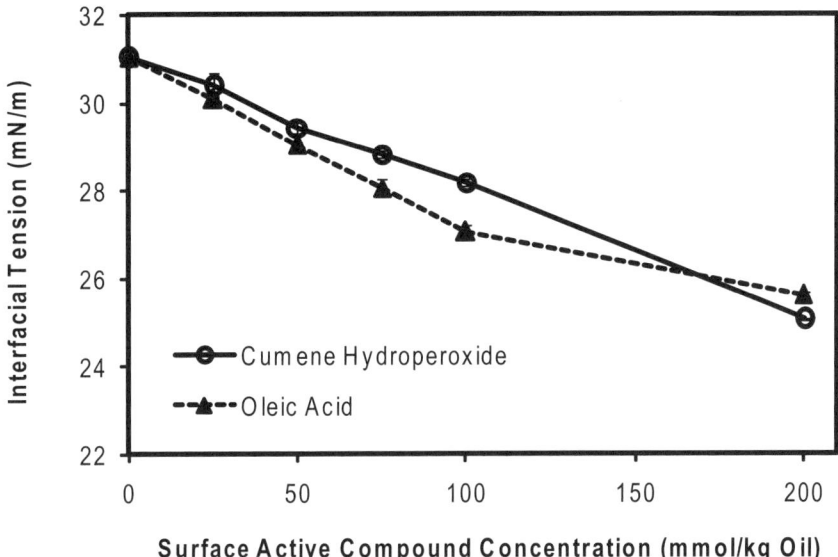

**Fig. 9.9.** Influence of cumene hydroperoxide and oleic acid (0-200 mmol/kg lipid) on interfacial tension of stripped corn oil as determined by drop shape analysis at room temperature after 20 min of equilibration.

interfacial tension (Fig. 9.10). The total concentration of surface active compounds in commercial oils is greater than the levels needed to saturate the water-oil interface of association colloids. This can be observed by the fact that additional water can be added to oils without the oil becoming cloudy (an indication of formation of an emulsion instead of association colloids) or forming oil-water bilayers. Since commercial oils probably contain excess levels of surface-active components, an additional factor that would impact the concentration and type of surface-active compounds in association colloids would be competition among the surface-active molecules for the water-oil interface. Interfacial tension can be used to determine if the surface-active minor components commonly found in commercial oils will migrate to the water-oil interface. These experiments were performed by adding oleic acid (25 mmol oleic acid/kg oil) or phosphatidylcholine (0.1 mmol phosphatidylcholine/kg oil) to stripped corn oil followed by addition of cumene hydroperoxide over a concentration range of 0-100 mmol/kg oil. This model was used to determine if a lipid hydroperoxide, the least surface active of all minor components tested, would be able to migrate to the water-oil interface in the presence of other surfactants where it could be decomposed by aqueous phase prooxidant metals and result in accelerated lipid oxidation. In this system, cumene hydroperoxide was able to significantly decrease ($p<0.05$) interfacial tension of the water-oil bilayer in the presence of either oleic acid or phosphatidylcholine (Fig. 9.11) indicating that it could gain access to aqueous phase prooxidative metals in the presence of other surface-active compounds.

The ability of surface-active minor components in oils to concentrate at the oil-water interface could also be influenced by oil polarity. Increasing δ-tocopherol concentrations from 25 to 75 mmol/kg oil did not significantly decrease (p>0.05) interfacial tension in stripped corn oil suggesting that there were not enough driving forces for δ-tocopherol to migrate to the water-oil interface (Fig. 9.12). However, the further increases in δ-tocopherol concentrations to 100 and 200 mmol/kg oil resulted in significant decreases (p<0.05) in interfacial tension compared to control. Compared to previous findings, δ-tocopherol as low as 1 mmol/kg hexadecane was able to decrease interfacial tension of water-hexadecane bilayer (Chaiyasit et al., 2005). The polarity difference is greater between hexadecane and δ-tocopherol than stripped corn oil and δ-tocopherol, thus allowing δ-tocopherol to migrate to the water-hexadecane interface more effectively than to the water-stripped corn oil interface. This could be another reason why the effectiveness of antioxidants varies in different food oils.

## Conclusions

Most lipids in food exist as dispersions or emulsions. Lipid dispersions contain lipid-water interfaces whose physical properties greatly impact lipid oxidation chemistry. The physical properties that can impact lipid oxidation include surface charge that can

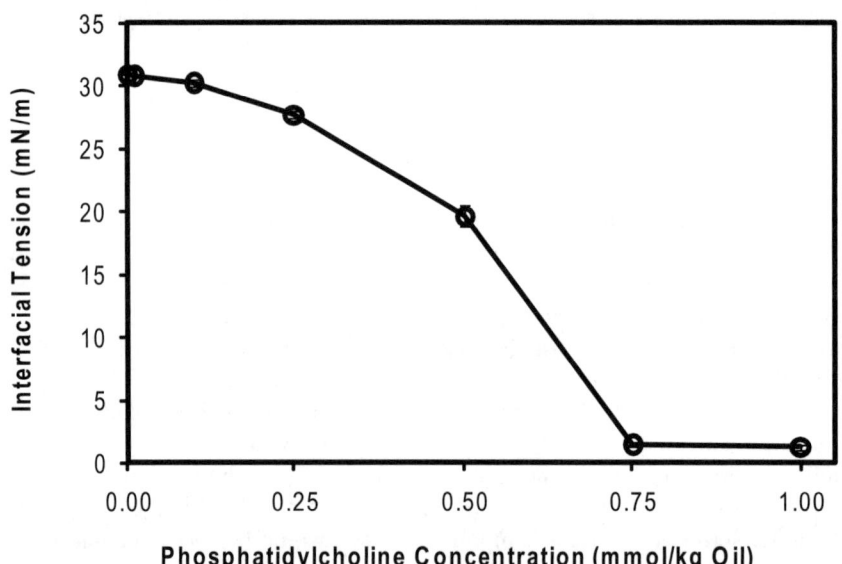

**Fig. 9.10.** Influence of phosphatidylcholine dioleyl (0-1 mmol/kg lipid) on interfacial tension of stripped corn oil as determined by drop shape analysis at room temperature after 20 min of equilibration.

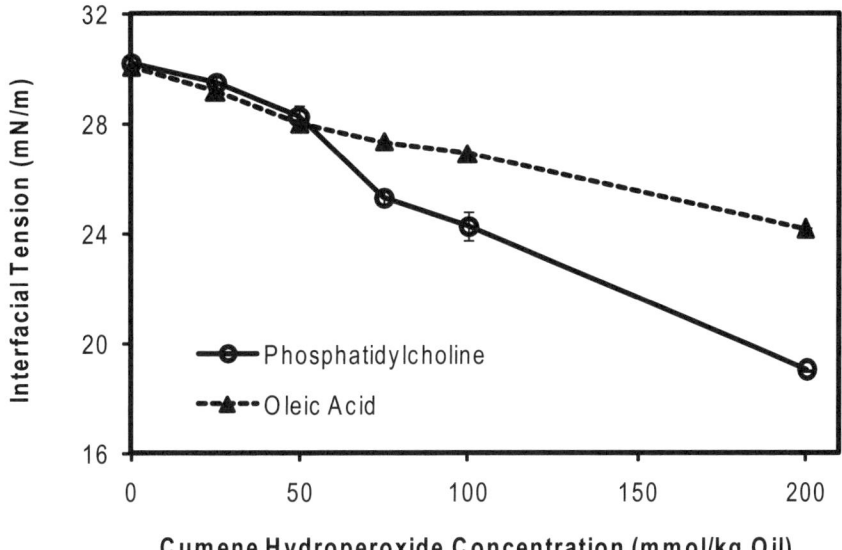

**Fig. 9.11.** Influence of cumene hydroperoxide (0-200 mmol/kg lipid) on interfacial tension of stripped corn oil containing oleic acid (25 mmol/kg lipid) or phosphatidylcholine dioleyl (0.1 Mmol/kg lipid) as determined by drop shape analysis at room temperature after 20 min of equilibration.

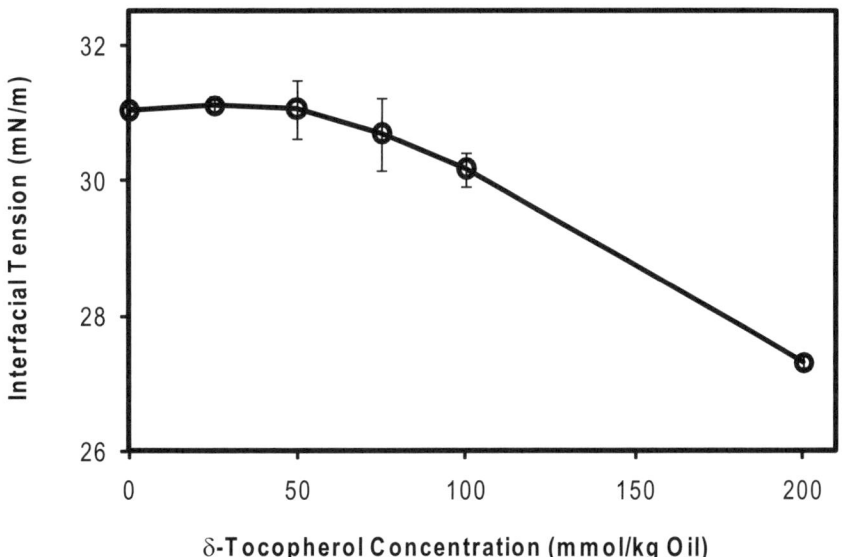

**Fig. 9.12.** Influence of δ-tocopherol (0-200 mmol/kg lipid) on interfacial tension of stripped corn oil as determined by drop shape analysis at room temperature after 20 min of equilibration.

attract or repel prooxidant metals and interfacial thickness that can inhibit interactions between aqueous phase prooxidants and lipids. Physical structures also seem to play an important role in the oxidative stability of bulk oils. The ability of minor components to migrate and concentrate in associated colloids in commercial oils could help explain why free fatty acids are prooxidative, phospholipids are antioxidative, and polar antioxidants are more effective than non-polar antioxidants. Understanding how the physical properties of food dispersions impacts lipid oxidation could lead to the development of novel antioxidant technologies that help improve the oxidative stability of oils containing increased concentrations of polyunsaturated fatty acids.

## Acknowledgement

This research was supported in part by grant 2007-02650 from the NRI competitive grants program of the United States Department of Agriculture.

## References

Abdalla, A. E.; and J.P. Roozen. Effect of Plant Extracts on the Oxidative Stability of Sunflower Oil and Emulsion. *Food Chem.* **1999**, *64*, 323-329.

Cercaci, L.; M.T. Rodriguez-Estrada; G. Lercker; and E.A. Decker. Phytosterol Oxidation in Oil-in-Water Emulsions and Bulk Oil. *Food Chem.* **2007**, *102*, 161-167.

Chaiyasit, W.; R.J. Elias; D.J. McClements; and E.A. Decker. Role of Physical Structures in Bulk Oils on Lipid Oxidation. *Crit. Rev. Food Sci. Nutr.* **2007**, *47*, 299-317.

Chaiyasit, W.; D.J. McClements; and E.A. Decker. The Relationship Between the Physicochemical Properties of Antioxidants and Their Ability to Inhibit Lipid Oxidation in Bulk Oil and Oil-in-Water Emulsions. *J. Agric. Food Chem.* **2005**, *53*, 4982-4988.

CRC Press. *CRC Handbook of Chemistry and Physics. 63$^{rd}$ Edition.* Weast, R.C.; and Astle, M.J. Eds., CRC Press: Boca Raton, Florida, 1982, p. E54.

Dana, D.; and I.S. Saguy. Mechanism of Oil Uptake during Deep-Fat Frying and the Surfactant Effect-Theory and Myth. *Adv. Colloid Int. Sci.* **2006**, *128*, 267-272.

Decker, E.A.; and D.J. McClements. Transition Metal and Hydroperoxide Interactions: An Important Determinant in the Oxidative Stability of Lipid Dispersions. *Inform.* **2001**, *12*, 251-255.

Diaz, M.; C.M. Dunn; D.J. McClements; and E.A. Decker.. Use of Caseinophosphopeptides as Natural Antioxidants in Oil-in-Water Emulsions. *J. Agric. Food Chem.* **2003**, *51*, 2365-2370.

Dickinson, E. *An Introduction to Food Colloids.* Oxford University Press: Oxford, 1992, p. 2.

Dickinson, E.; and D.J. McClements.. *Advances in Food Colloids.* Blackie Academic and Professional: Glasgow, 1995, p. 15-17.

Dickinson, E.; and G. Stainsby. *Colloids in Foods.* Applied Science Publishers: London, 1982, p. 15.

Frankel, E.N. *Lipid Oxidation.* The Oily Press: Dundee, Scotland, 1998, pp. 166-185.

Frankel, E.N.; S.W. Huang; J. Kanner; and J.B. German. Interfacial Phenomena in the Evaluation of Antioxidants-Bulk Oils vs Emulsions. *J. Agric. Food Chem.* **1994**, *42*, 1054-1059.

Fritsch, C.W. Lipid Oxidation-The Other Dimensions. *Inform.* **1994**, *5*, 423-436.

Garti, N.; and M.E. Leser. Emulsification Properties of Hydrocolloids. *Polym. Adv. Tech.* **2001** *12,* 123-135.

Gupta, R.; H.S. Muralidhara; and H.T. Davis. Structure and Phase Behavior of Phospholipids-Based Micelles in Nonaqueous Media. *Langmuir.* **2001,** *17(17),* 5176-5183.

Halliwell, B.; M.A. Murcia; S. Chirico; and O.I. Aruoma. Free Radicals and Antioxidants in Food and in vivo: What They Do and How They Work. *Crit. Rev. Food. Sci. Nutr.* **1995** *35,* 7-20.

Hu, M.; D.J. McClements; and E.A. Decker. Impact of Whey Protein Emulsifiers on the Oxidative Stability of Salmon Oil-in-Water Emulsions. *J. Agric. Food Chem.* **2003a,** *51,* 1435-1439.

Hu, M.; D.J. McClements; and E.A. Decker. Lipid Oxidation in Corn Oil-in-Water Emulsions Stabilized by Casein, Whey Protein Isolate, and Soy Protein Isolate. *J. Agric. Food Chem.* **2003b,** *51,* 1696-1700.

Koga, T.; and J. Terao. Phospholipids Increase Radical-Scavenging Activity of Vitamin E in a Bulk Oil Model System. *J. Agric. Food Chem.* **1995,** *43,* 1450-1454.

Mancuso, J.R.; D.J. McClements; and E.A. Decker. The Effects of Surfactant Type, pH, and Chelators on the Oxidation of Salmon Oil-in-Water Emulsions. *J. Agric. Food Chem.* **1999,** *47,* 4112-4116.

McClements, D.J. *Food Emulsions: Principles, Practice and Techniques*, CRC Press: Boca Raton, Florida, 1999, p. 167.

McClements, D.J. *Food Emulsions: Principles, Practice and Techniques, 2nd Edition,* CRC Press: Boca Raton, Florida, 2004, p. 3.

McClements, D.J.; and E.A. Decker.. Lipid Oxidation in Oil-in-Water Emulsions: Impact of Molecular Environment on Chemical Reactions in Heterogeneous Food Systems. *J. Food Sci.* **2000,** *65(8),* 1270-1282.

Naz, S.; H. Sheikh; R. Siddiqi; and S.A. Sayeed. Deterioration of Olive, Corn and Soybean Oils Due to Air, Light, Heat and Deep-Frying. *Food Res. Int.* **2005,** *38,* 127-134.

Porter; W.L. Recent Trends in Food Applications of Antioxidants. In *Autoxidation in Food and Biological Systems;* Simic, M.G.; and Karel, M. Eds., Plenum Press: New York, 1980, pp. 295-365.

Silvestre, M.P.C.; W. Chaiyasit; R.G. Brannan; D.J. McClements; and E.A. Decker. Ability of Surfactant Headgroup Size to Alter Lipid and Antioxidant. *J. Agric. Food Chem.* **2000,** *48,* 2057-2061.

Tong, L.M.; S. Sasaki; D.J. McClements; and E.A. Decker. Antioxidant Activity of Whey in a Salmon Oil Emulsion. *J. Food Sci.* **2000a,** *65(8),* 1325-1329.

Tong, L.M.; S. Sasaki; D.J. McClements; and E.A. Decker. Mechanisms of the Antioxidant Activity of a High Molecular Weight Fraction of Whey. *J. Agric. Food Chem.* **2000b,** *48,* 1473-1478.

# 10

# Antioxidant Evaluation Strategies
Leif H. Skibsted
Food Chemistry, Department of Food Science, University of Copenhagen, Rolighedsvej 30, DK-1958 Frederiksberg C, Denmark

## Introduction

Throughout the world societies are recognized where life expectancy is longer than in comparable societies. Longevity has been associated with availability and choice of fresh and good food. Certain bacteria from fermented milk having the potential to colonize the human intestine have been suggested to be crucial for mountaineers ageing with no loss of physical and mental capabilities. In coastal societies a diet rich in fish has been identified as important for the absence of certain cardiovascular diseases. Moreover it has often been speculated whether one single dietary factor could be identified as the "secret" for human health, such as a specific lipid source, a mineral present in the soil, or certain herbs.

Around the Mediterranean, the diet is rich in vegetables, nuts, fruits, the bread is whole grain, and fish is preferred for meat, which is eaten in moderation. Antioxidants has been suggested as the health factor common for citrus fruits, tomatoes, nuts, spices, whole grain bread, and the red wine typical for the Mediterranean meal and that accompany fish fried in olive oil.

In relation to food and human nutrition, antioxidants were originally defined as "substrates that in small quantities are able to prevent or greatly retard the oxidation of easily oxidizable nutrients such as fats" (Chipault, 1962). Antioxidants could be active at three stages in relation to human nutrition. Antioxidants can prevent oxidative damage to food during processing, storage, and preparation of the meal; in effect limiting formation of toxic oxidation products in the food eaten. Without being absorbed, antioxidants can protect sensitive tissue in the human digestive tract against aggressive prooxidants formed during digestion. Antioxidants may have protective effects in the human body following absorption, and as such may be important nutrients (Vulcain et al., 2005). In order to evaluate antioxidants it is important to recognize at what stage antioxidative protection is needed or expected and to adjust the evaluation protocol accordingly (Becker et al., 2004). Different protocols should be applied to evaluate health effects of antioxidants and to evaluate protection of food against oxidative deterioration. A methodological shortcoming has been identified since antioxidants are multifunctional in biological systems but the methods traditionally used for evaluation are only "one-dimensional" (Frankel and Meyer, 2000). The multifunctionality of antioxidants has also led to more general or broad definitions of antioxidants, including functions as regulators of gene expression (Rice-Evans, 2004).

As for the health effects of antioxidants, epidemiological studies have suggested beneficial effects of antioxidants especially from fruits and vegetables. The result of human intervention studies are, however, less clear and supplementation with antioxidants like vitamin E have shown little if any protective effect on cardiovascular diseases and on cancer (Vivekananthan et al., 2003; Lonn et al., 2005). In contrast to supplementation, a change in diet will involve a changed pattern of interaction between the numerous bioactive compounds present in plant food with the perspective of synergistic effects between the individual antioxidants not obvious following supplementation with single or a few compounds. Further development in antioxidant evaluation methods should accordingly focus on antioxidant interaction. Synergistic effects of antioxidants were recently demonstrated for dietary protection against age-related macular degeneration (van Leeuwen et al., 2005).

Antioxidant protective effects have been demonstrated much more convincingly for food quality than for human health. Based on dietary recommendations, attempts have been made to change the lipid profile of pork and poultry and even bovine milk to a higher degree of unsaturation by changing the feeding regime. Such products with improved nutritive value are in general vulnerable to oxidation (Sandström et al., 2000). However, increasing the dietary level of vitamin E for the production animal has been found to more than compensate for the oxidative instability introduced by the higher unsaturation in the meat products (Jensen et al., 1998).

Possible effects of non-absorbed antioxidants on human health through activity in the digestive tract have only lately been recognized and deserve further attention (Kanner and Lapidot, 2001). Protection by plant-based antioxidants may even be more relevant in relation to iron-fortification of certain foods, in which the increased availability of iron may create iron-based prooxidants during digestion.

## Evaluation Protocols

A four step strategy for antioxidant evaluation was proposed (Becker et al., 2004). This strategy was based on the assumption that lipid oxidation is initiated through free radical mechanisms and that phenolic compounds are the main group of primary antioxidants. The strategy included the following steps to evaluate potential sources of antioxidants, such as protein hydrolysates and plant extracts:

i. Quantification and possible identification of phenolic compounds

ii. Quantification of radical-scavenging activity using radicals with relevant life-time and reactivity for biological systems

iii. Evaluation of the activity to inhibit or halt lipid oxidation in model biological systems

iv. Depending on the use of the antioxidants:

   a. Storage studies using actual antioxidants incorporated in the food product of relevance to evaluate food stability

   b. Human intervention studies using relevant markers for oxidative status and oxidative stress to evaluate health effects.

Step (i) is an initial screening of potential sources of antioxidants and may also be used to optimize extraction procedures and to select extraction solvents. Step (ii) evaluates the chain-breaking capacity of the extract recognizing that lipid oxidation is a free radical chain reaction. In step (iii) an oxidation substrate is introduced; the capability of the potential antioxidant to limit oxygen consumption or formation of secondary lipid oxidation products is determined and expressed as a prolongation of the lag-phase or a decrease in rate. Many natural compounds are scavengers of especially short-lived free radicals without being effective at protecting lipids against oxidation, and evaluation including only step (i) and/or step (ii), as is often seen in literature, may give false positive results. Conclusions from step (iii) may, however, still be invalid to protect real foods or to predict health effects, and a complete evaluation of antioxidants should always include step (iv). There is no short cut from determining total phenolic content, determining radical scavenging activity, or even determining effects on lipid oxidation in model systems, to real food systems. The same reservation applies to antioxidants and health effects; again radical-scavenging screening and determination should be followed by intervention studies prior to dietary recommendations. Anthocyanines for example are efficient radical scavengers but have extremely low bioavailability from the digestive tract. There are many claims for protective effects on vision by anthocyanines as antioxidants which could not be confirmed in intervention studies (Canter and Ernst, 2004). Still, anthocyanines may have positive effects as antioxidants in the digestive tract without being absorbed by deactivating prooxidants from meat (Kanner and Lapidot, 2001). Accordingly the proposed 4 step evaluation has to be used with care and the initial step should primarily be applied to screen new sources of natural antioxidants. When the lipid system of step (iii) is changed from a homogeneous lipid phase to heterogeneous systems of increasing structural organization, antioxidant synergism may also be recognized during screening (Becker et al., 2007).

## Early Events in Lipid Oxidation

Oxidation is initiated by irradiation (including visible light), by enzymes, and by metal catalysis. Heat and pressure enhance these effects. It should be recognized that optimal antioxidant protection is obtained by preventing oxidation initiation rather than by limiting propagation of lipid oxidation when the initial oxidative damage has occurred. Thus for food products with long shelf-life, damage during production leading to a high level of lipid hydroperoxides holds the potential of severe oxidative damage during subsequent storage of the product. From Fig. 10.1, three major reaction paths for lipid oxidation may be identified:

I. Oxygen activation by metal catalysis, including activation by oxidoreductases to yield superoxide and hydroxyl radicals, like in the serum phase of milk. The hydroxyl radicals and chlorine radicals from disinfectants will attack unsaturated lipids and initiate chain reactions.

II. Lipoxygenase activity that incorporates oxygen in unsaturated lipids to yield lipid hydroperoxides like in vegetables and chicken meat.

III. Photosensitized oxidation as a result of light exposure that either activates ground state oxygen to form singlet oxygen by physical quenching of the excited state sensitizer or forms radicals through chemical quenching of the excited state sensitizer. Singlet oxygen may add to unsaturated lipids to form lipid hydroperoxides as for lipoxygenase activity (II). Substrate radicals may subsequently enter the free radical reaction sequences described for metal catalysis (I).

Antioxidant evaluation corresponding to the initial three steps of the four step evaluation protocol will have to be adjusted according to the oxidative stress expected for the food or the biological system under consideration for protection by antioxidants. The reactions or reaction sequences marked with capital letters in Fig. 10.1 show the targets for antioxidant protection at early stages of lipid oxidation. Efficient radical scavengers will prevent initiation of oxidation by radicals at A, while metal chelators may prevent cleavage of lipid hydroperoxides to initiate further chain reactions at B. Protection against light-induced lipid oxidation depends on interaction with the photosensitizer or on interaction with singlet oxygen (C). Inactivation of lipoxygenase, as in blanching of vegetables, will not be discussed any further (D). Initiation of oxidation by lipoxygenases or by light results in the formation of lipid hydroperoxides. Metal catalysis often depends on the existence of preformed lipid hydroperoxides (LOOH) and has been termed "LOOH-dependent oxidation". Differentiation between "LOOH-independent" lipid oxidation initiated by oxygen activation and metal catalysis:

$$Fe^{2+} + H_2O_2 \rightarrow Fe^{3+} + HO\bullet + OH^- \quad [1]$$
$$Fe^{3+} + H_2O_2 \rightarrow FeO^{2+} + HO\bullet + H^+ \quad [2]$$

and "LOOH-dependent" lipid oxidation based on initial enzymatic or photochemical generation of lipid hydroperoxides:

$$Fe^{2+} + LOOH \rightarrow Fe^{3+} + LO\bullet + OH^- \quad [3]$$
$$Fe^{3+} + LOOH \rightarrow FeO^{2+} + LO\bullet + H^+ \quad [4]$$

is usually difficult (Carlsen et al., 2005). However, heat and pressure enhances the cleavage of peroxides to initiate new chain reactions for both types of lipid oxidation. High-pressure processing has been introduced for new types of meat and dairy products for which antioxidant protection will also be required. For such products high pressure must be considered together with heat as an enhancer of lipid oxidation in the later steps of the four step procedure (Bragagnolo et al., 2005).

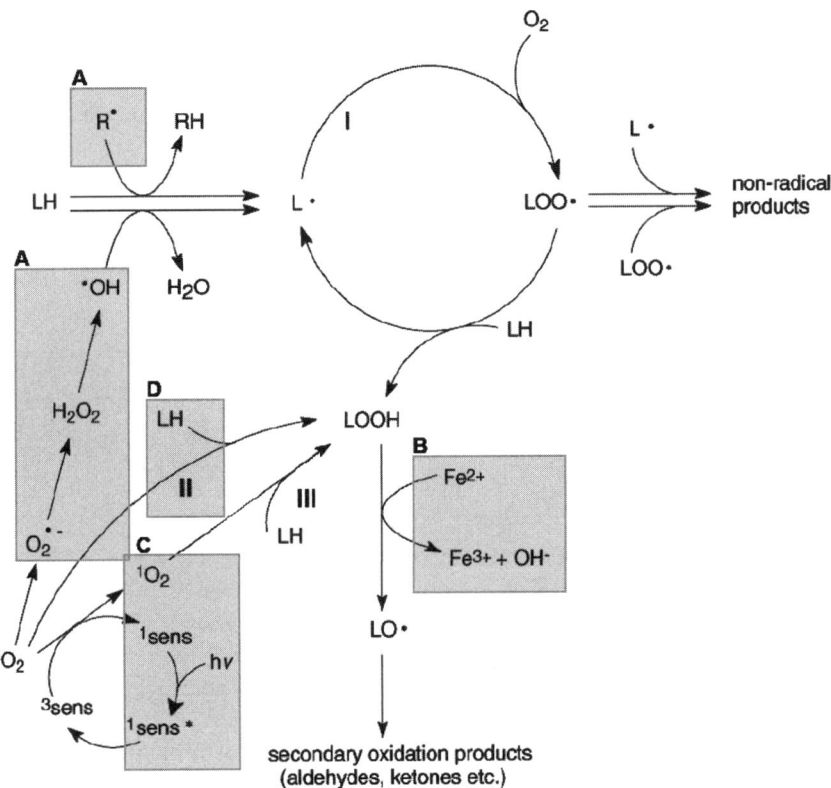

**Fig. 10.1.** Lipid oxidation depends on three reaction paths: (I) Free radical chain reaction initiated by oxygen activation to yield hydroxyl radicals by radical formation; (II) Lipoxygenase activity to yield lipid hydroperoxides; or (III) Photosensitized formation of singlet oxygen or free radicals. Areas A, B, C, and D show where protection against early events in lipid oxidation should be targeted.

## Protection Against Light

Under most conditions lipid oxidation is induced through sensitized reactions rather than by direct absorption of light by the lipid substrate (Bekbölet, 1990). Important photosensitizers in foods are riboflavin, chlorophylls, partly degraded heme pigments, and certain synthetic colorants like erythrosine. For certain biological structures, like the eye, exposure to light with a strong UV-component may involve direct excitation of aromatic side chains in the proteins leading to cross-linking. In order to optimize antioxidant protection of food exposed to light, it is important to distinguish between two limiting mechanisms as shown in Fig. 10.2 for riboflavin. Photosensitizers like riboflavin are characterized by efficient intersystem crossing to yield triplet states from the initially populated excited singlet states. The triplet excited state may be deactivated chemically by electron transfer to an oxidation substrate or by physical quenching by oxygen (Li et al., 2000). Protection against light-induced oxidation can be based on one or more of the following four principles:

i. Inner-filter effects.

ii. Quenching of triplet state photosensitizer.

iii. Quenching of singlet oxygen for Type II mechanism (Fig. 10.2).

iv. Scavenging of substrate radicals formed by chemical quenching of triplet photosensitizer by primary oxidation substrate for Type I mechanism (Fig. 10.2).

To protect by inner-filter effects, the light is absorbed by another compound than the photosensitizer. For a dairy spread, β-carotene was found to protect lipids against oxidation by riboflavin-sensitized oxidation through such an inner-filter effect, as light was found to be absorbed preferentially by β-carotene rather than by riboflavin (Hansen and Skibsted, 2000). Protection by direct quenching of the triplet state photosensitizer was observed for ascorbic acid, plant phenols, Trolox, and tocopherols, while carotenoids somewhat surprising were found inactive (Jung et al., 2007; Becker et al., 2005; Cardoso et al., 2007). In contrast, carotenoids are efficient physical quenchers of singlet oxygen, and the most effective are the carotenoids with the longest conjugated systems (Bradley and Min, 1992). As for the last principle of protection, the mechanism is similar to the mechanism for protection against thermal lipid oxidation and the radicals formed by chemical quenching of the photosensitizer will initiate chain reaction if not scavenged by an antioxidant (Fig. 10.1). Notably, a number of plant phenols known to be efficient antioxidants in thermal oxidation of lipids, have recently been found to have a dual function in light-induced lipid oxidation as they are efficient quenchers of triplet-excited state riboflavin besides being efficient scavengers of any radicals formed by reaction of oxidation substrates with triplet-

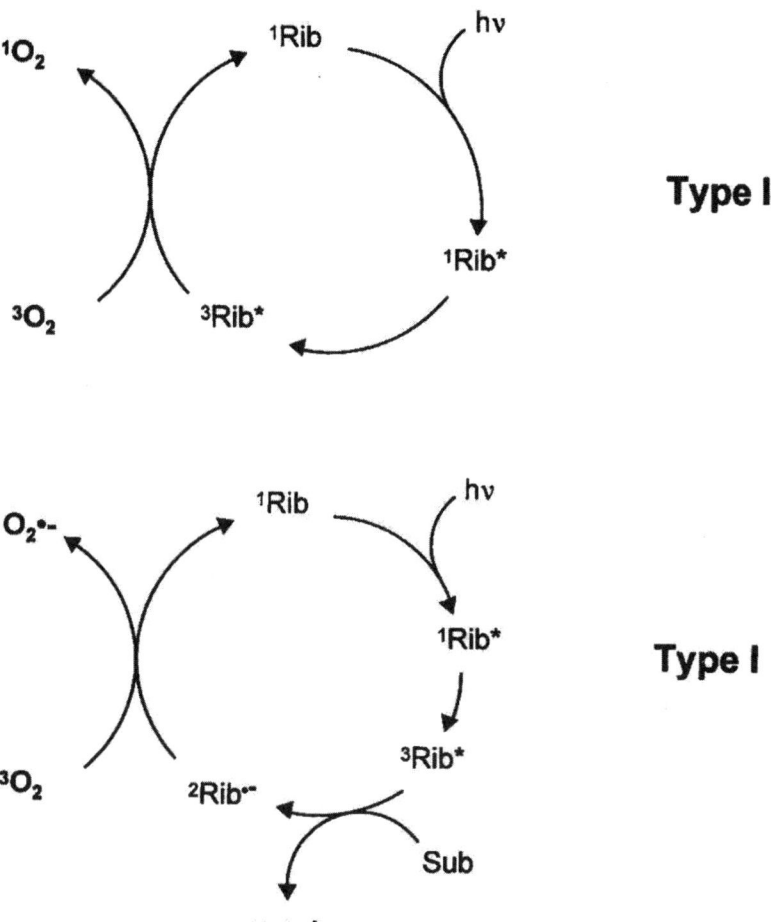

**Fig. 10.2.** Two limiting mechanisms for photosensitized initiation of lipid and protein oxidation with riboflavin (Vitamin B2) as sensitizer: Type I involves direct radical formation through chemical quenching of triplet riboflavin by a substrate, while Type II depends on physical quenching of triplet riboflavin by oxygen to yield singlet oxygen.

state riboflavin (Becker et al., 2005). In Table 10.1 second-order rate constants for a number of quenchers relevant to dairy products are collected together with activation parameters. The quenching rate approaches the diffusion limit. Although the values of the activation parameters for the quenching reaction, have a large variation, they confirm a common deactivation mechanism since they show so-called isokinetic behavior ($\Delta H^{\#}$ depends linearly on $\Delta S^{\#}$) together with purine bases. The mechanism by which these compounds protect dairy products, beer, and other foods and beverages against light-induced oxidation in agreement with the isokinetic behavior has been described as a bimolecular diffusion-controlled encounter with electron (or hydrogen atom) transfer as the rate-determining step.

**Table 10.1.** Second-Order Rate Constants at 25°C for Quenching of Triplet-Excited State Riboflavin by Compounds of Interest for Protection of Food Against Light-Induced Oxidation.[a]

| Quencher | Solvent | $k_2$ (l × mol$^{-1}$ × s$^{-1}$) | $\Delta H^{\#}$ (kJ × mol$^{-1}$) | $\Delta S^{\#}$ (J × mol$^{-1}$ × K$^{-1}$) |
|---|---|---|---|---|
| Ascorbate | Water, pH=6.4 | 2.0 × 10$^9$ | 184 | 551 |
| Trolox | Water, pH=6.4 | 2.6 × 10$^9$ | 14 | -19 |
| α-Tocopherol | Tween-20 Emulsion | 1.1 × 10$^8$ | 43 | 53 |
| Caffeic Acid | Acetonitrile/Water | 2.2 × 10$^9$ | 27 | 66 |
| Rutin | Acetonitrile/Water | 1.0 × 10$^9$ | | |
| (+)-Catechin | Acetonitrile/Water | 1.4 × 10$^9$ | | |

[a]From Becker et al., 2005, Cardoso et al., 2006, and Cardoso et al., 2007.

To optimize protection against light-induced oxidation in food and beverages, the four step strategy originally formulated as a general procedure for antioxidant protection should be modified:

i. Analysis should also include other types of compounds, like terpenes and carotenoids, in the plant material or extract which could serve as inner-filters or singlet oxygen quenchers

ii. Radical-scavenging experiments should be replaced by quenching studies of the actual photosensitizer present in the product in combination with protective quenchers. Such quenching studies could be based on dynamic, real-time methods using laser flash photolysis or on static methods based on the Stern-Volmer formalism (Cardoso et al., 2006).

iii. Experiments in which rates for oxygen consumption or formation of secondary lipid oxidation products are determined for a specific oxidation substrate should be replaced by photochemical experiments in which quantum yields, based on chemical actinometry, are determined for relevant wavelengths for the biological tissue or for the beverage or food product (Hansen and Skibsted, 2000).

iv. For the actual product formulated according to the results obtained in the previous steps, storage experiments with controlled light exposure seem mandatory. Effects on product quality should be evaluated by chemical analysis or by sensory evaluation. A combination of the wavelength dependence of quantum yield for the oxidative process, the spectral characteristics of the light source and the absorption spectrum of the photosensitizer in the product may together provide a so-called action spectrum.

For milk-based beverages, the finding that plant phenols are efficient quenchers (in step iii) have led to the suggestion that plant extracts should be explored to protect fruit flavored products, and tested in step (iv) (Becker et al., 2005). Riboflavin quenchers have been found to inhibit lightstruck flavor formation in beer (Goldsmith et al., 2005). However terpenes, like eucalyptol, present in some spices used as flavors in beer have not been found to quench triplet riboflavin and will accordingly not yield any direct protection (Cardoso et al., 2006).

In conclusion, it is important to know whether photoxidation is occurring by a Type I or a Type II mechanism or by both mechanisms at the same time in competition (Jung et al., 2007). Besides inner-filter protection, carotenoids will thus yield protection against photosensitization by the Type II mechanism but not against photosensitization by the Type I mechanism. In milk products, physical quenching of triplet riboflavin by oxygen to yield singlet oxygen will not compete efficiently with chemical quenching by peptides and free amino acids even in air-saturated milk (Cardoso et al., 2004). Accordingly peptides and amino acids will be oxidized by a Type I mechanism. In milk with high uric acid content, uric acid will deactivate triplet riboflavin in competition with the peptides and amino acids, in effect yielding optimal protection of milk against light. The uric acid level in milk depends on the feeding regime of the dairy cows.

## Protection Against Metal Catalysis

In food and other biological systems, transition metal ions are chelated by various ligands. The standard reduction potential valid for the iron couple

$$Fe(H_2O)_6^{3+} + e^- \rightarrow Fe(H_2O)_6^{2+} \quad [5]$$

is in the presence of flavonoids, like quercetin, which bind iron(II) more strongly than iron(III), higher than $E^\circ = 0.77$ V, which is the value for the hexaqua ions (Ferrali et al., 1997). Thus Iron(II) becomes less reducing, and quercetin may prevent the Fenton reaction (Eqn. 1) as an initiator of lipid oxidation. As for the general antioxidant evaluation procedure, it should be realized that when lipid oxidation is expected to be metal-ion dependent, as in precooked meat and other processed foods, antioxidants should be tested as scavengers of metal-related prooxidants rather than of semi-stable radicals like DPPH• or others frequently used in standard tests. Heme pigments are a source of iron in meat and prooxidative activity has been related to reaction cycles similar to the pseudoperoxidase cycle shown in Fig. 10.3. Deactivation of the hypervalent iron pigment perferryl and ferryl by potential antioxidants is relevant (Andersen et al., 2003). For an increasing number of plant phenols from sources like green tea, fruits and vegetables, scavenging rate constants have been determined. In Table 10.2, a few examples are compared with rate constants for deactivation of the superoxide anion radical and lipid peroxyl radicals. Notably, phenols are rather effective deactivating hypervalent iron compared to vitamin antioxidants, like ascorbate, and plant phenols are also effective in preventing simple iron ions from becoming prooxidative. Deactivation of hypervalent iron by plant phenols has been suggested as an important mechanism for protection of sensitive tissue in the digestive tract against oxidative stress (Kanner and Lapidot, 2001; Vulcain et al., 2005). There is a clear need for standard procedures to evaluate the efficiency of plant extracts as protectors of metal catalysis in lipid oxidation, including both heme pigments, partly degraded heme pigments, and the simpler "free" iron ions. Attempts have been made to define such evaluation methods, but they still depend on relatively advanced instrumentation (Carlsen et al., 2003).

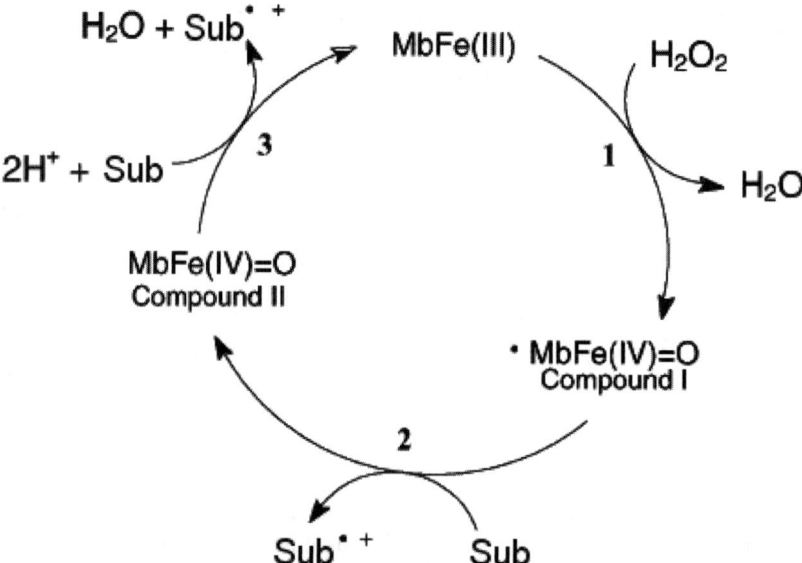

**Fig. 10.3.** Heme pigment may act as prooxidants through activation by hydrogen peroxide (1) or Lipid Hydroperoxides. The hypervalent metmyoglobin (or hemoglobin) species will initiate lipid or protein oxidation as shown in step 2 and step 3.

Table 10.2. Second-Order Rate Constants at 25°C for Scavenging of Prooxidants by Compounds of Interest for Protection of Food Against Metal-Catalyzed Lipid Oxidation.[a]

| Scavenger | $k_{ferryl}$ (l × mol$^{-1}$ × s$^{-1}$)[b] | $k_{inh}$ (l × mol$^{-1}$ × s$^{-1}$)[c] | kOO- (l × mol$^{-1}$ × s$^{-1}$)[b] |
|---|---|---|---|
| Ascorbate | 16 | | |
| Trolox | 20 | | $5.8 \times 10^3$ |
| α-Tocopherol | | $3.5 \times 10^6$ | |
| Caffeic Acid | 65 | | |
| Rutin | 23 | | $5.1 \times 10^4$ |
| Catechin/Epicatechin | 20 | $\sim 4 \times 10^5$ | $\sim 2 \times 10^4$ |

[a]From Andersen et al., 2003. [b]In water. [c]n t-butyl alcohol, $k_{inh}$ is the effective inhibitory rate constant for deactivating lipid peroxyl radicals.

## Protection Against Heat and Pressure

Most methods currently in use to determine antioxidative capacity of plant material, including food products, are based on combinations of phenol analysis and assays based on scavenging of stable or short-lived radicals (Becker et al., 2004). Such methods relate to the oxidative stress introduced in food systems or in the human body by normal thermal processes (in contrast to photochemical reactions). As for methods based on scavenging stable radicals like ABST•+, DPPH•, Fremy's salt, and the galvinoxyl radical, a relevant criticism is that none of these radicals are of biological importance or present in any food systems. For short-lived radicals protection of specific probes against free radicals, like in the ORAC assay, does not mimic biological systems. Attempts have been made to use more moderate reactive radicals like the 1-hydroxyethyl radical in combination with the ESR spin-trapping technique for antioxidant evaluation. Promising results were obtained, since such methodologies provide a detailed picture of the balance between prooxidative and antioxidative effects, which normally not are available from the standard methods (Rødtjer et al., 2006).

The general four step procedure described previously, however, does work for thermally processed food and is illustrated by results from a recent study (Racanicci et al., 2007). Chicken meat is becoming increasingly important world-wide and in countries like Brazil, production is increasing rapidly. Among different waste materials from local vegetable production in Brazil, mate (dried leaves of *Ilex paraguariensis*) was selected as a potential source for antioxidants to pre-cooked chicken-meat products. In step (i) of the four step procedure, extracts were made with water, methanol, ethanol, and 70% aqueous acetone. Water (and 70% aqueous acetone) was superior as an extraction solvent compared to methanol and ethanol, and in step (ii) the radical-scavenging capacity of the aqueous extract showed that each phenolic equivalent corresponded to one mole of Fremy's salt. Fremy's salt scavenging method accordingly is similar to a titration of the plant phenol. In step (iii), heme pigment initiation of lipid oxidation was used in order to mimic the conditions in the pre-cooked meat. Aqueous mate extract was found to be efficient in halting lipid oxidation, and a dose/respond curve could be constructed. In the final testing step (iv), mate aqueous extract was added to chicken meat balls prior to cooking; in a comparison with an unprotected product and meat balls protected by rosemary as reference, mate was found to give equal or better protection than rosemary. Notably, both mate and rosemary were found to protect vitamin E in the product during storage. The four step procedure accordingly leads directly to practical application of a new herbal source for antioxidants to meat products, including optimization of extraction and adjustment of dose. The evaluation procedure described is recommended for other sensitive dairy and meat products in combination with local plant sources with a GRAS-status.

Less investigated is antioxidative protection of high-pressure processed food products, a protection which is clearly needed. Pressure-induced lipid oxidation in muscle systems, has been assigned to two main factors, iron released from heme proteins, or membrane disruption. However, iron release could not be detected following high-pressure processing of chicken meat for thermal processes (Orlien et al., 2000). Accordingly protection should be targeted towards enzymatic formation of radicals in disrupted membranes, and rosemary extract was found to scavenge such radicals efficiently (Bragagnolo et al., 2005). Pressure effects on lipid oxidation have been expressed as volume of activation analogously with energy of activation for thermal process (Orlien et al., 2000). However evaluation procedures to protect high-pressure processed food needs further investigation related to effects of antioxidants on volume of activation in order to use this parameter in practical work. New imaging techniques based on ESR-spectroscopy should also be developed in order to locate oxidation initiation related to design of optimal protection.

## Kinetic versus Thermodynamic Control

Antioxidant synergism is getting increasing attention, and synergism is a result of kinetic control of antioxidant interaction with respect to primary antioxidants. In homogeneous solution, antioxidants are expected to regenerate each other according to the ordering of standard reduction potentials of the corresponding free radicals; accordingly ascorbate should regenerate plant phenols, tocopherols, and other antioxidants as the most reducing in agreement with thermodynamic product control. Ascorbate has thus been shown to regenerate tocopherols from their one-electron oxidized forms in a process of physiological importance in agreement with the predictions from thermodynamics. In lipid systems of increasing structural organization, synergism between carotenoids and tocopherols were clearly demonstrated indicating kinetic control of the antioxidant interaction (Becker et al., 2007). For the four groups of antioxidants shown in Table 10.3, six possible types of interaction are possible and hold the potential of synergistic effects. Antioxidant synergism becomes possible when a better antioxidant, like the tocopherols, is regenerated by flavonoids or other plant phenols. In meat systems, plant antioxidants have thus been found to protect vitamin E, probably through a regeneration mechanism, indicating kinetic control related to phase separation of the antioxidants (Racanicci et al., 2007). Synergism is also possible in systems where the less efficient antioxidants, such as flavonoids or carotenoids, are sacrificially oxidized, in effect protecting the better antioxidant like tocopherols and again constitutes an example of kinetic control. The barely studied interaction between carotenoids and plant phenols seems of importance for synergistic interaction and should be further investigated (Han et al., 2007).

Table 10.3. Four Groups of Antioxidants

|  | Hydrophilic | Lipophilic |
|---|---|---|
| Vitamins | Ascorbic Acid | Tocopherols |
| Non-Vitamins | Polyphenols | Carotenoids |

In homogeneous solution, the distribution between lipid peroxyl radicals and carbon-centered lipid radicals may be important for antioxidant interaction. Tocopherols will be used up by the reaction of the tocopheryl radical with peroxyl radicals under high oxygen conditions, while flavonoids and other plant phenols may regenerate tocopherol in competition with a less efficient reaction between the tocopheryl radical and carbon-centered radicals during oxygen depletion.

A comparison between high ascorbic blackcurrant juice and high phenolic black chokeberry in relation to antioxidant effects in combination with α-tocopherol, a very interesting observation has been made. The chokeberry juice was, in contrast to the blackcurrant juice, found to protect vitamin E efficiently against oxidation. This must clearly be an example of kinetic control, since ascorbic acid is the more reducing antioxidant compared to both the tocopherol and especially the flavonoids (Graversen et al., 2007).

## Conclusions

Lipid oxidation is often investigated in food systems without considering oxidation of other components. However protein oxidation is getting increased attention in relation to food quality and to optimal biological function. In meat, protein oxidation is known to decrease eating quality by reducing tenderness and juiciness and by enhancing discoloration and flavor deterioration (Xiong, 2000). Evaluation protocols should be established for proteins similar for those being used for lipids. Such antioxidant evaluation should also include the effect on a possible interaction between lipid and protein oxidation. Some methods adapted for evaluation against light-induced oxidation in foods is already being used for proteins like in milk products (Cardoso et al., 2004). For meat proteins, oxidation reactions are leading to cross-linking and carbonyl formation. Antioxidants like rosemary, very effective in inhibiting lipid oxidation in meat, seem to have little if any effect on protein oxidation (Lund et al., 2007). Protein and lipid oxidation may be uncoupled in such products and design of dual acting antioxidant systems present a real challenge. In tissue, further complications to protect proteins with antioxidants will arise, since oxidized lipids may modify proteins through coupling reactions between lipid-derived aldehydes and protein amino side-chain groups.

## References

Andersen, M.L.; R.K. Lauridsen; and L.H. Skibsted. Optimizing the Use of Phenolic Compounds in Food. In *Phytochemical Functional Foods,* Johnson, I.; and Williamson, G. eds., Woodhead: CRC, Cambridge, 2003, pp. 315-346.

Becker, E.M.; D.R. Cardoso; and L.H. Skibsted. Deactivation of Riboflavin Triplet-Exited State by Phenolic Antioxidants: Mechanism Behind Protective Effects in Photooxidation of Milk-Based Beverages. *Eur. Food Res. Tech.* **2005,** *221,* 382-386.

Becker, E.M.; L.R. Nissen; and L.H. Skibsted. Antioxidant Evaluation Protocols: Food Quality or Health Effects. *Eur. Food Res. Technol.* **2004,** *219,* 561-571.

Becker, E.M.; G. Ntouma; and L.H. Skibsted. Synergism and Antagonism Between Quercetin and Other Chain-Breaking Antioxidants in Lipid Systems of Increasing Structural Organisation. *Food Chem.* **2007,** *103,* 1288-1296.

Bekbölet, M. Light Effects on Food. *J. Food Prot.* **1990,** 53, 430-440.

Bradley, D.G.; and D.B. Min. Singlet Oxygen Oxidation of Foods. *Crit. Rev. Food Sci. Nutr.* **1992,** *31,* 211-236.

Bragagnolo, N.; B. Danielsen; and L.H. Skibsted. Effect of Rosemary on Lipid Oxidation in Pressure-Processed, Minced Chicken Breast during Refrigerated Storage and Subsequent Heat Treatment. *Eur. Food Res. Tech.* **2005,** *221,* 610-615.

Cardoso, D.R.; D.W. Franco; K. Olsen; M.L. Andersen; and L.H. Skibsted. Reactivity of Bovine Whey Proteins, Peptides and Amino Acids Toward Triplet Riboflavin as Studied by Laser Flash Photolysis. *J. Agric. Food Chem.* **2004,** *52,* 6602-6606.

Cardoso, D.R.; K. Olsen; J.K.S. Møller; and L.H. Skibsted. Phenol and Terpene Quenching of Singlet and Triplet Excited States of Riboflavin in Relation to Light-Struck Flavor Formation in Beer. *J. Agric. Food Chem.* **2006,** *54,* 5630-5636, eratum 54, 9278.

Cardoso, D.R.; K. Olsen; and L.H. Skibsted. Mechanism of Deactivation of Triplet-Excited Riboflavin by Ascorbate, Carotenoids and Tocopherols in Homogeneous and Heterogeneous Aqueous Food Model System. *J. Agric. Food Chem.* **2007,** *55,* 6285-6291.

Carlsen, C.U.; J.K.S. Møller; and L.H. Skibsted. Heme Iron in Lipid Oxidation. *Coord. Chem. Rev.* **2005,** *249,* 485-498.

Carlsen, C.U.; I.M. Skovgaard; and L.H. Skibsted. Pseudoperoxidase Activity of Myoglobin. Kinetics and Mechanism of the Peroxidase Cycle of Myoglobin with $H_2O_2$ and 2,2-Azino-*bis*(3-Ethylbenzthiazoline-6-Sulfonate) as Substrates. *J. Agric. Food Chem.* **2003,** *51,* 5815-5823.

Carter, P.H.; and E. Ernst. Anthocyanosides of *Vaccinium myrtillus* (Bilberry) for Night Vision–A Systematic Review of Placebo-Controlled Trials. *Surv. Ophthalmol.* **2004,** *49,* 38-50.

Chipault, J.R. Antioxidants for Food Use. In *Autoxidation and Antioxidants,* Lundberg, W.O. Ed., Wiley: New York, 1962, pp. 477-542.

Ferrali, M.; C. Signorini; N. Caciotti; L. Sugherini; L. Cicoli; D. Giachetti; and M. Comporti. Protection Against Oxidative Damage of Erythrocyte Membrane by the Flavonoid Quercetin and Its Relation to Iron Chelating Activity. *FEBS Letters* **1997,** *416,* 123-129.

Frankel, E.N.; and A.S. Meyer. The Problem of Using One-Dimensional Methods to Evaluate Multifunctional Food and Biological Antioxidants. *J. Sci. Food Agricult.* **2000,** *80(13),* 1925-1941.

Goldsmith, M.R.; P.J. Rogers; N.M. Cabral; K.P. Ghiggens; and F.A. Roddick. Riboflavin Triplet Quenchers Inhibit Lightstruck Flavor Formation in Beer. *J. Am. Soc. Brew. Chem.*

2005, *63*, 177-184.

Graversen, H.B.; E.M. Becker; L.H. Skibsted; and M.L. Andersen. Antioxidant Synergism Between Fruit Juice and α-Tocopherol. A Comparison Between High Phenolic Black Chokeberry (*Aronia melanocarpa*) and High Ascorbic Blackcurrant (*Ribes nigrum*). *Eur. Food Res. Technol.* **2008**, *226*, 737-743.

Han, R-M.; Y-X. Tian; E.M. Becker; M.L. Andersen; J-P. Zhang; and L.H. Skibsted. Puerarin and Conjugate Bases as Radical Scavengers and Antioxidants. Molecular Mechanism and Synergism with β-Carotene. *J. Agric. Food Chem.* **2007**, *55*, 2384-2391.

Hansen, E.; and L.H. Skibsted. Light Induced Oxidative Changes in a Model Dairy Spread. Wavelength Dependence of Quantum Yields. *J. Agric. Food Chem.* **2000**, *48*, 3090-3094.

Jensen, C.; M. Flensted-Jensen; L.H. Skibsted; and G. Bertelsen. Effects of Dietary Rape Seed Oil, Copper(II) Sulphate and Vitamin E on Drip Loss, Colour and Lipid Oxidation of Chilled Pork Chops Packed in Atmospheric Air or in a High Oxygen Atmosphere. *Meat Sci.* **1998**, *50*, 211-221.

Jung, M.Y.; Y.S. Oh; D.K. Kim; H.J. Kim; D.B. Min. Photoinduced Generation of 2,3-Butanedione from Riboflavin. *J. Agric. Food Chem.* **2007**, *55*, 170-174.

Kanner, J.; and T. Lapidot. The Stomach as a Bioreactor: Dietary Lipid Peroxidation in the Gastric Fluid and the Effect of Plant-Derived Antioxidants. *Free Rad. Biol. Med.* **2001**, *31*, 1388-1395.

Li, T.; J.M. King; and D.B. Min. Quenching Mechanism and Kinetics of Carotenoids in Riboflavin Photosensitized Singlet Oxidation of Vitamin D2. *J. Food Biochem.* **2000**, *24*, 477-492.

Lonn, E.; J. Bosch; S. Yusuf; P. Sheridan; J. Pogue; J.M.O. Arnold; C. Ross; A. Arnold; P. Sleight; J. Probstfield; et al. Effects of long-term vitamin E supplementation on cardiovascular events and cancer: a randomized controlled trial. *JAMA* **2005**, *293(11)*, 1338-1347.

Lund, M.L.; M.S. Hviid; and L.H. Skibsted. The Combined Effect of Antioxidants and Modified Atmosphere Packaging on Protein and Lipid Oxidation in Beef Patties during Chill Storage. *Meat Sci.* **2007**, *76*, 226-233.

Orlien, V.; E. Hansen; and L.H. Skibsted. Lipid Oxidation in High-Pressure Processed Chicken Breast Muscle during Chill Storage. Critical Working Pressure in Relation to Oxidation Mechanism. *Eur. Food Res. Technol.* **2000**, *211*, 99-104.

Racanicci, A. M.C.; B. Danielsen; and L.H. Skibsted. Mate (*Ilex paraguariensis*) as a Source of Water Extractable Antioxidant for Use in Chicken Meat. *Eur. Food Res. Technol.* **2007**, accepted for publication.

Rice-Evans, C. Flavonoids and Isoflavonoids: Absorption, Metabolism, and Bioactivity. *Free Rad. Biol. Med.* **2004**, *36*, 827-828.

Rødtjer, A.; L.H. Skibsted; and M.L. Andersen. Antioxidative and Prooxidative Effects of Extracts Made from Cherry Liqueour Promace. *Food Chem.* **2006**, *99*, 6-14.

Sandström, B.; S. Bügel; C. Lauridsen; F. Nielsen; C. Jensen; and L.H. Skibsted. Cholesterol Lowering Potential in Man of Fat from Pigs Fed Rapeseed Oil. *Brit. J. Nutr.* **2000**, *84*, 143-150.

van Leeuwen, R.; S. Boekhoorn; J.R. Vingerling; J.C.M. Witteman; C.C.W. Klaver; A. Hofman; and P.T.V.M. de Jong. Dietary Intake of Antioxidants and Risk of Age-Related Macular Degeneration. *JAMA* **2005**, *294(24)*, 3101-3107.

Vivekananthan, D.P.; M.S. Penn; S.K. Sapp; A. Hsu; and E.J. Topol. Use of Antioxidant Vitamins for the Prevention of Cardiovascular Disease: Meta-Analysis of Randomised Trials. *Lancet* **2003**, *361*, 2017-2030.

Vulcain, E.; P. Goupe; C. Caris-Veyrat; and O. Dangles. Inhibition of the Metmyoglobin-Induced Peroxidation of Linoleic Acid by Dietary Antioxidants: Action in the Aqueous vs. Lipid Phase. *Free Rad. Res.* **2005,** *39,* 547-563.

Xiong, Y.L. Protein Oxidation and Implications for Muscle Food Quality. In *Antioxidants in Muscle Foods,* Decker, E.; and Faustman, E. Eds., Chichester, John Wiky and Sons, 2000, pp. 85-111.

# Index

## A

Acrolein/crotonaldehyde, co-oxidation of proteins and, 204–206
Adiposity, CLA isomer effects on, 77
Age pigment formation, co-oxidation of proteins and, 233–240
Aldehydes
  and oxidation of EPA/DHA, 55
  ozone reaction and, 44
  ozone/lipid reaction and, 46
α-linolenic acid, and oxidation of EPA/DHA, 53–54
Alkanals, CLA oxidation and, 99
Alkoxyl radicals, 35–36
Amino acid losses, co-oxidation of proteins and, 212–220
Antioxidants. *See also* Carotenoids
  and autoxidation of CLA studies, 85–86
  carotenoids, 158–159
  and comparison of oxidation rates of CLA/LA isomers, 89
  evaluation strategies
    early lipid oxidation events, 293–295
    heat/pressure protection, 301–302
    kinetic versus thermodynamic control, 302–303
    light protection, 296–299
    metal catalysis protection, 299–301
    overview, 291–292, 303
    protocols, 292–293
  and levels of EPA/DHA in marine animal tissue, 56
  lipid oxidation and, 52–53
  oil-in-water emulsions, 279–281
  tocopherol concentration
    antioxidant mechanism of α-tocopherol, 128–129
    effect of temperature, 138
    overview, 127–128
    pro-oxidant mechanism of α-tocopherol, 129, 131–138
Arachidonic acid, and oxidative stability of PUFA, 52–53
Aromatic compounds, Type II reaction of Sensitizer* and, 8
Ascorbic acids, as oxygen quenching mechanisms, 21–22
Atherosclerosis, CLA isomer effects on, 77
α-tocopherol. *See also* Tocopherols
  antioxidant mechanism of, 128–129
  chemical structures of, 131
  effect of temperature on antioxidant activity, 138
  pro-oxidant mechanism
    autoinitiation from hydroperoxide decomposition, 134
    chain transfer reactions by radical, 132, 134
    pro-oxidant effects due to products, 134–138
    Yanishlieva-Marinova model, 131–132
  rates constants of autoxidation reactions, 130
Autoxidation

conjugated linoleic acid (CLA)
furan FA/secondary product
formation, 97, 99–100
hydroperoxide formation, 89–93
oligomer formation, 93–98
rate/routes, 88–89
studies on CLA as free acid or
ester, 78–87
rates constants involving LH/TH,
130
Azide, as quencher of singlet oxygen
reaction, 10

## B

β-carotene (CarH). *See also* Carotenoids
hydroxyl radicals and, 33–34
peroxyl radicals and, 38–40
structure/nomenclature of, 144–145
Bifunctional saturated aldehydes, co-
oxidation of proteins and, 199–202
Biological function alteration/loss,
co-oxidation of proteins and, 241,
243–244
Bis-allylic hydrogens, and oxidation of
PUFA, 62, 70
Bonito oil, lipids levels and, 56, 58
Bulk oils. *See* Oil-in-water emulsions

## C

Campesterol, as sterol oxidation
product, 113
Cancer, CLA isomer effects on, 77
Carbonyl oxidation, 156
Carotenes
hydroxyl radicals and, 33–34
peroxyl radicals and, 38–40
singlet oxygen quenching
mechanisms and, 23
Carotenoids
chemical quenching of
photosensitizers/singlet oxygen by,
153–156
light absorption, 149–150
oxidation reactions, 143–144, 170
as oxygen quenching mechanisms,
21–24
peroxyl radical reactions
antioxidant activity, 158–159
enzyme-catalyzed co-oxidation
with UFAs, 161–162
kinetics/mechanisms, 155,
157–158
oxidation products, 159–161
physical quenching of
photosensitizers/singlet oxygen by,
152–153
pro-oxidant effects
effect of heat, 166–169
triplet molecular oxygen reactions,
163–165
as quencher of singlet oxygen
reaction, 10
reactivity
configuration of cyclic ends,
147–148
degree of conjugation, 145–146
geometrical isomerism, 147
oxygenated groups at cyclic ends,
148–149
structure/nomenclature of, 144–145
and synergism with tocopherols in
lipid oxidation reactions, 168, 170
Cellular lipids, oxidation of long-chain
PUFA in, 63–65, 70–71
Chemical quenching. *See* Oxygen
quenching mechanisms
Chemical quenching of photosensitizers/
singlet oxygen by carotenoids,
153–156
Chemical traps, singlet oxygen and,
9–10
Chemiluminence, singlet oxygen and,
12
Cholesterol
oxidation
kinetics of sterol autoxidation,
119–122
products/mechanisms of sterol
autoxidation, 114–118
products/mechanisms of sterol
photoxidation, 118–119
sterol/stanol structures, 111–114
singlet/triplet oxygen oxidation and,

10–11
Cholesterol oxidation products (COP), 111–114
Conjugated DHA (CDHA), and formation of conjugated trienoic acids in oils with PUFA, 69, 71
Conjugated EPA (CEPA)
 and formation of conjugated trienoic acids in oils with PUFA, 69, 71
Conjugated linoleic acid (CLA)
 autoxidation
  furan FA/secondary product formation, 97, 99–100
  hydroperoxide formation, 89–93
  oligomer formation, 93–98
  overview, 102
  rate/routes, 88–89
  studies on CLA as free acid or ester, 78–87
 and formation of conjugated trienoic acids in oils with PUFA, 65–66, 71
 oxidation
  overview, 77–78, 102, 105
 singlet oxygen oxidation
  primary product formation, 100–104
  secondary product formation, 104–105
Conjugated linolenic acid (CLN), and formation of conjugated trienoic acids in oils with PUFA, 65–69, 71
Conjugation and carotenoids, 145–146
Co-oxidation of carotenoids, 161–162
Co-oxidation of proteins
 lipid epoxide reactions, 192–197
 lipid hydroperoxide reactions, 189–192
 macromolecular damage overview, 181–183
 molecular damage reactions
  abnormal function/immunochemistry/disease contributions, 244–248
  amino acid losses, 212–220
  biological function alteration/loss, 241, 243–244
  change: solubility/hydrophobicity/conformation/aggregation, 220, 222
  crosslinking, 222–230. *See also* Crosslinking of proteins
  enzyme activity inhibition, 240–243
  fluorescent adduct/age pigment formation, 233–240
  fragmentation, 231–233
  free radical transfer from lipids to proteins, 220–221
 secondary product reactions
  4-hydroxy-2-alkenals/4-oxo-2-alkenals, 206–209
  acrolein/crotonaldehyde, 204–206
  bifunctional saturated aldehydes, 199–202
  free radical oxidation of unsaturated aldehydes, 210
  monofunctional alkanals, 197–199
  physiological γ-ketoaldehydes/levuglandins, 209–211
  unsaturated aldehydes, 202–204
 transfer of lipid radicals to proteins
  consequences, 188–189
  evidence, 184–188
Crosslinking of proteins
 carbonyl-mediated
  Michael addition of thiols/amines, 226–227, 229–230
  pyridinium crosslink formation, 229–230
  pyrrole link formation, 227–229
  Schiff base amine condensation, 225, 227, 229–230
 effects of, 222–223
 free radical crosslinking (polymerization), 223–225
Crude oils, and formation of conjugated trienoic acids in oils with PUFA, 65

# D

DABCO, as quencher of singlet oxygen reaction, 10
Deuterium oxide, singlet oxygen detection and, 9

DHA
  conjugated DHA (CDHA), 69, 71
  oxidation overview, 51–53, 69–71
  oxidative stability in liposomes, 61–63
  oxidative stability in phospholipids of, 55–60
  and oxidative stability of PUFA, 52–55
  structures of, 53–54
Dioxetanes as product of CLA oxidation, 101, 103
Dioxines as product of CLA oxidation, 101, 103
Disease contributions, co-oxidation of proteins and, 244–248
Dismutation, hydrogen peroxide and, 43–44
Doering's diradical, 163–165
Double bonds, singlet oxygen quenching mechanism rates, 24
DPBF (1,3-diphenylisobenzofuran), singlet oxygen and, 9

E
Emulsions. *See* Oil-in-water emulsions
Endoperoxide oxidation, 156
Energy levels of triplet/singlet oxygen, 4
Enzyme activity inhibition, co-oxidation of proteins and, 240–243
Enzyme-catalyzed co-oxidation with UFAs, carotenoids and, 161–162
EPA
  conjugated EPA (CEPA), 69, 71
  oxidation overview, 51–53, 69–71
  oxidative stability in phospholipids of, 55–60
  and oxidative stability of PUFA, 52–55
  structures of, 53–54
Epoxide reactions, co-oxidation of proteins and, 192–197
ESR (electron spin resonance) spectroscopy, singlet oxygen and, 12–14
Esters, studies on CLA as free acid or, 78–87

F
Fatty acids
  CLA oxidation
    furan FA/secondary product formation, 97, 99–100, 104–105
    reaction with singlet oxygen, 13–17
Fenton reaction, hydroxyl radicals and, 32–33
Fish oil. *See also* Long-chain polyunsaturated fatty acids (PUFA)
  and oxidative stability of EPA/DHA, 55–60
  triacylglycerol (TAG) as main lipid class of, 58
Flavor
  PUFAs and, 51
  reversion flavor, 20–21, 68, 71
Flavor stability, effects of light on, 5–6
Fluorescent adduct/age pigment formation
  alkenals and, 239
  bifunctional aldehydes and, 237–238
  lipid hydroxyperoxide and, 239–240
  monofunctional aldehydes/alkanals and, 235, 237
Food dispersions
  oil-in-water emulsions
    bulk oils, 280, 282–286
    emulsifier roles, 276–279
    free radical scavenging antioxidant impact, 279–281
    lipid oxidation, 274–275
    overview, 273–274, 286–288
    protein impact, 275–276
Fragmentation of proteins, 231–233
Free acids
  studies on CLA as ester or, 78–87
Free fatty acids
  marine animal tissue and, 59
Free radicals. *See also* Long-chain polyunsaturated fatty acids (PUFA)
  and antioxidant mechanism of α-tocopherol, 128–129
  and autoxidation of CLA, 92
  CLA isomer effects on, 77
  ESR spectroscopy and, 12–14
  oil-in-water emulsions/food

dispersions and, 279–281
and oxidation of unsaturated aldehydes, 210
singlet oxygen oxidation and, 13–14
sterol autoxidation and, 121
and transfer from lipids to proteins, 220–221
Furan fatty acid formation, CLA oxidation and, 97, 99–100, 104–105

## G

Gas chromatography (GC), and oxidation of EPA/DHA, 55
γ-ketoaldehydes/levuglandins
co-oxidation of proteins and, 209–211

## H

Haber-Weiss reaction
hydroxyl radicals and, 33
singlet oxygen formation and, 4–5
superoxide anion radicals and, 43
Half-lives of reactive oxygen species (ROS), 31
Hammond postulate, and diastereoselectivity of CLA autoxidation, 92
Herring roe lipids, 59–60
Histidine, as quencher of singlet oxygen reaction, 10
Hund's rule, and elecron configuration of triplet/singlet oxygen, 1, 3
Hydrocarbons, and oxidation of EPA/DHA, 55
Hydrogen peroxide
chemistry/reactions of, 43–44
ozone/lipid reaction and, 44, 46
Hydroperoxides (ROOH)
alkoxyl radicals formation by, 35
α-tocopherol antioxidant activity and, 131–132
autoinitiation from decomposition of, 134
autoxidation of CLA methyl ester and, 89–93
co-oxidation of proteins and, 189–192
as product of CLA oxidation, 101, 103–104
PV measurements and CLA autoxidation, 78
singlet/triplet oxygen oxidation and, 16–17
Hydroperoxy epidioxides, and oxidation of PUFA, 53–54
Hydroperoxyl radicals, 35, 37
Hydrophilic antioxidants, 303
Hydrophobicity/conformation/ aggregation/solubility
co-oxidation of proteins and, 220, 222
4-hydroxy-2-alkenals/4-oxo-2-alkenals, co-oxidation of proteins and, 206–209
Hydroxyl radical scavenger, 10
Hydroxyl radicals, chemistry/reactions of, 31–34

## I

Immunochemistry/disease contributions, co-oxidation of proteins and, 244–248
Insulin resistance, CLA isomer effects on, 77
Isomers, geometrical, 147
Isotopes, deuterium oxide effect and, 9

## J

Jablonski diagram, 5–6
Japan, and CLN content in soybean oil, 66–67

## K

Ketoaldehydes/levuglandins, co-oxidation of proteins and, 209–211
Kinetic control of reactions, 302–303

## L

Laser deflection calorimetry, singlet oxygen detection and, 13
Levuglandins, co-oxidation of proteins and, 209–211

Lifetime of singlet oxygen, 3
Light absorption of carotenoids, 149–155
Linoleate, singlet/triplet oxygen oxidation rates and, 17
Linoleic acid
   hydroxyl radical initiated oxidation of, 33–34
   and oxidation rate of CLA isomers, 88–89
   PUFAs of, 51
   and reversion flavor in soybean oil, 21
   singlet oxygen and formation of 2-pentyl furan, 18
Linolenate, singlet/triplet oxygen oxidation rates and, 17
Linolenic acid
   and formation of conjugated trienoic acids in oils with PUFA, 65–69
   and oxidation of EPA/DHA, 53–54
   PUFAs of, 51
   and reversion flavor in soybean oil, 21
   singlet oxygen and formation of 2-pentyl furan, 18
Lipid epoxide reactions, co-oxidation of proteins and, 192–197
Lipid hydroxyperoxide reactions, co-oxidation of proteins and, 189–192
Lipid oxidation. *See also* Co-oxidation of proteins; Oxidation; Tocopherols
   carotenoid/tocopherol synergism in, 168, 170
   mechanism, 52–53
Lipid radicals. *See* Co-oxidation of proteins
Lipophilic antioxidants, 303
Liposomes
   oxidative stability of DHA in, 61–63, 71
Long-chain polyunsaturated fatty acids (PUFA)
   DHA
      oxidative stability in liposomes, 61–63
   EPA/DHA
      oxidation, 52–55
      oxidation overview, 51–53, 69–71
      oxidative stability in phospholipids, 55–60
   formation of conjugated trienoic acids in oils containing, 65–69
   oxidation in cellular lipids of, 63–65

# M

Methyl esters, and autoxidation of CLA, 89–93
Methyl linoleate oxidation, 132–133
MHP (monohydroperoxide) isomers
   and oxidation of PUFA esters, 62–63
   and oxidation of PUFA in cellular lipids, 64–65
Michael addition of thiols/amines, and carbonyl-mediated crosslinking of proteins, 226–227, 229–230
Middle chain fatty acid (MCT), beany/grassy flavor and, 68–69
Molecular damage. *See* Co-oxidation of proteins
Molecular orbital theory and triplet/singlet oxygens, 1–4
Monofunctional alkanals, co-oxidation of proteins and, 197–199

# N

NADPH, superoxide anion radicals and, 42

# O

Oil-in-water emulsions
   food dispersions
      bulk oils, 280, 282–286
      emulsifier roles, 276–279
      free radical scavenging antioxidant impact, 279–281
      lipid oxidation, 274–275
      overview, 273–274, 286–288
      protein impact, 275–276
Oleate, singlet/triplet oxygen oxidation rates and, 17
Olefins, Type II reaction of Sensitizer*

and, 8
Oleic acid
  hydroxyl radical addition reaction to, 33–34
  ozone reaction with, 44–45
Oligomer formation
  and polymerization from propagation reactions, 93–95
  and polymerization from termination reactions, 95–97
Olive oil, singlet oxygen photooxidation and, 20
Oxidation
  carotenoid classification degree of, 165
  carotenoids and, 159–161
  cholesterol/phytosterols
    kinetics of sterol autoxidation, 119–122
    products/mechanisms of sterol autoxidation, 114–118
    products/mechanisms of sterol photoxidation, 118–119
    sterol/stanol structures, 111–114
  singlet oxygen and, 17–20
  singlet oxygen/fatty acid reaction and, 13–17
Oxidized product formation ($\emptyset AO_2$), quantum yield of, 25–27
Oxygen consumption during oxidation of fish lipids, 59–60
Oxygen quenching mechanisms
  carotenoids, 22–24
  determination, 24–27
  rates, 23–24
  tocopherols, 24
Oxygen radicals. See Reactive oxygen species (ROS)
Oxygen species. See Reactive oxygen species (ROS)
Ozone, 44–46
Ozonide, 44

# P

Pauli's exclusion principle, and elecron configuration of triplet/singlet oxygen, 1, 3

2-pentyl furan
  formation from linoleic acid by singlet oxygen, 18
  formation from linolenic acid by singlet oxygen, 19
  and reversion flavor in soybean oil, 21
Perepoxides, as product of CLA oxidation, 101–103
Peroxide value (PV)
  and autoxidation of CLA, 78–86
  and oxidation of EPA/DHA, 54–57
Peroxyl radical reactions
  and evaluation strategies, 303
Peroxyl radicals
  carotenoids
    antioxidant activity, 158–159
    enzyme-catalyzed co-oxidation with UFAs, 161–162
    kinetics/mechanisms, 155, 157–158
    oxidation products, 159–161
  chemistry/reactions of, 37–41
  sterol autoxidation and, 115, 117
Phosphatidylcholine (PC), oxidative stability of DHA and, 61–62
Phospholipids
  and levels of EPA/DHA in marine animal tissue, 56
  marine animal tissue and, 59
Photooxidation. See also Oxidation; Oxygen quenching mechanisms
  and autoxidation of CLA studies, 80
  of CLA methyl esters/ML, 104
  singlet oxygen
    olive oil, 20
    soybean oil, 20
  of sterols, 118–119
Photosensitization
  chemical quenching by carotenoids, 153–156
  deuterium oxide effect and, 9
  physical quenching by carotenoids, 150–153
  singlet oxygen formation and, 5
Physical quenching of photosensitizers/ singlet oxygen by carotenoids,

152–153
Physiological γ-ketoaldehydes/
levuglandins
  co-oxidation of proteins and,
    209–211
Phytosterol oxidation products (POP),
  111–114
Phytosterols
  oxidation
    kinetics of sterol autoxidation,
      119–122
    products/mechanisms of sterol
      autoxidation, 114–118
    products/mechanisms of sterol
      photoxidation, 118–119
    sterol/stanol structures, 111–114
Polymerization
  oligomer formation and, 93–97
  protein crosslinking and, 223–225
Polyunsaturated fatty acids, singlet
  oxygen oxidation and, 13–14
Pro-oxidant. See α-tocopherol
Propanal, formation during oxidation of
  fish lipids, 59–60
Proteins, co-oxidation of. See Co-
  oxidation of proteins
PUFA. See Long-chain polyunsaturated
  fatty acids (PUFA)
Pyridinium crosslink formation, and
  carbonyl-mediated crosslinking of
  proteins, 229–230
Pyrrole link formation, and carbonyl-
  mediated crosslinking of proteins,
  227–229

## Q
Quantum yield of oxidized product
  formation (ØAO$_2$), 25–27
Quenching by carotenoids
  chemical, 153–156
  physical, 152–153

## R
Radical compounds
  alkoxyl radicals, 35–36
  hydrogen peroxide, 43–44
  hydroperoxyl radicals, 35, 37
  hydroxyl radicals, 31–34
  ozone, 44–46
  peroxyl radicals, 37–41
  superoxide anion radicals, 42–43
  triplet oxygen as, 4
Rancidity, low-molecular-weight
  compounds and, 1
Rapeseed oil, stigmasterol
  transformation to stigmasterol oxide
  in, 121–122
Rate of initiation (RI), and effect of
  temperature on tocopherols, 138
Reactive oxygen species (ROS)
  alkoxyl radicals, 35–36
  hydrogen peroxide, 43–44
  hydroperoxyl radicals, 35, 37
  hydroxyl radicals, 31–34
  overview, 31
  ozone, 44–46
  peroxyl radicals, 37–41
  superoxide anion radicals, 42–43
Reversion flavor
  oxidation of long-chain PUFA and,
    71
  and PV of soybean oil, 68
  and singlet oxygen, 20–21
  soybean oil and, 20–21
Riboflavin (RF), superoxide anion
  radicals and, 42–43

## S
Salmon roe lipids, 59–60
Schiff base amine condensation
  and carbonyl-mediated crosslinking
    of proteins, 225, 227, 229–230
Sensitizer*
  conversion to triplet state, 6
  riboflavin (RF)
    superoxide anion radicals and,
      42–43
  singlet oxygen quenching
    mechanisms and, 22
  Type I/II mechanism of, 6–8
Singlet oxygen. See also Sensitizer*
  2-pentyl furan formation, 18–19

and chemical quenching by
  carotenoids, 150–153
cholesterol oxidation products and,
  10–11
CLA oxidation
  primary product formation,
    100–104
  secondary product formation,
    104–105
detection
  chemical traps, 9–10
  chemiluminence, 12
  deuterium oxide effect, 9
  ESR spectroscopy, 12–14
  quenchers, 10
electron configuration of, 1–4
formation
  chemical/photochemical/biological
    methods, 4–6
and formation of tetraoxy
  intermediate, 137–138
formed from α-tocopherol radicals,
  135–137
lifetime in solution of, 8
natural decay of, 8
oxidation in foods, 17–20
oxygen quenching mechanisms
  carotenoids, 21–24
  determination, 24–27
  tocopherols, 24
and physical quenching by
  carotenoids, 153–156
reaction with fatty acids
  hydroperoxides formed, 16–17
  olefins/dienes, 14–16
  polyunsaturated, 13–14
and reversion flavor in soybean oil,
  20–21
Sitosterol, as sterol oxidation product,
  113
Solubility/hydrophobicity/
  conformation/aggregation
  co-oxidation of proteins and, 220,
    222
Soybean oil
  and formation of conjugated trienoic
    acids in oils with PUFA, 65–69

reversion flavor in, 20–21
singlet oxygen, photooxidation, 20
Spin multiplicity, and elecron
  configuration of triplet/singlet oxygen,
  3
Squid lipids, and levels of EPA/DHA,
  56–58
Sterols
  autoxidation kinetics of, 119–122
  autoxidation of, 114–118
  campesterol oxidation products of,
    113
  cholesterol oxidation products of,
    112
  dehydration products of, 118–119
  marine animal tissue and, 59
  photooxidation of, 118–119
  sitosterol oxidation products of, 113
  sterol oxides in food, 114
  stigmasterol oxidation products of,
    113, 121–122
  structure of, 111
Stigmasterol, 113, 121–122
Superoxide anion radicals, chemistry/
  reactions of, 42–43
Superoxide anion reactions, deuterium
  oxide effect and, 9

## T

Temperature
  effect on carotenoid oxidation and
    cleavage, 166–168
  effect on tocopherol antioxidant
    activity, 138
  heat/pressure protection (evaluation),
    301–302
  singlet oxygen oxidation and, 13
Tertiary amines, as quencher of singlet
  oxygen reaction, 10
Thermodynamic control of reactions,
  302–303
Thiobarbituric acid reactive substances
  (TBARS), and oxidative stability of
  EPA/DHA, 55–60
Time-resolved singlet oxygen detection,
  13

Tocopherols
  alkoxyl radicals and, 35–36
  antioxidant efficacy due to concentration
    antioxidant mechanism of α-tocopherol, 128–129
    effect of temperature, 138
    overview, 127–128
    pro-oxidant mechanism of α-tocopherol, 129, 131–138
  hydroxyl radicals and, 34
  lipid content from marine organisms and, 58
  marine animal tissue and, 56, 59
  and oxidation rate of CLA isomers, 89
  as oxygen quenching mechanisms, 21–22, 24
  peroxyl radicals and, 39–41
  as quencher of singlet oxygen reaction, 10
  superoxide anion radicals and, 43
Triacylglycerol (TAG)
  beany/grassy flavor of MCT and, 68–69
  as main lipid class of fish oils, 58
  sterol autoxidation and, 121
Trienoic acids, formation of conjugated, 65–69
Tripalmitin, stigmasterol transformation to stigmasterol oxide in, 121–122
Triplet oxygen. *See also* Singlet oxygen
  carotenoids and, 163–165
  electron configuration of, 1–4
  hydroperoxides formation and, 16–17
  hydroperoxyl radical formation and, 35, 37
  hydroxyl radicals and, 33
  oxidation in foods, 17–20
  oxidized product formation (ØAO$_2$) and, 25–27
  ozone production and, 44
  peroxyl radicals and, 37–38
  Sensitizer* conversion to, 6
Trout egg lipids, and levels of EPA/DHA, 58

Tuna lipids, and levels of EPA/DHA, 56–58
Type I/II mechanism of Sensitizer*, 6–8

## U

Unsaturated aldehydes, co-oxidation of proteins and, 202–204
Unsaturated fatty acids
  carotenoid enzyme-catalyzed co-oxidation with, 161–162
  rates constants of autoxidation reactions, 130
  sterol autoxidation and, 120–121

## W

Water. *See* Reactive oxygen species (ROS)

## X

Xanthene, superoxide anion radicals and, 42
Xanthophylls
  carotenoid structre/antioxidant activity of, 158–159
  singlet oxygen quenching mechanisms and, 23